# Take Note!

To accompany

# Principles of Human Anatomy

*Ninth Edition*

Gerard J. Tortora
*Bergen Community College*

JOHN WILEY & SONS, INC.

To order books or for customer service call 1-800-CALL-WILEY (225-5945).

---



ISBN 0-471-41328-3

Printed in the United States of America

10 9 8 7 6 5 4 3 2 1

Printed and bound by Courier Westford, Inc.

NOTES

**Figure 1.1** Levels of structural organization that compose the human body (page 3).

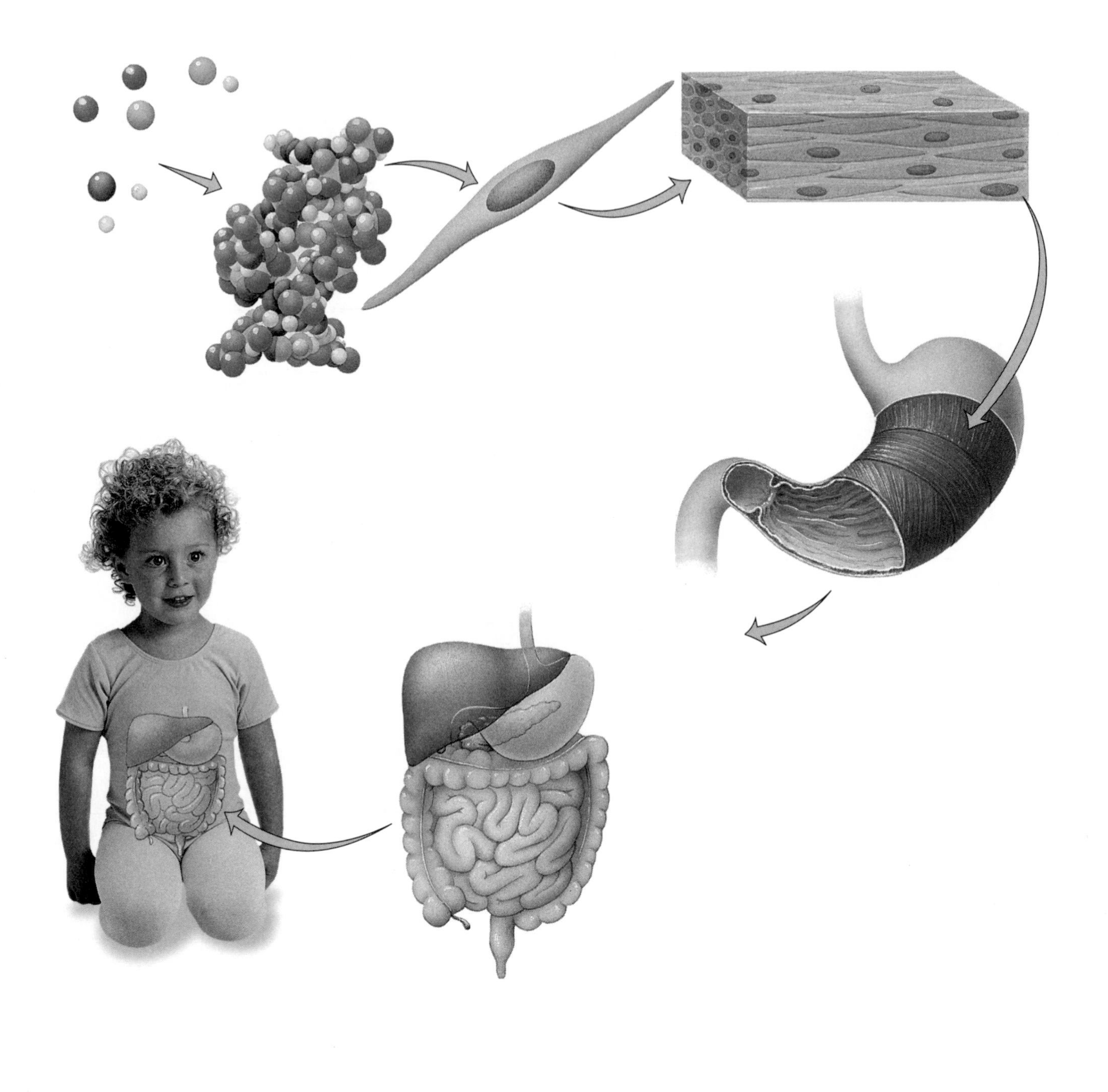

# NOTES

**Figure 1.2** The anatomical position (page 8).

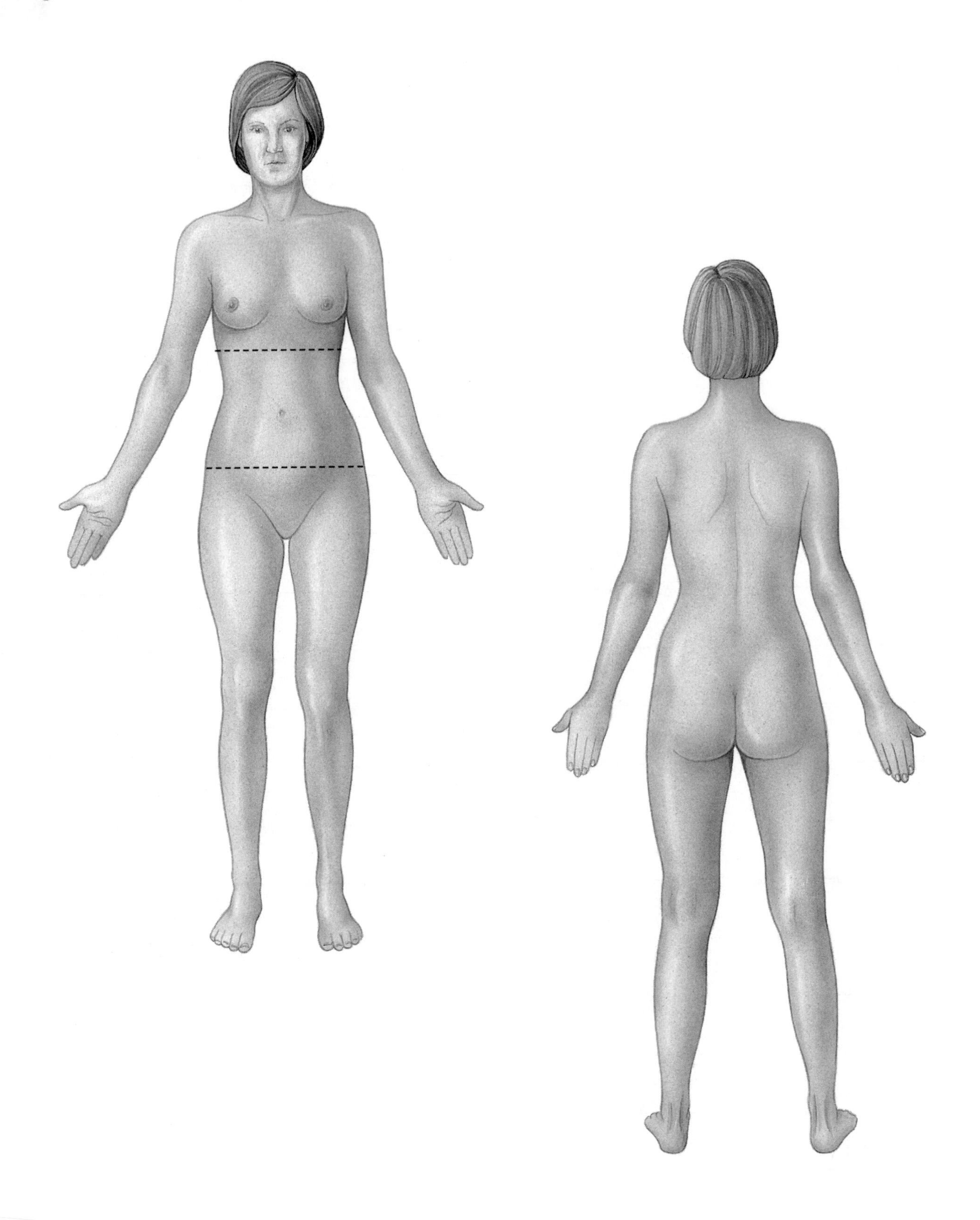

NOTES

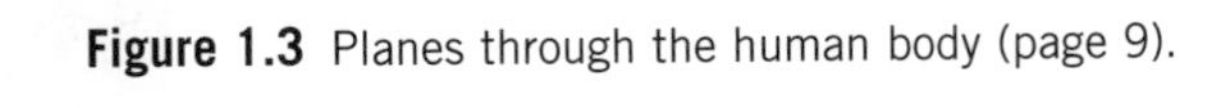

**Figure 1.3** Planes through the human body (page 9).

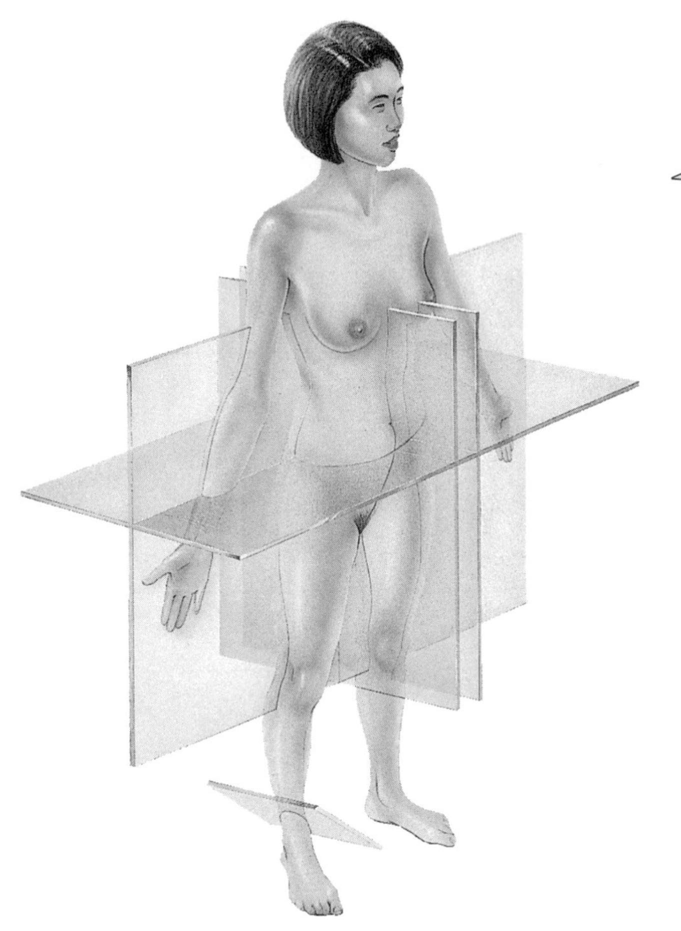

**Figure 1.4** Planes and sections through different parts of the brain (page 9).

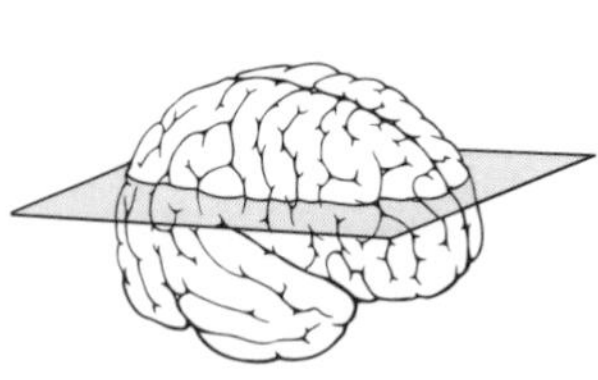

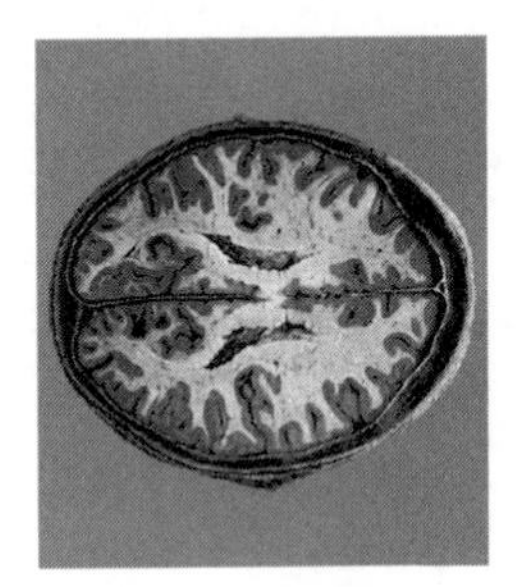

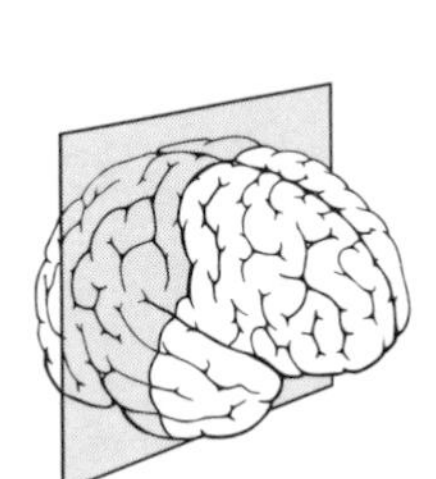

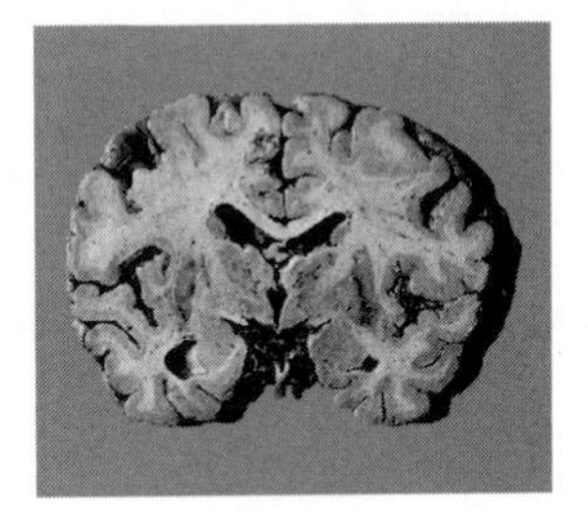

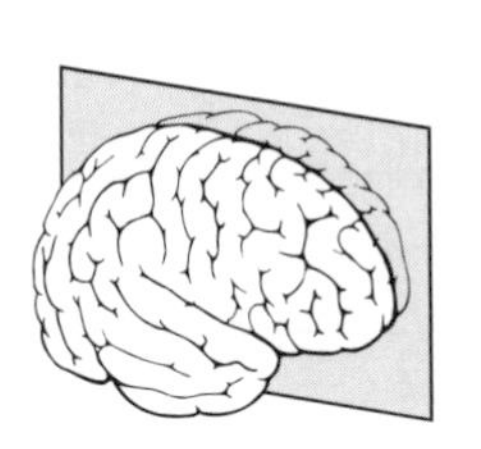

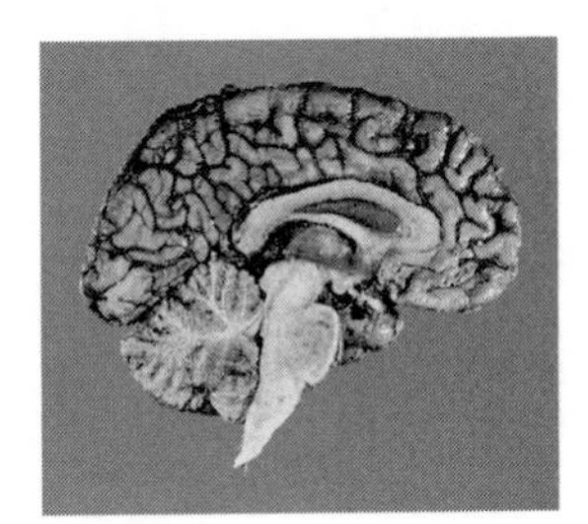

## NOTES

**Figure 1.5** Directional terms (page 11).

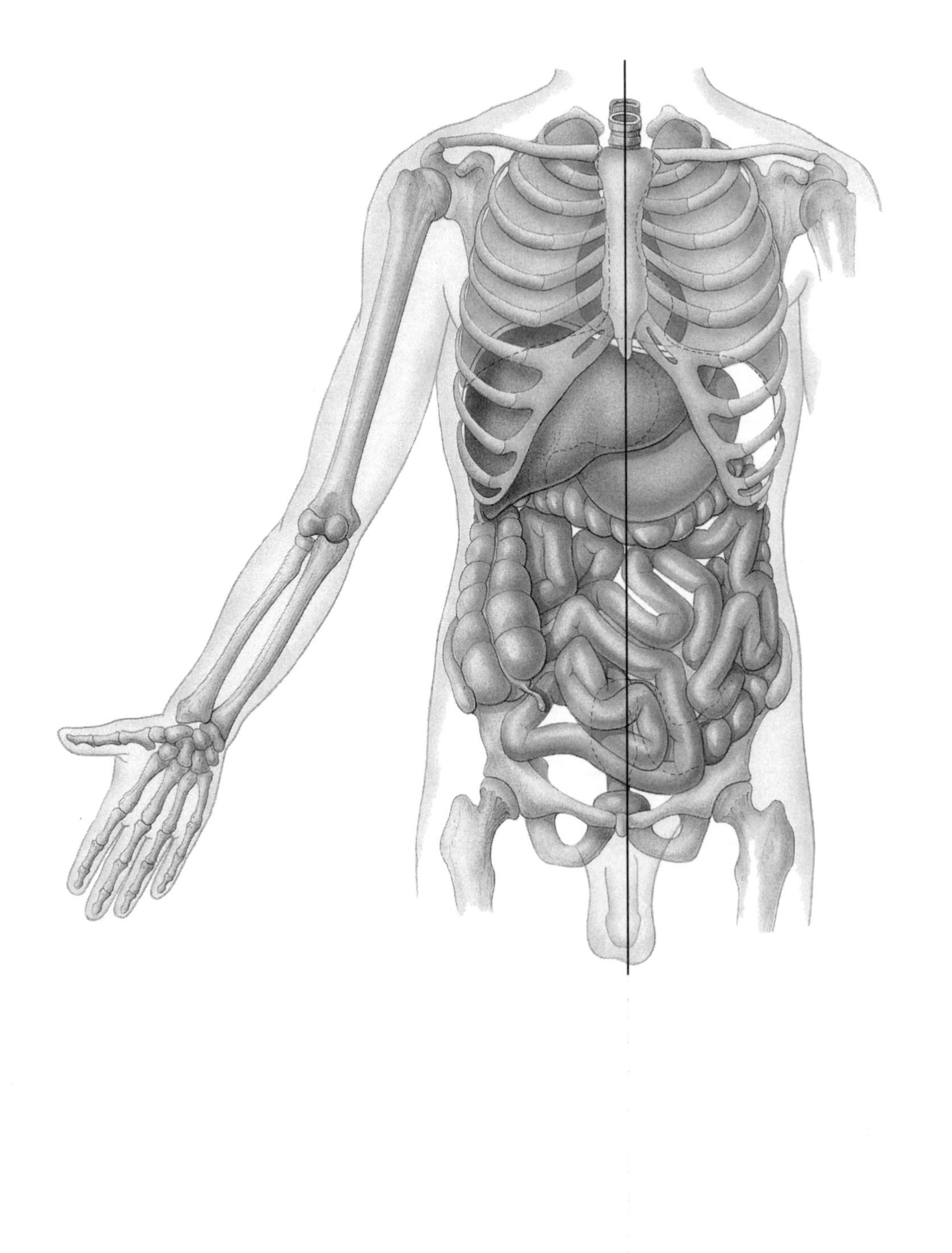

## NOTES

**Figure 1.6** Body cavities (page 12).

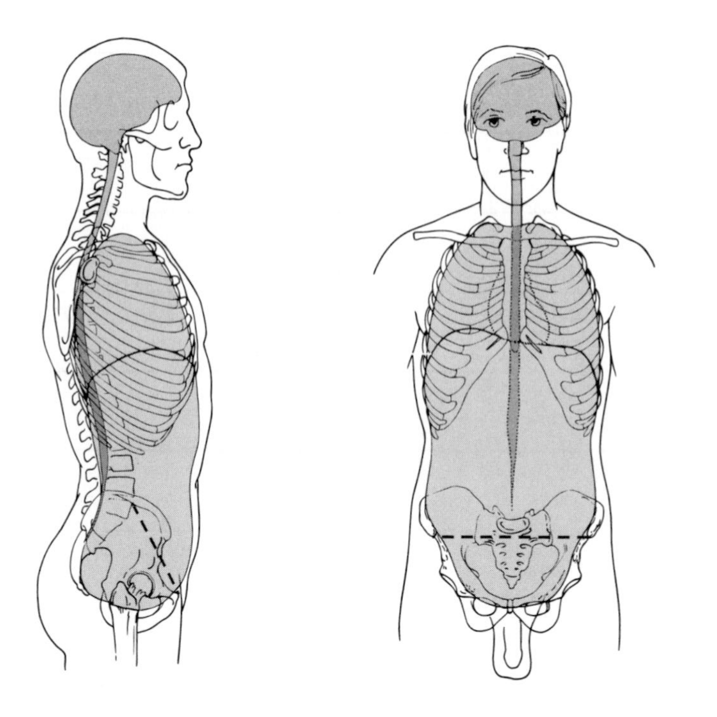

## NOTES

**Figure 1.7a,b** The thoracic cavity (page 13).

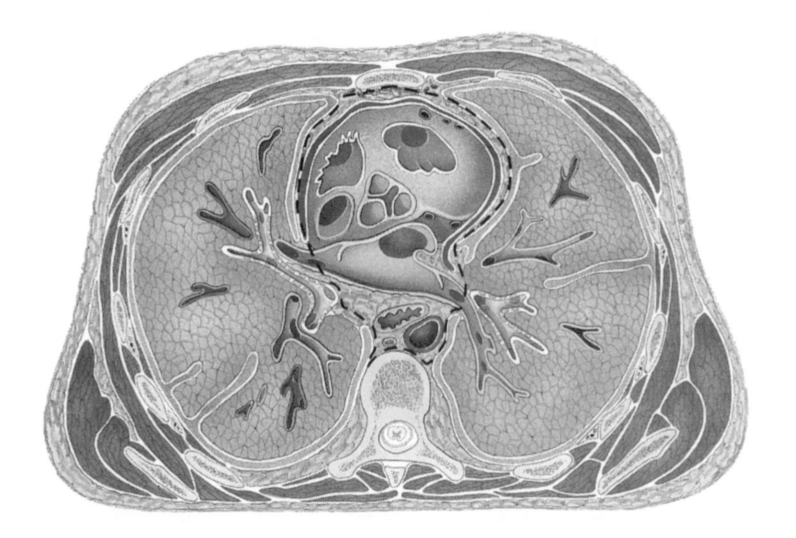

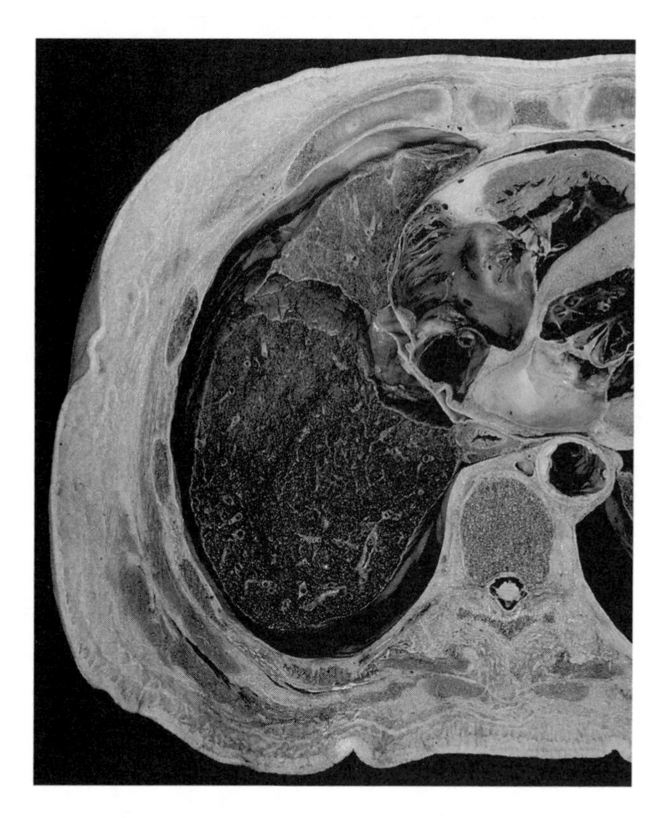

NOTES

**Figure 1.7c** The thoracic cavity (page 14).

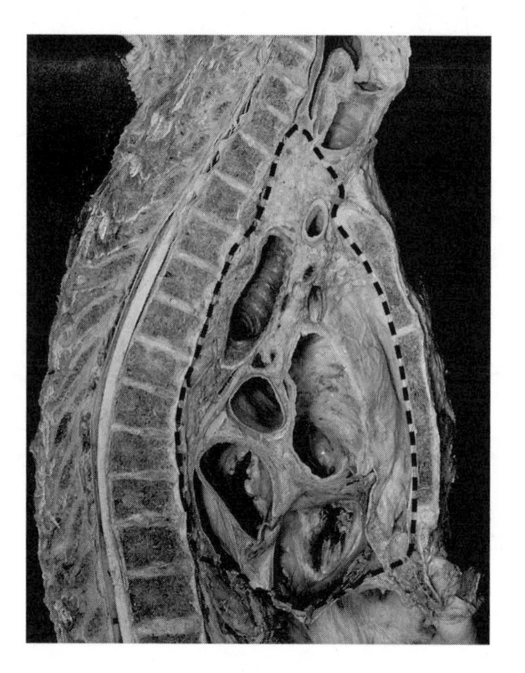

**Figure 1.8** The abdominopelvic cavity (page 15).

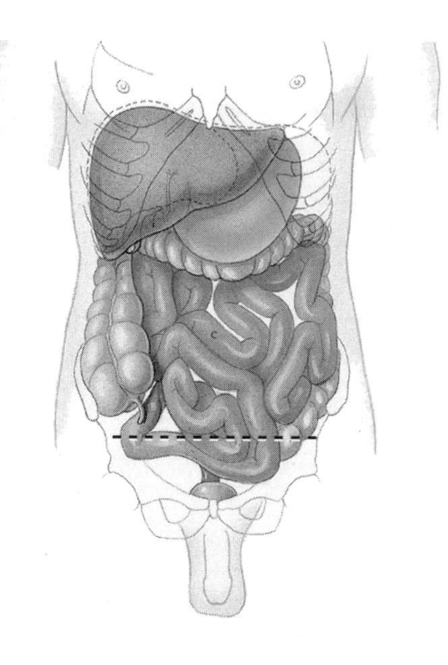

# NOTES

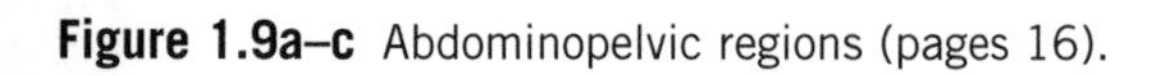
**Figure 1.9a–c** Abdominopelvic regions (pages 16).

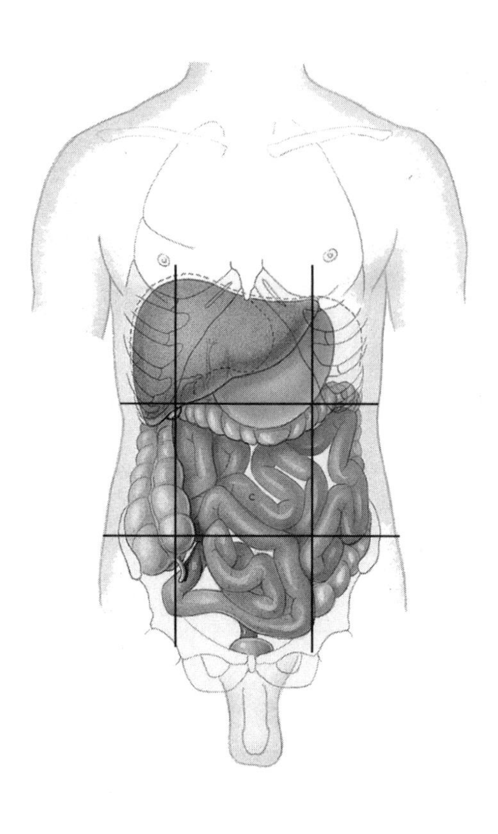

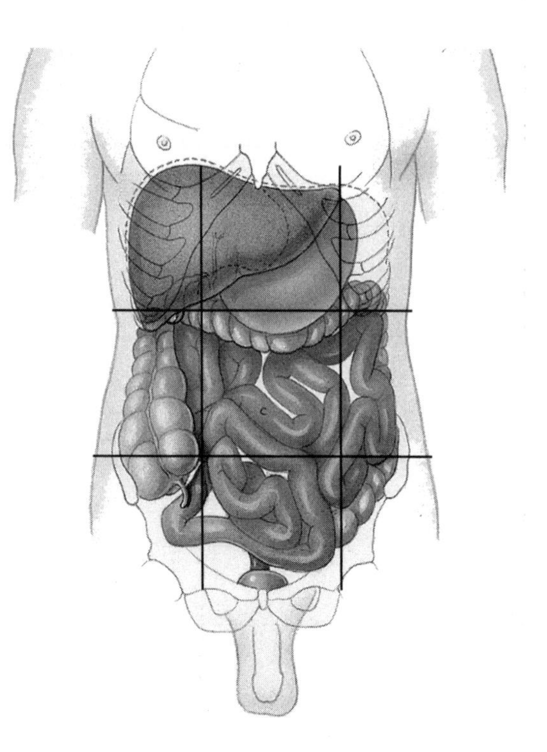

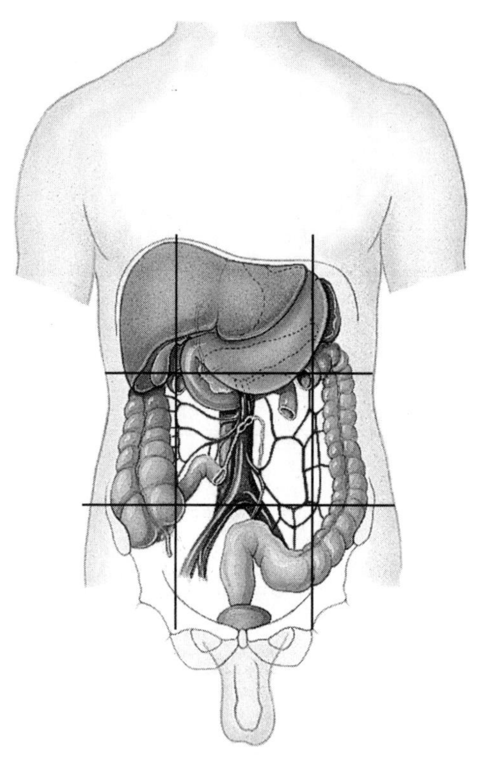

NOTES

**Figure 1.9d** Abdominopelvic regions (pages 17).

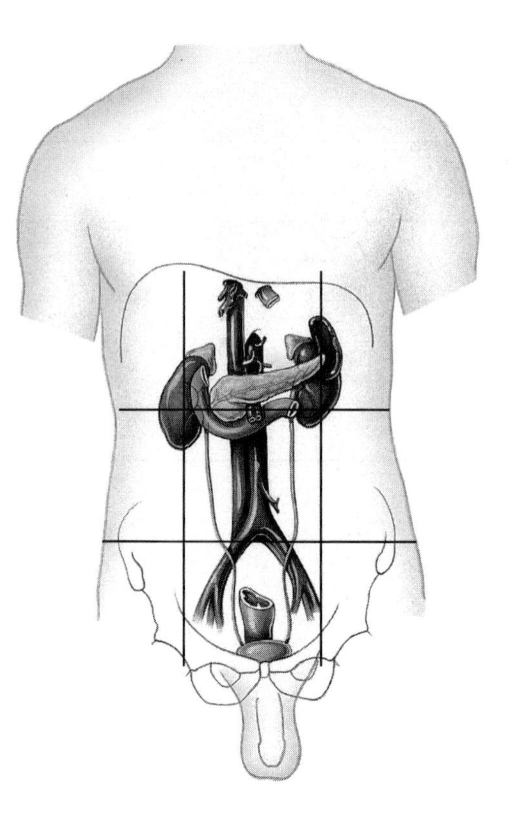

**Figure 1.10** Quadrants of the abdominopelvic cavity (page 19).

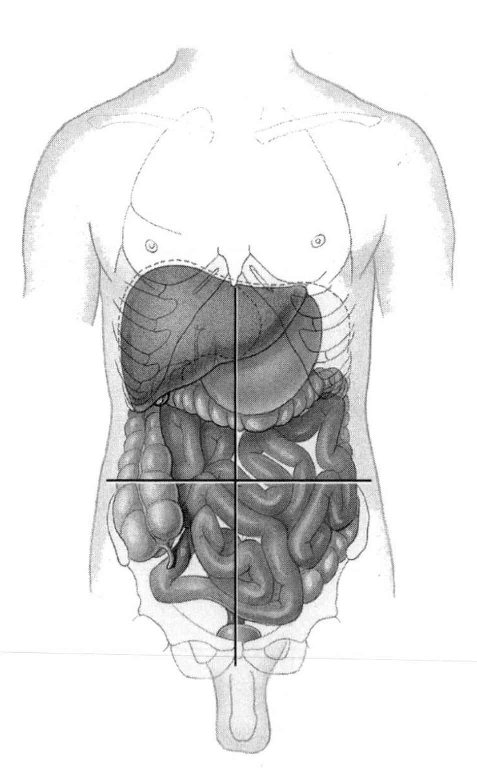

NOTES

**Figure 2.1** Generalized body cell (page 26).

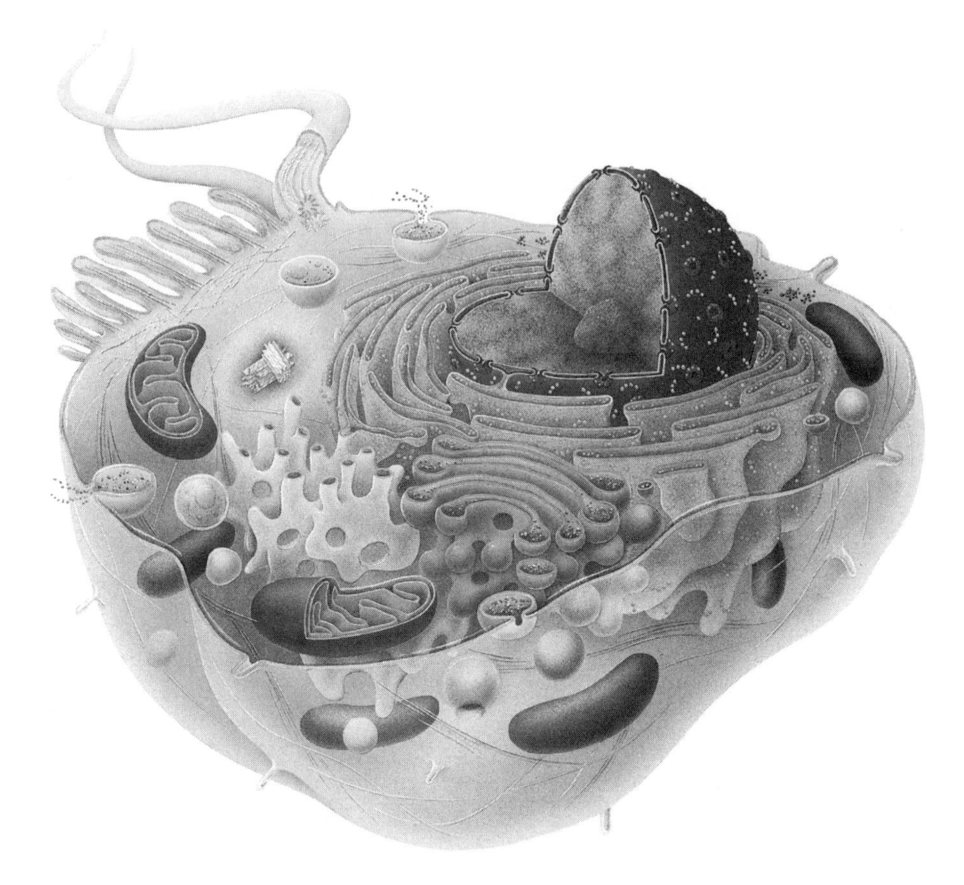

**Figure 2.2** Structure of the plasma membrane (page 27).

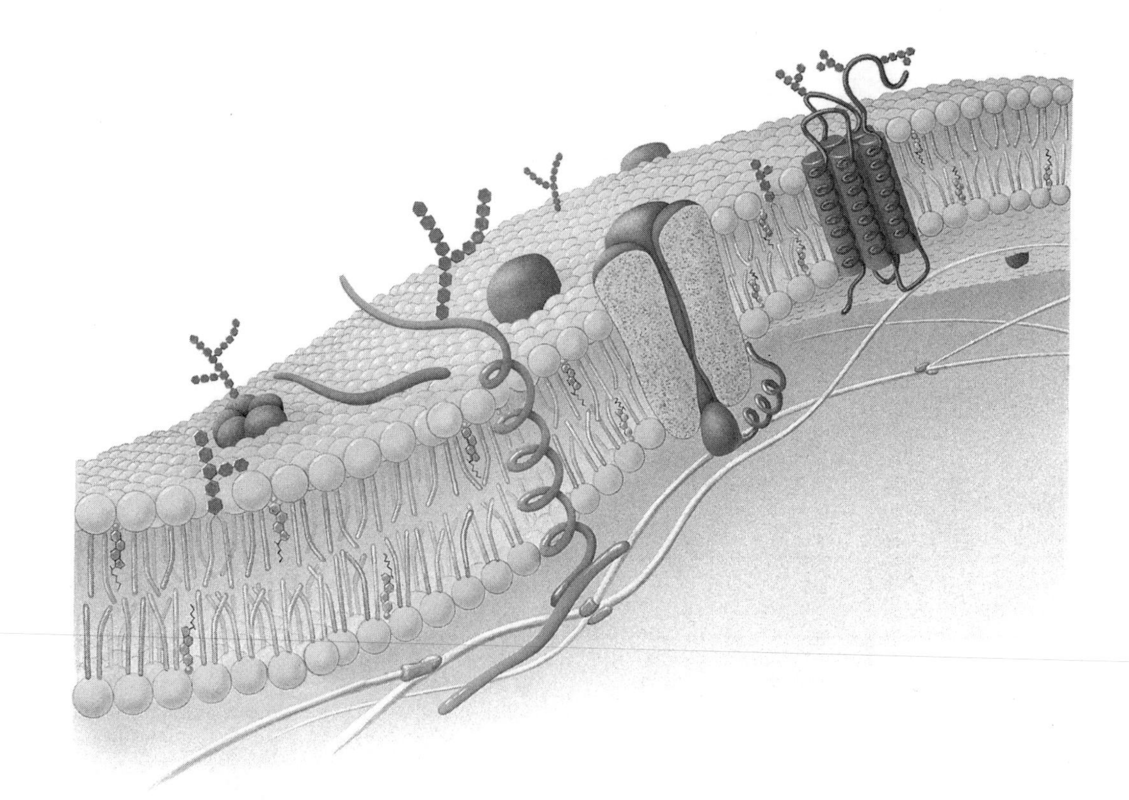

NOTES

**Figure 2.3** Body fluids (page 29).

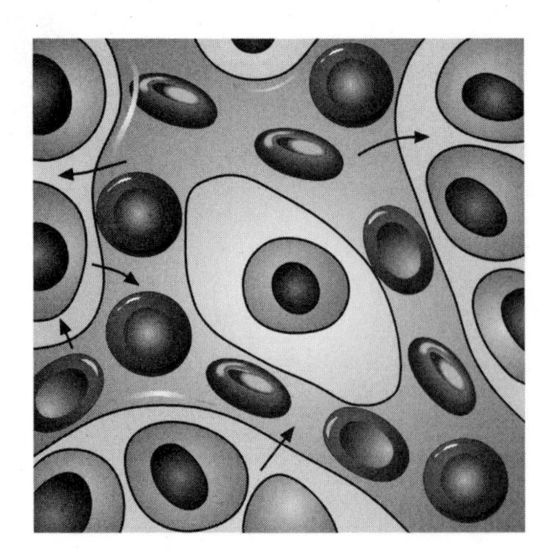

**Figure 2.4** Receptor-mediated endocytosis (page 30).

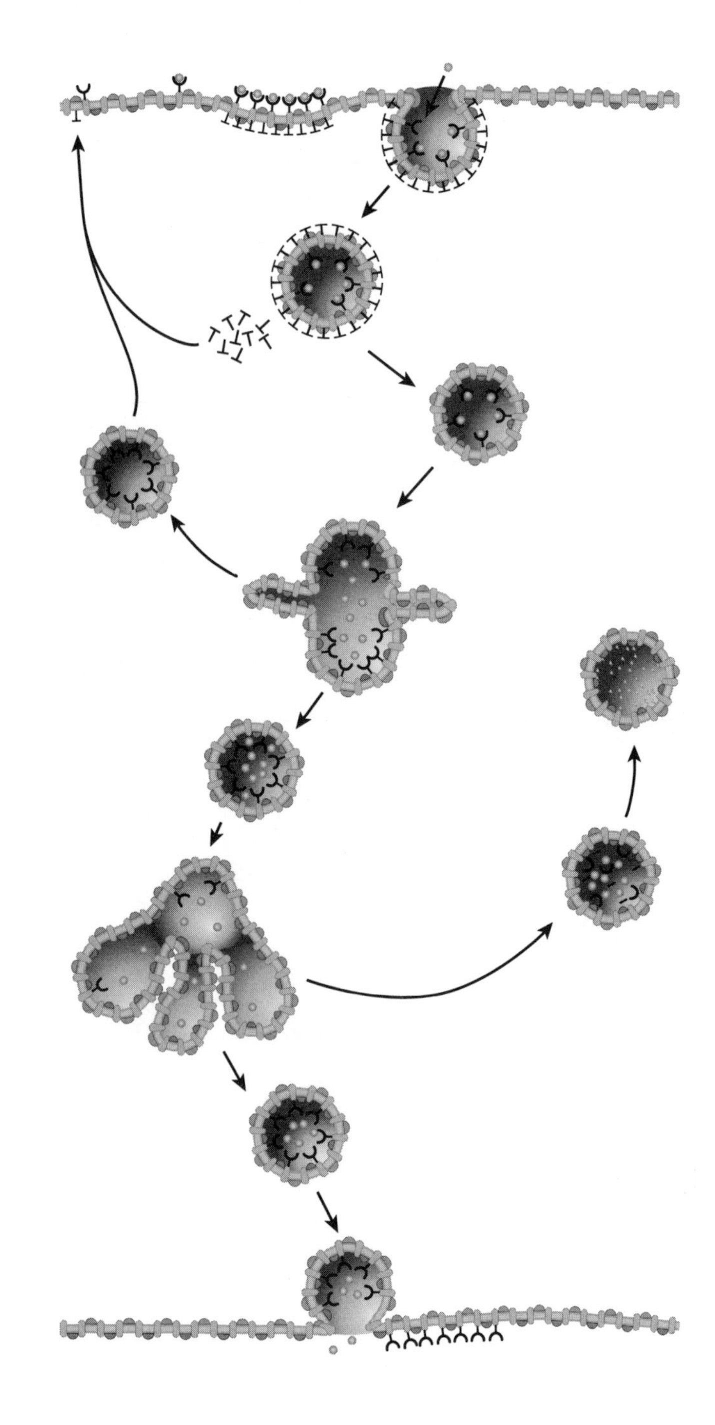

NOTES

**Figure 2.5a** Phagocytosis (page 31).

NOTES

2

**Figure 2.6** Cytoskeleton (page 33).

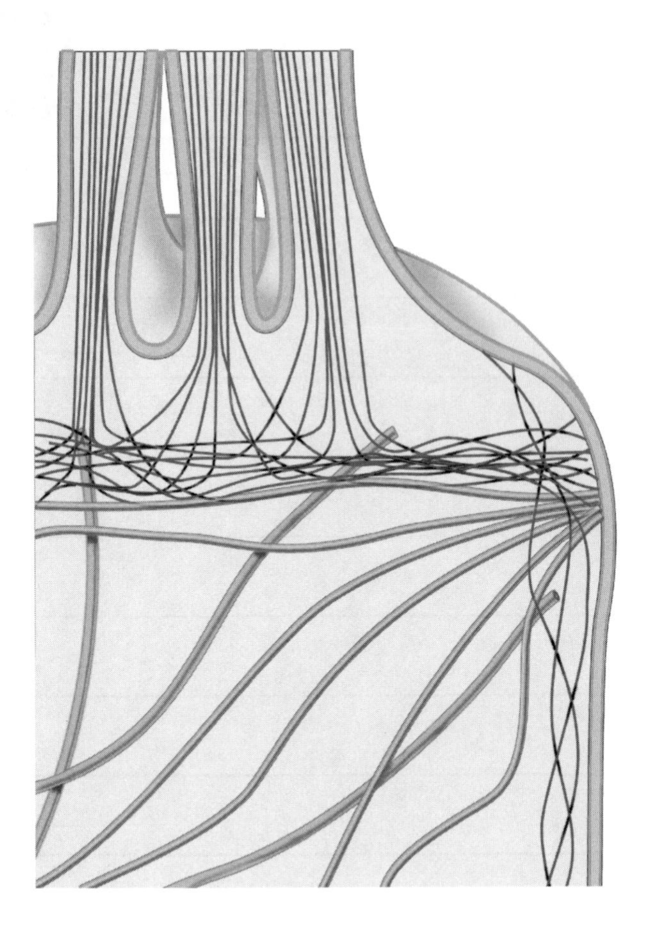

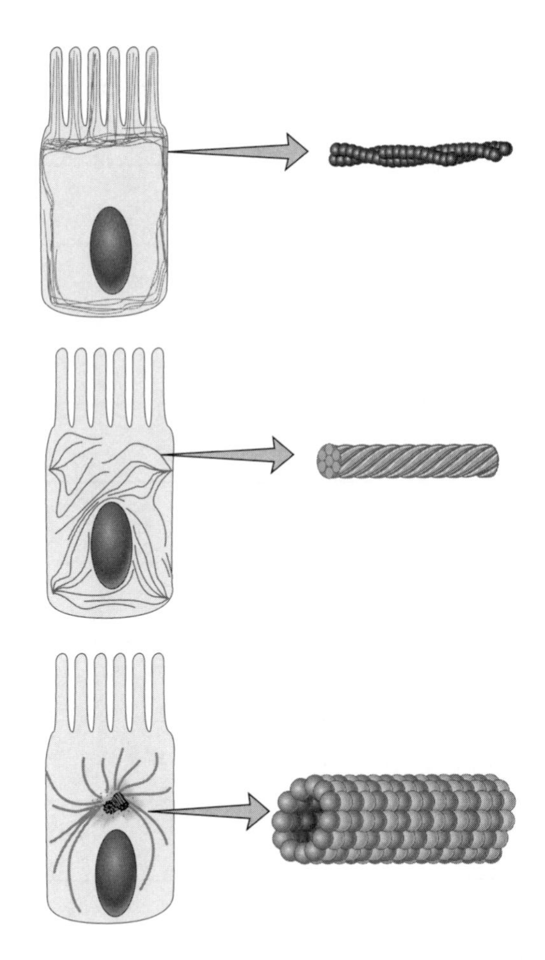

**Figure 2.7a** Centrosome (page 34).

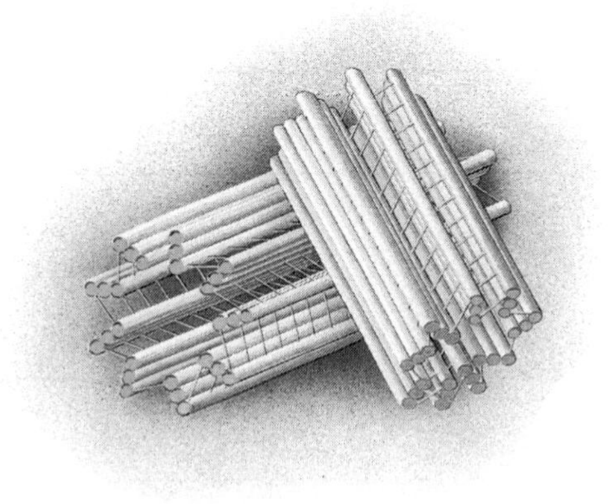

NOTES

2

**Figure 2.8** Cilia and flagella (page 35).

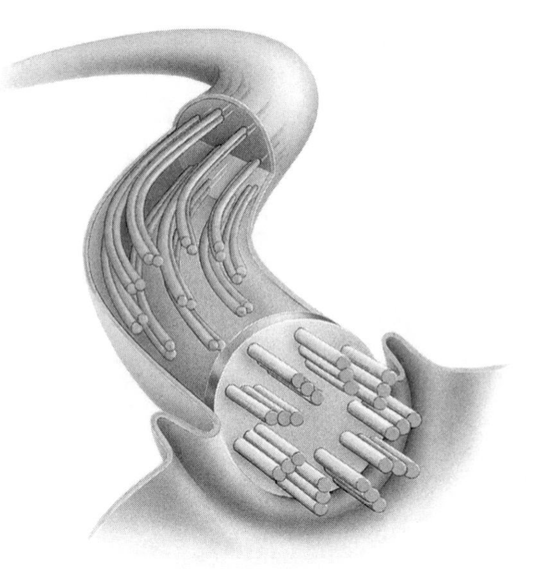

**Figure 2.9** Ribosomes (page 36).

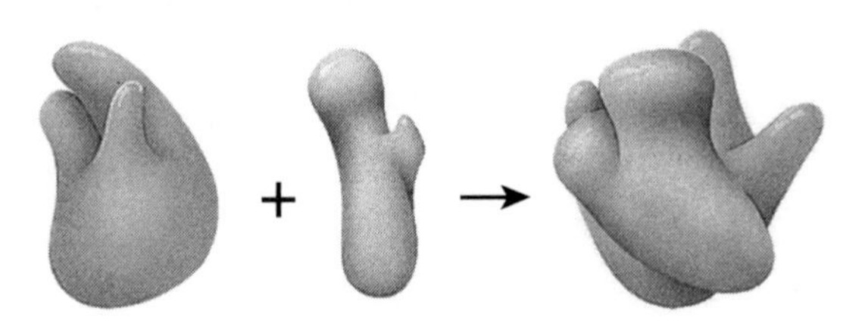

NOTES

**Figure 2.10a** Endoplasmic reticulum (page 36).

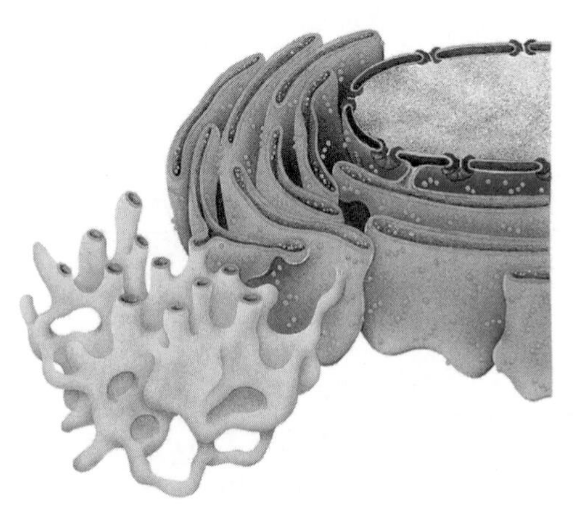

**Figure 2.11a** Golgi complex (page 37).

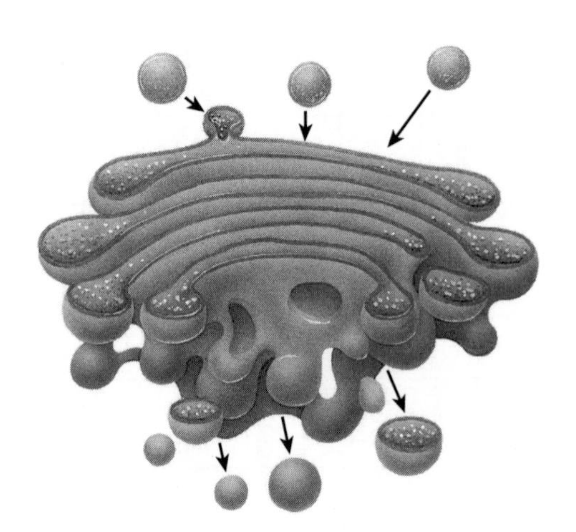

NOTES

2

**Figure 2.12** Packaging of synthesized proteins by the Golgi complex (page 38).

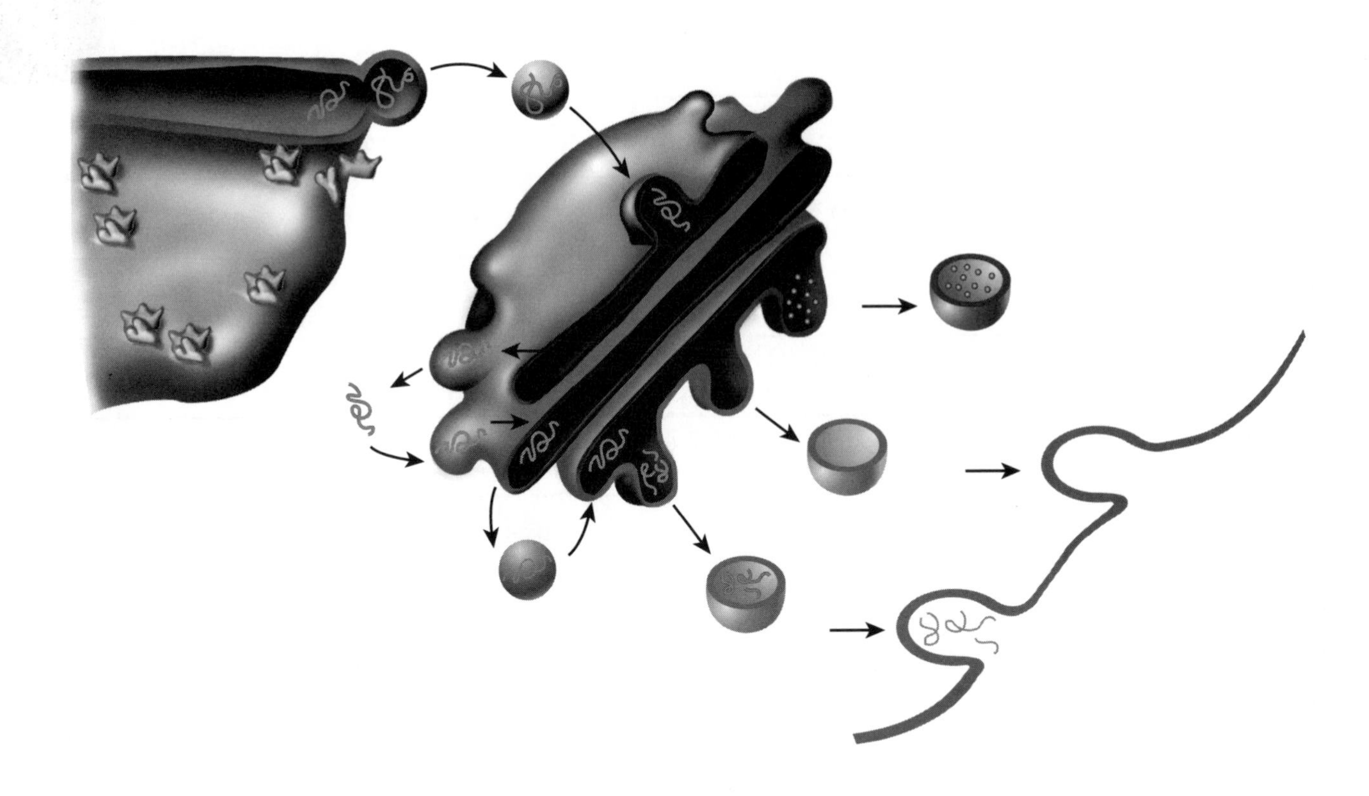

**Figure 2.13a** Lysosomes (page 39).

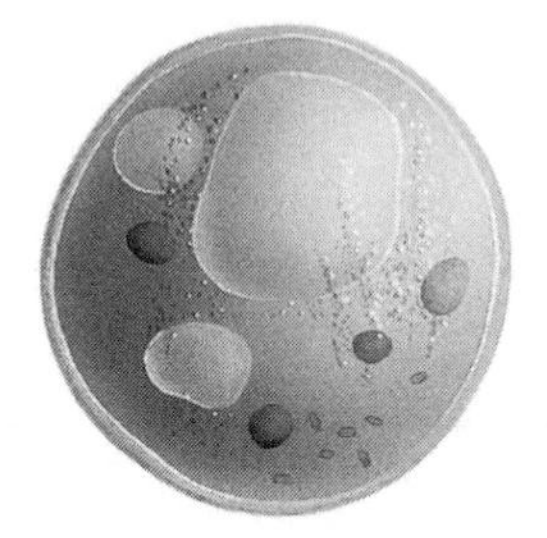

NOTES

**Figure 2.14a** Mitochondria (page 40).

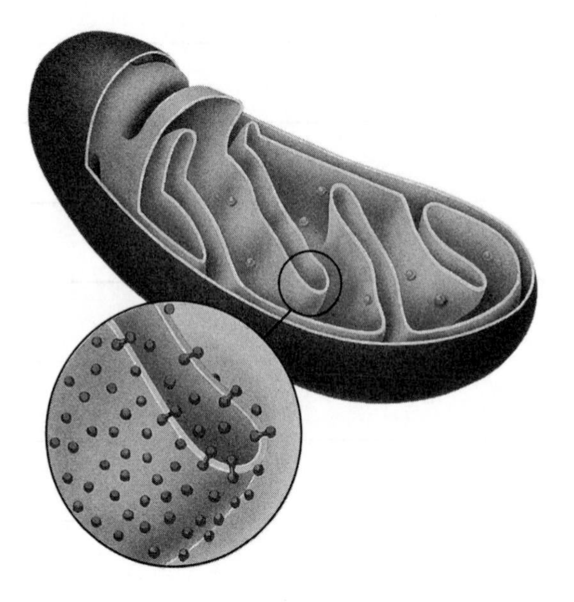

**Figure 2.15a** Nucleus (page 41).

NOTES

**Figure 2.16** Packing of the DNA into a chromosome in a dividing cell (page 42).

NOTES

**Figure 2.17** The cell cycle for a typical body cell (page 44).

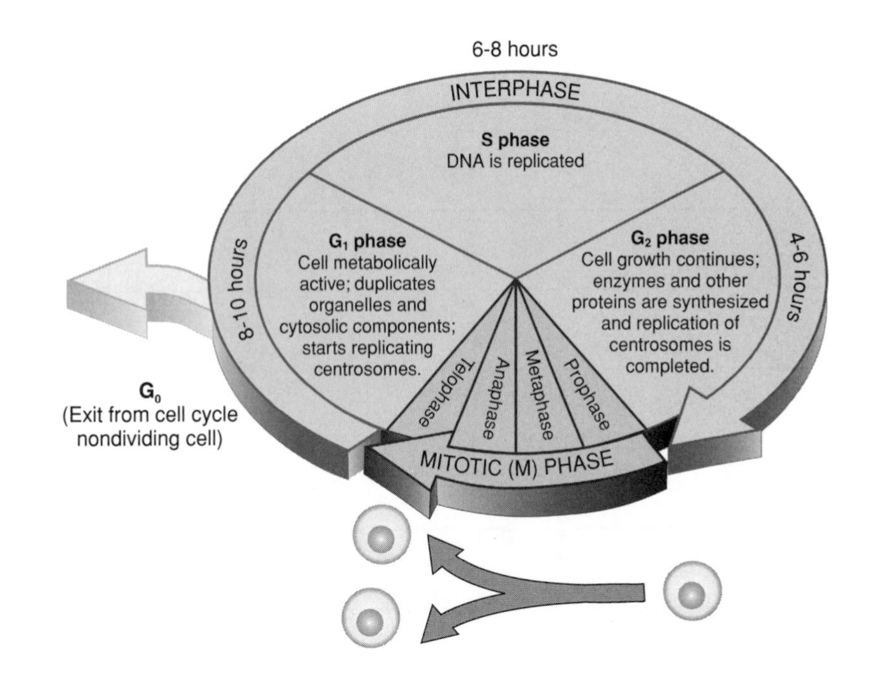


**Figure 2.18** Replication of DNA (page 44).

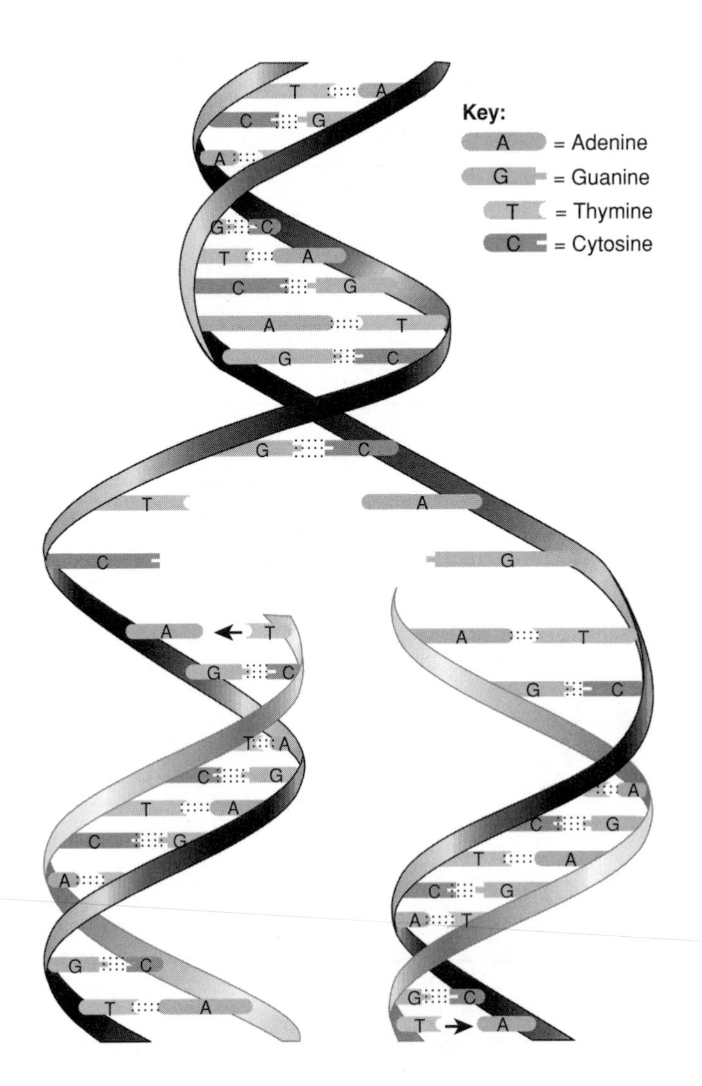

NOTES

**Figure 2.19** Cell division: mitosis and cytokinesis (page 46).

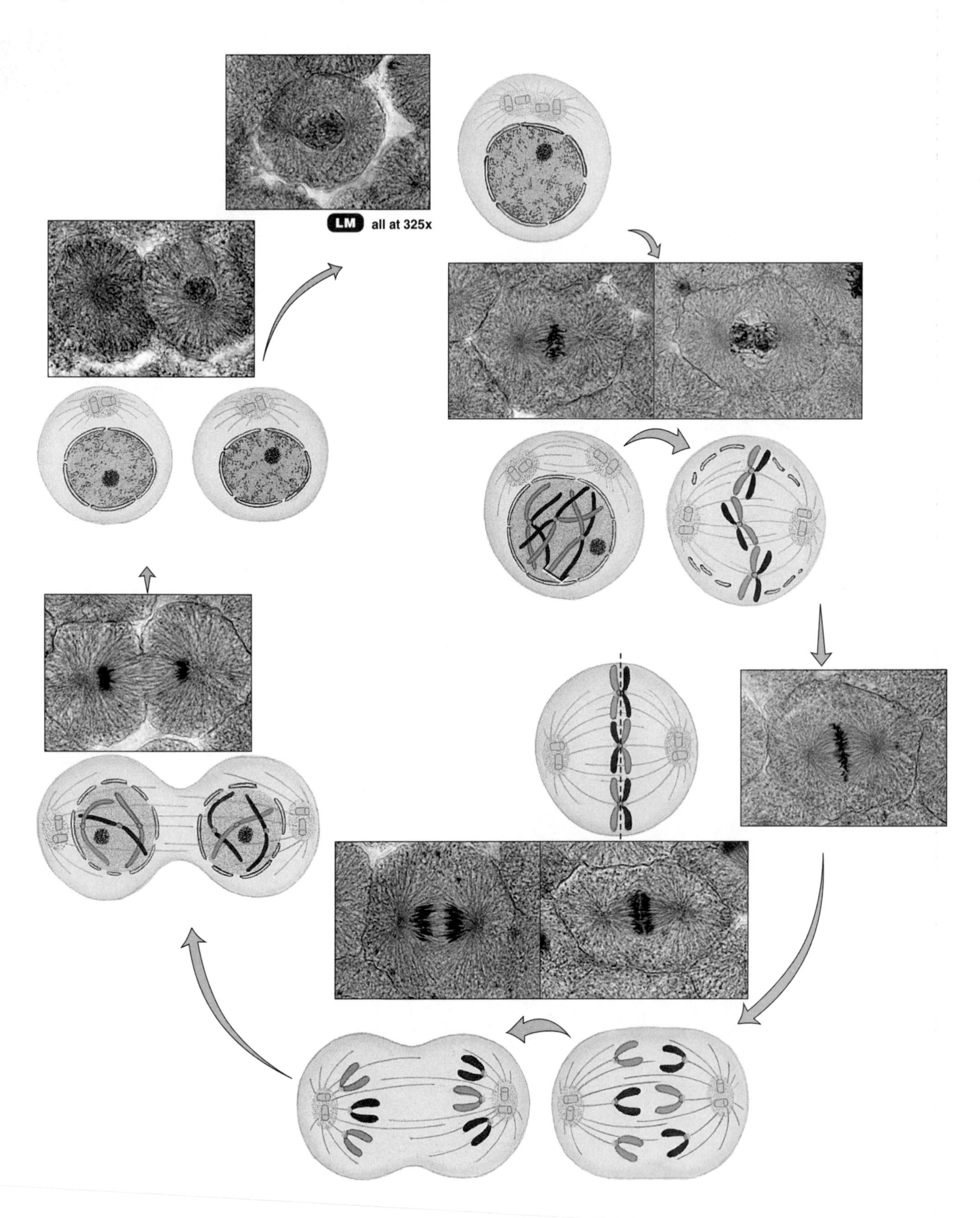

NOTES

**Figure 2.20** Diverse shapes and sizes of human cells (page 49).

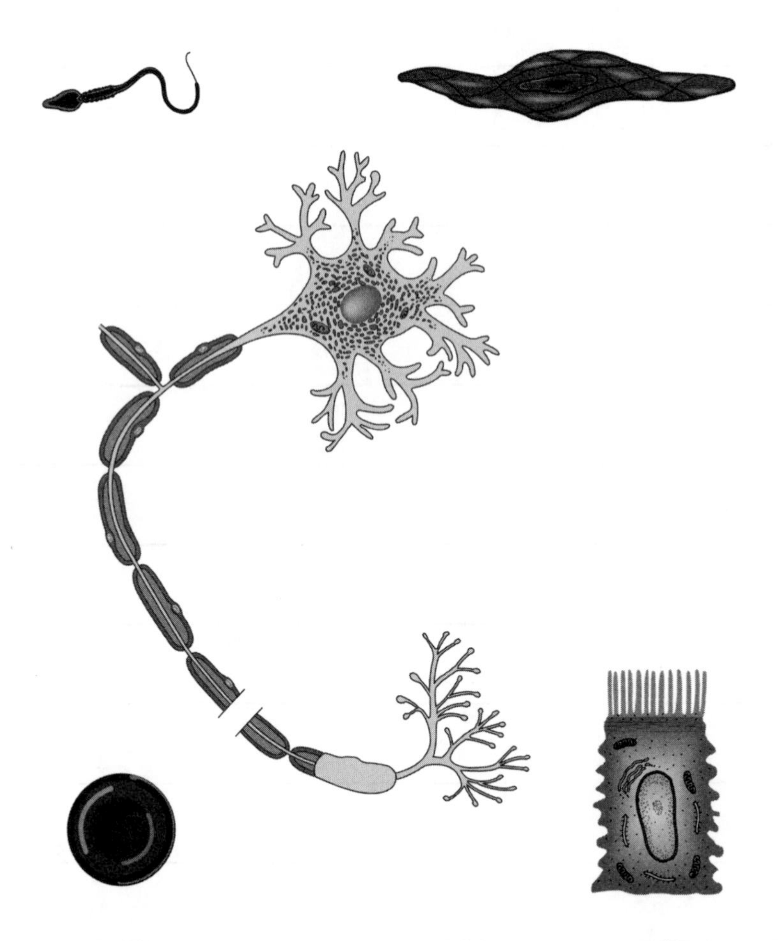

NOTES

2

**Figure 3.1** Cell junctions (page 57).

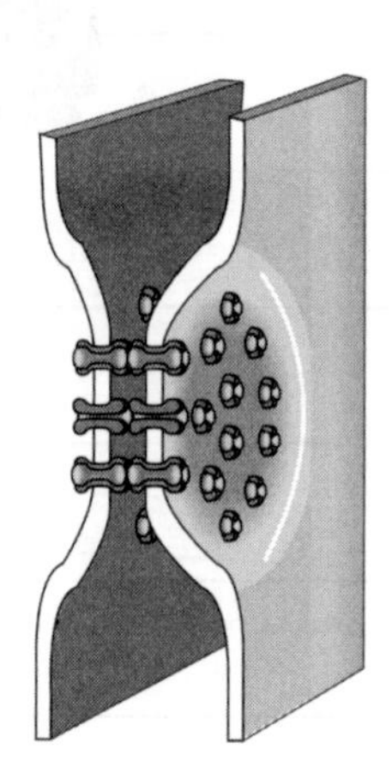

(f) Gap junction

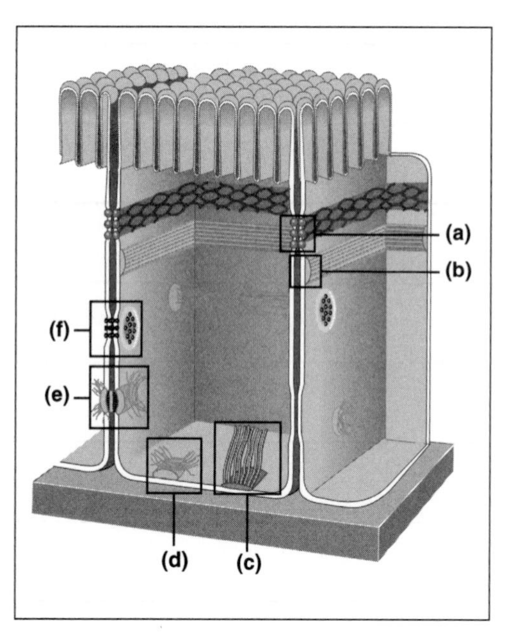


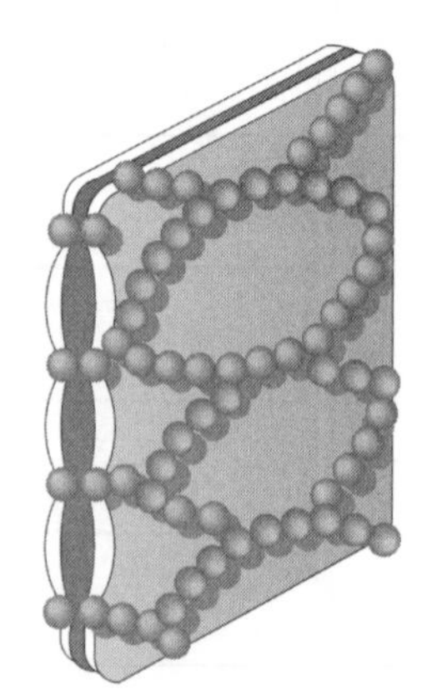

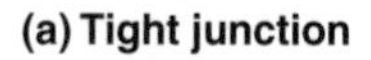

(a) Tight junction

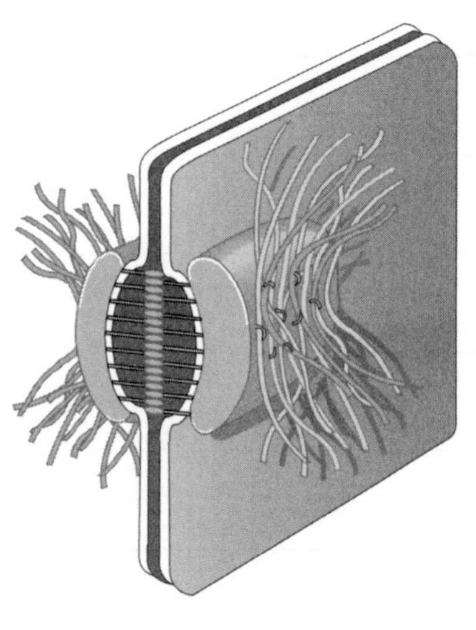

(e) Desmosome

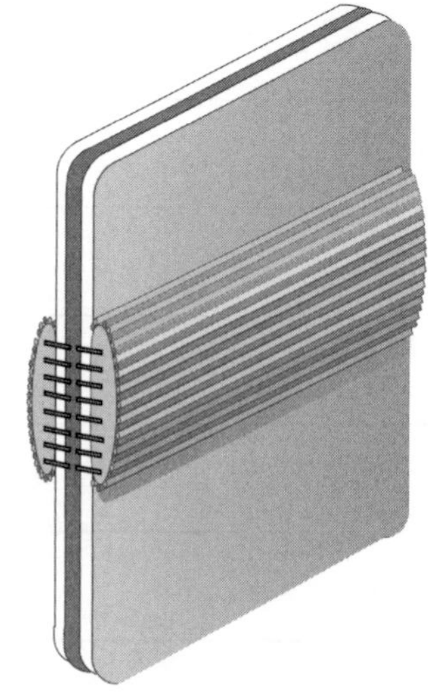

(b) Cell–to–cell adherens junction

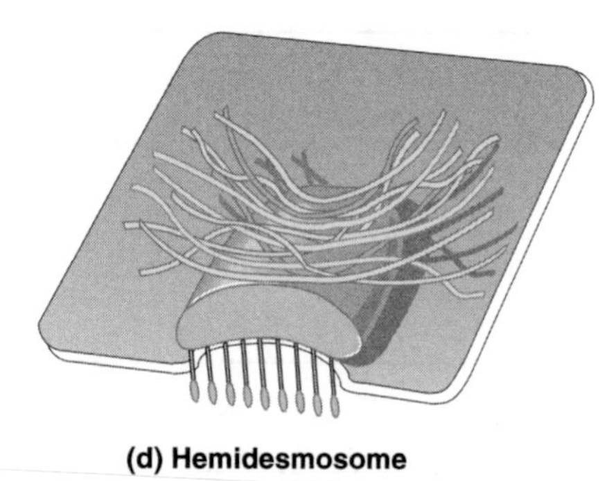

(d) Hemidesmosome

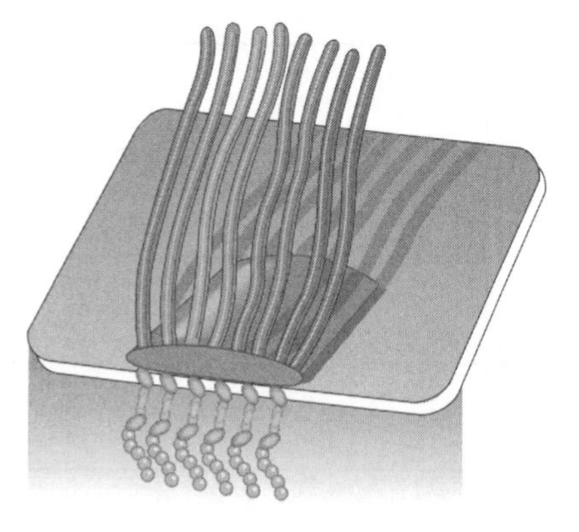

(c) Cell–to–extracellular material adherens junction (focal adhesion)

NOTES

3

**Figure 3.2** Surfaces of epithelial cells and the structure and location of the basement membrane (page 59).

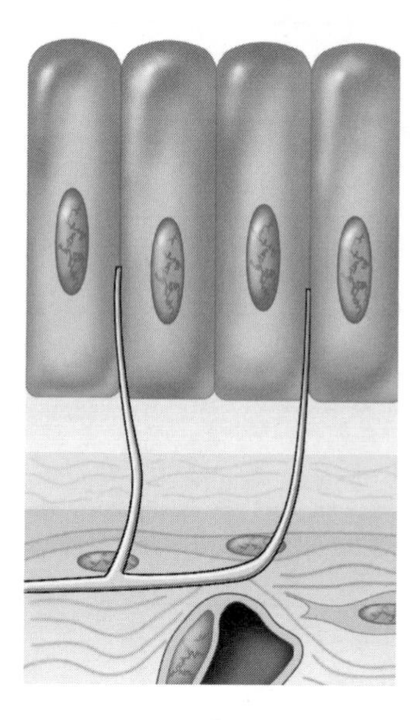

**Table 3.1a** Simple squamous epithelium (page 60).

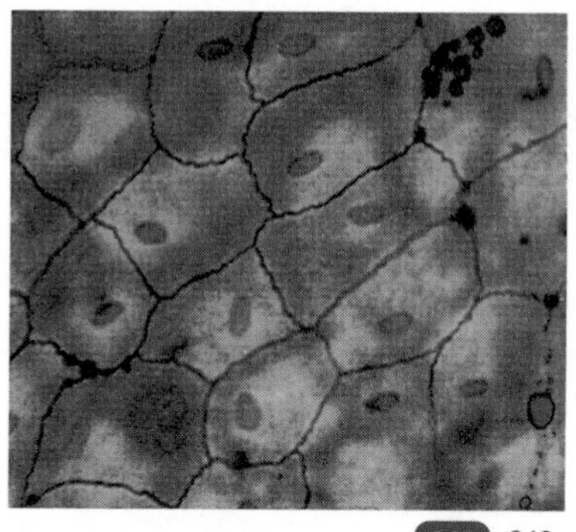


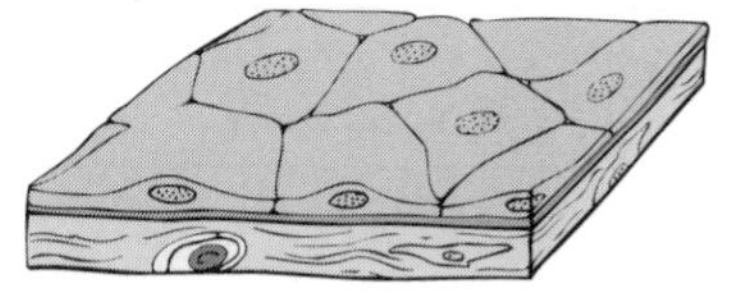

NOTES

3

**Table 3.1b** Simple cuboidal epithelium (page 61).

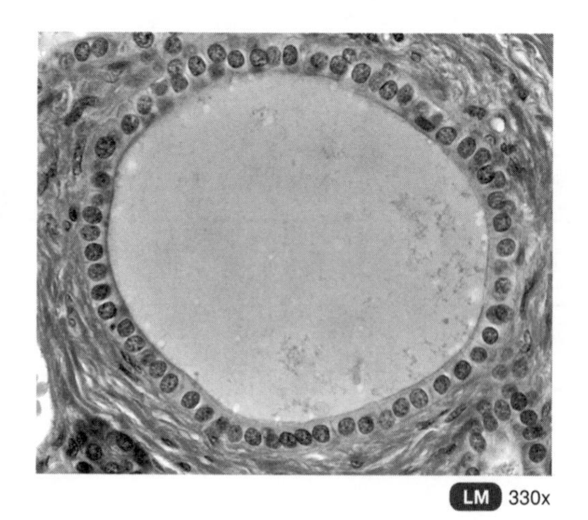


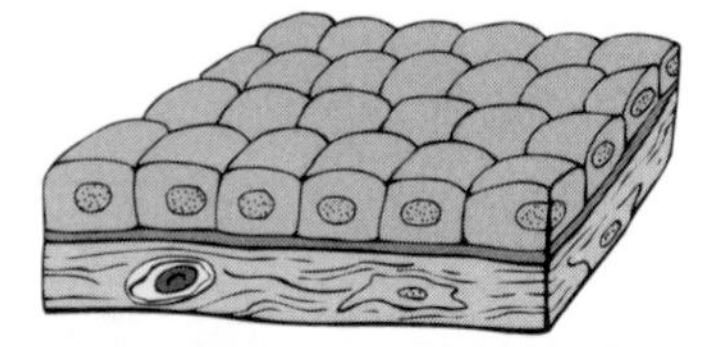

**Table 3.1c** Nonciliated simple columnar epithelium (page 62).

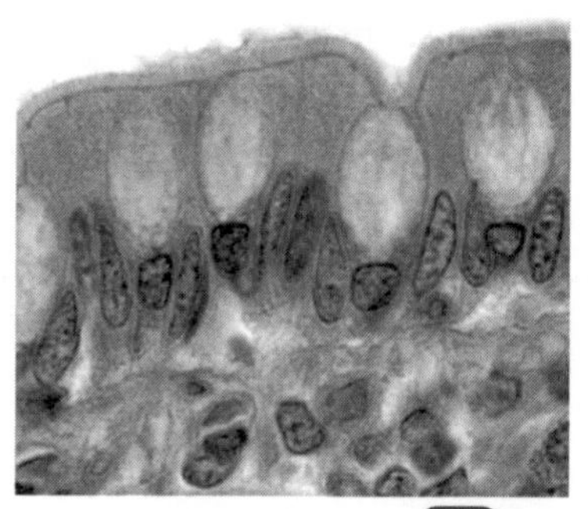


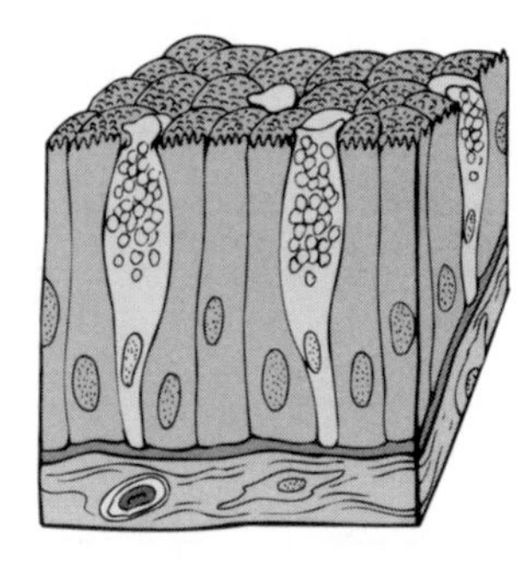

NOTES

3

**Table 3.1d** Ciliated simple columnar epithelium (page 62).

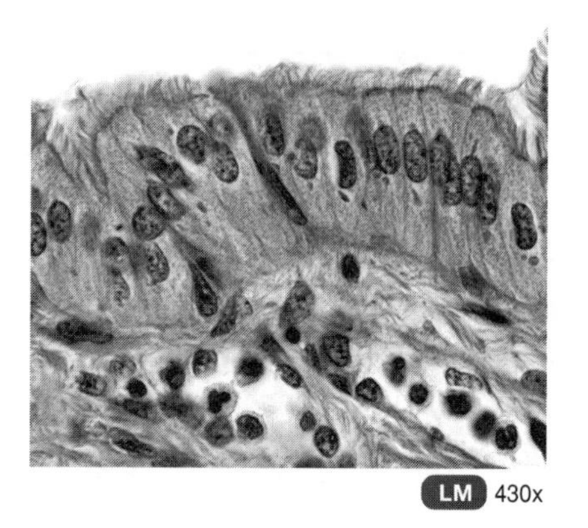


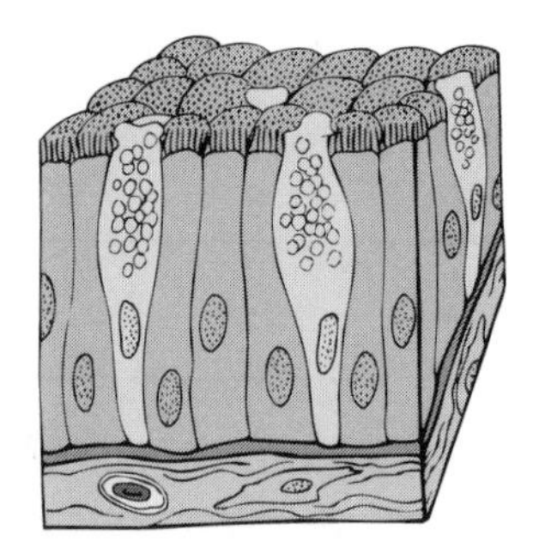

**Table 3.1e** Stratified squamous epithelium (page 63).

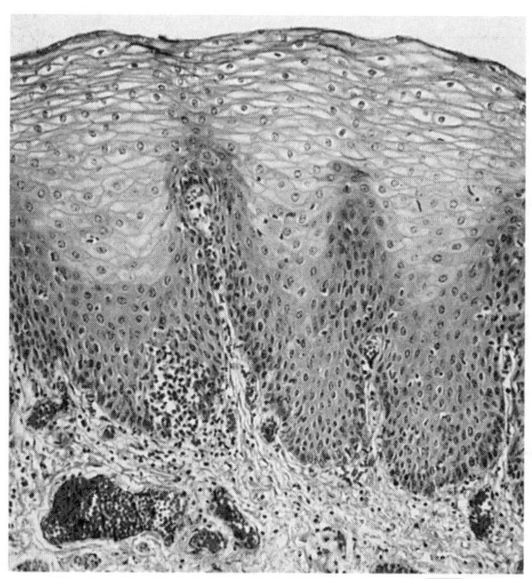


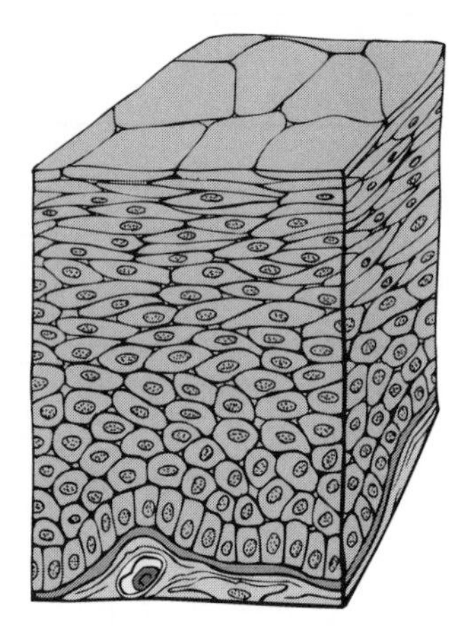

## NOTES

3

**Table 3.1f** Stratified cuboidal epithelium (page 63).

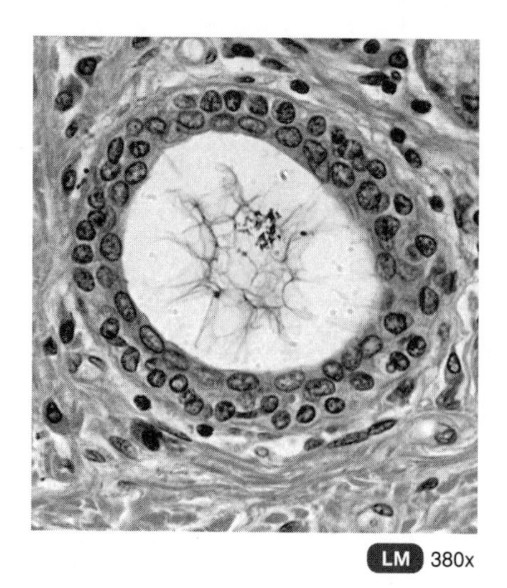


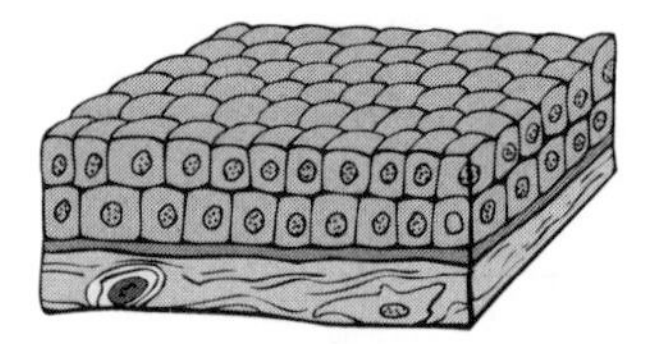

**Table 3.1g** Stratified columnar epithelium (page 64).

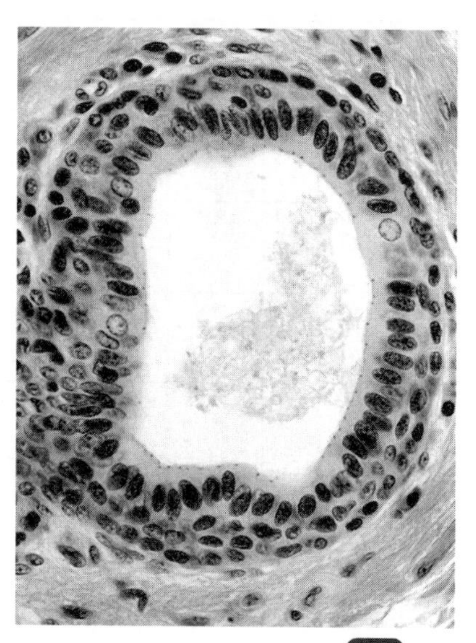


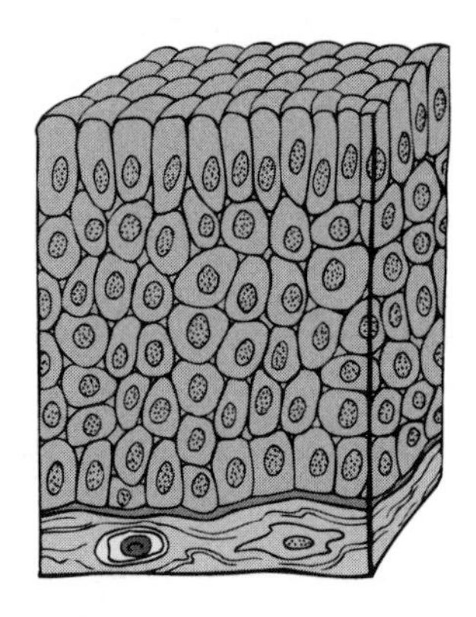

NOTES

3

**Table 3.1h** Transitional epithelium (page 64).

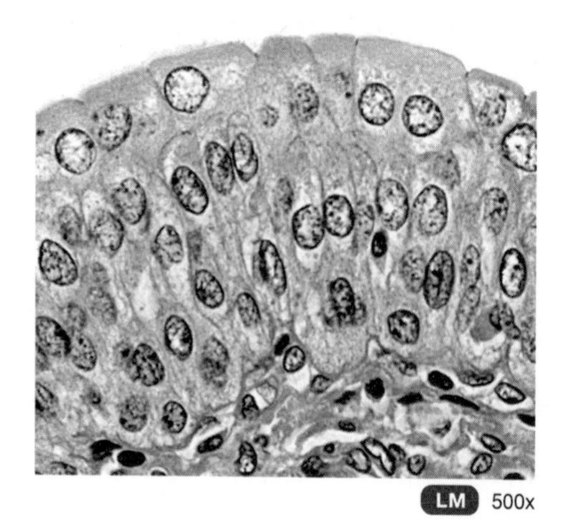


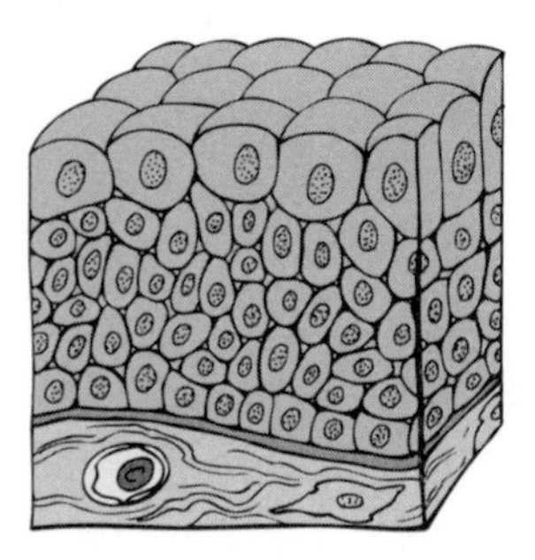

**Table 3.1i** Pseudostratified columnar epithelium (page 65).

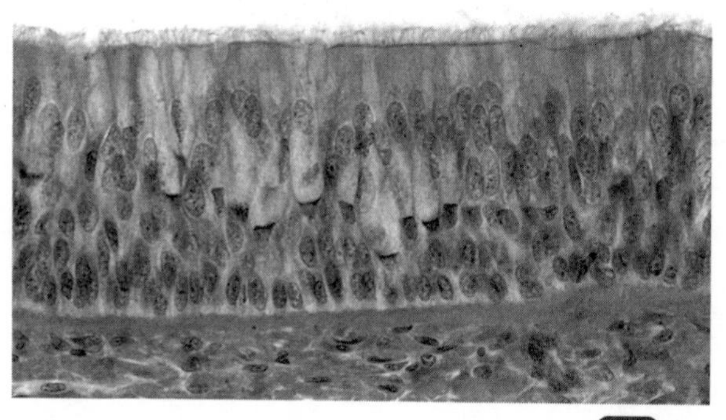


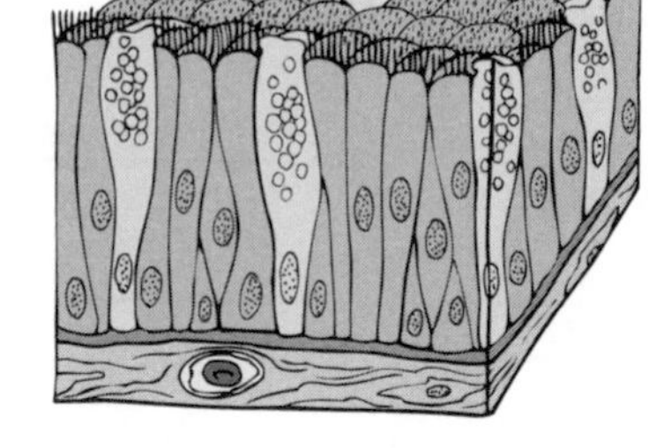

NOTES

3

**Table 3.1j** Endocrine glands (page 66).

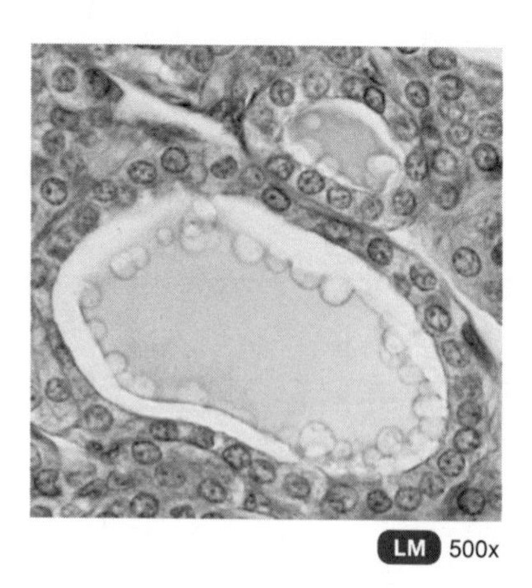


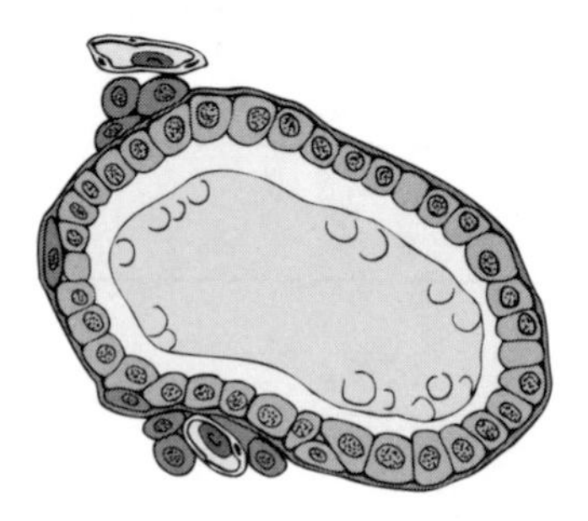

**Table 3.1k** Exocrine glands (page 66).

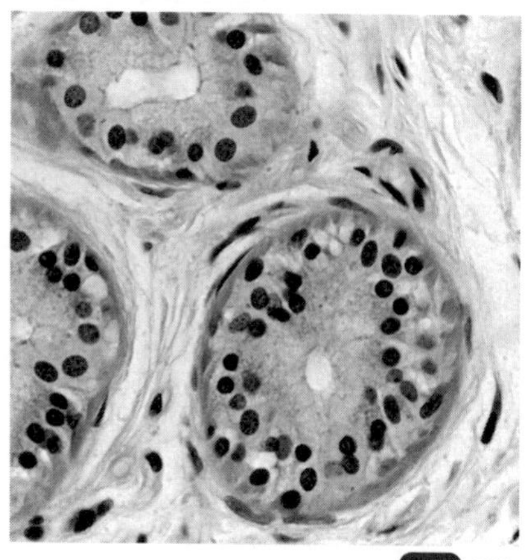


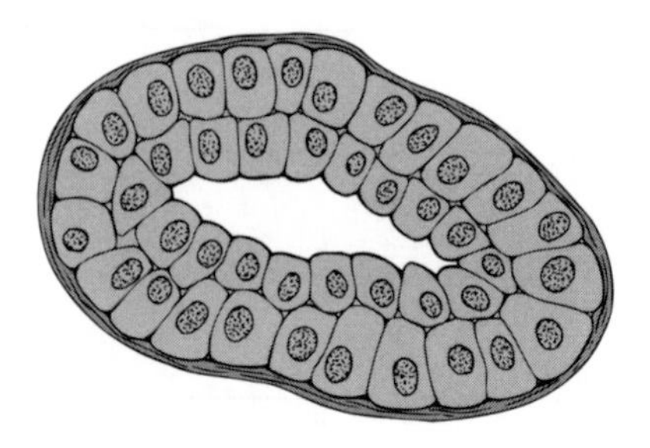

NOTES

3

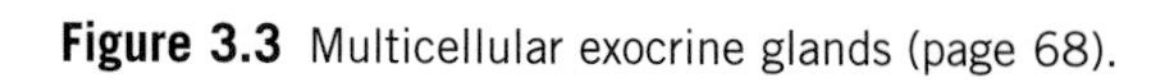

**Figure 3.3** Multicellular exocrine glands (page 68).

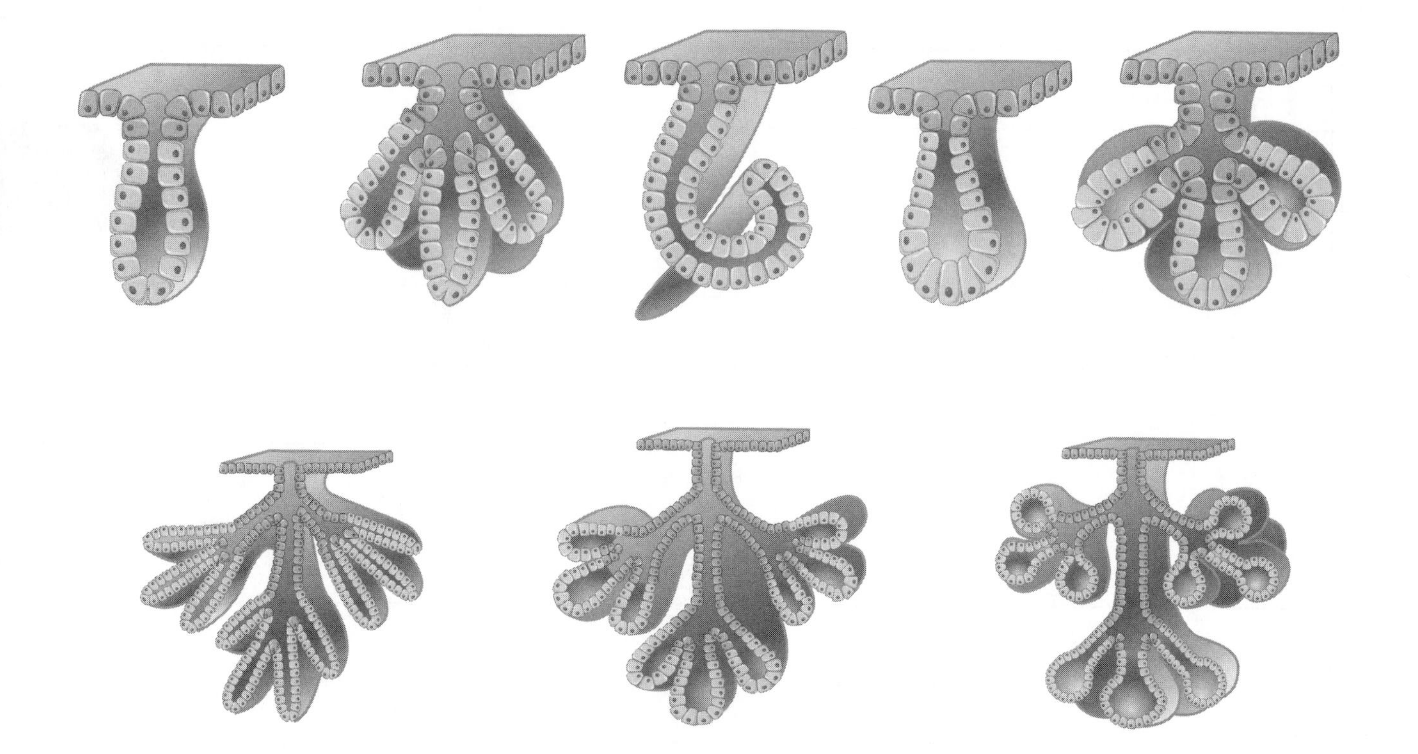

NOTES

3

**Figure 3.4** Functional classification of multicellular exocrine glands (page 69).

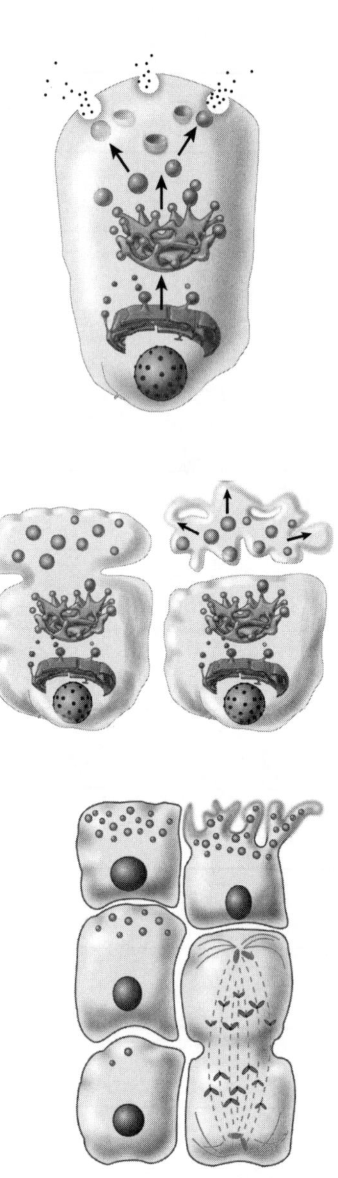

NOTES

**Figure 3.5** Representative cells and fibers present in connective tissues (page 70).

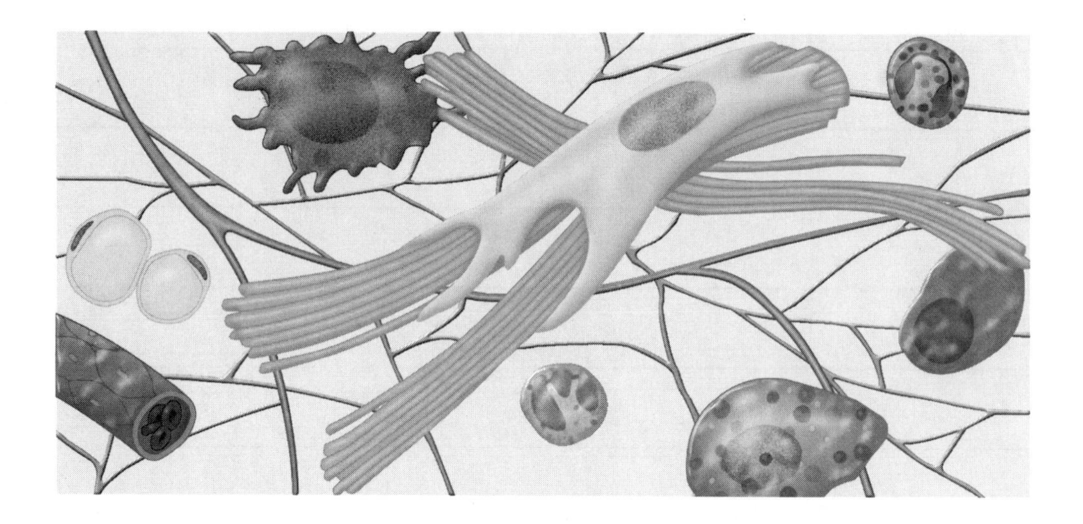

**Table 3.2a** Mesenchyme (page 72).

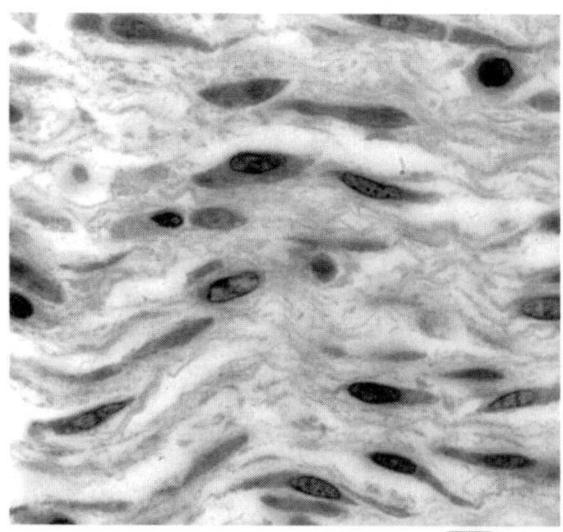


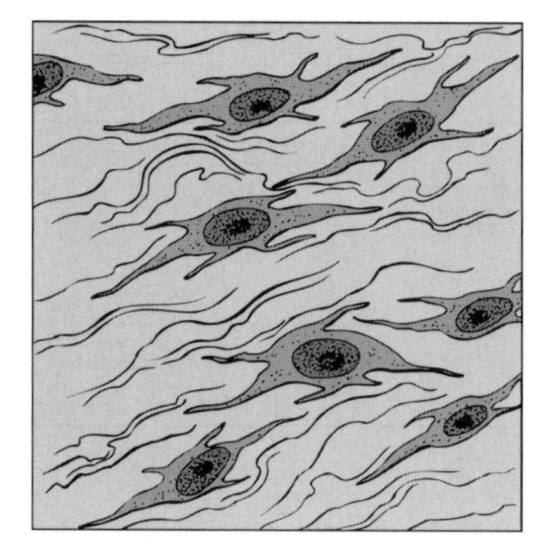

NOTES

3

**Table 3.2b** Mucus connective tissue (page 72).

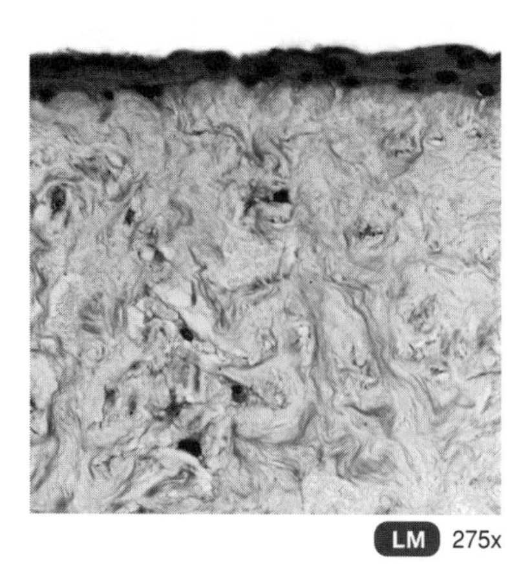


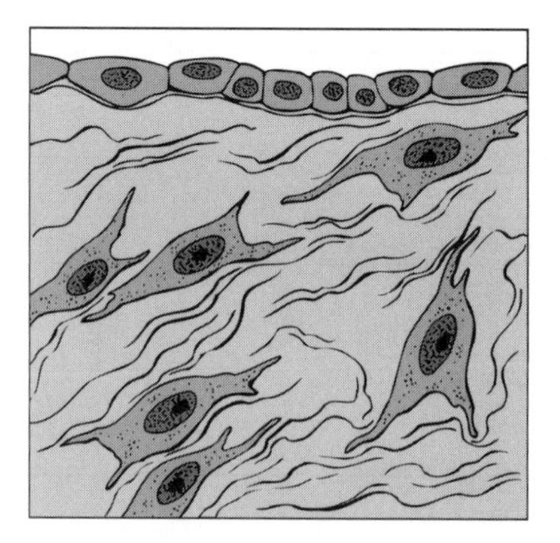

**Table 3.3a** Areolar connective tissue (page 73).

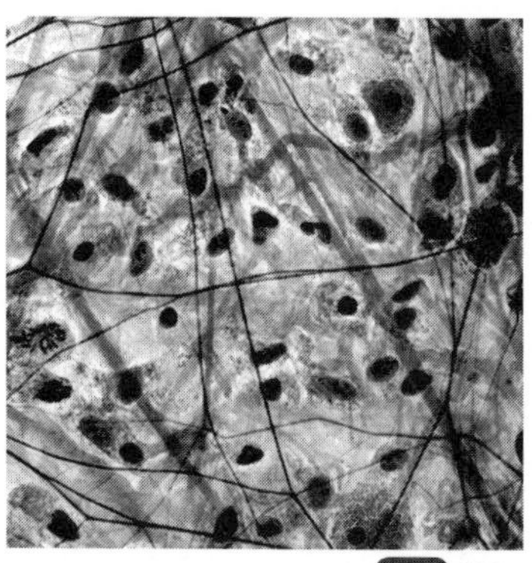


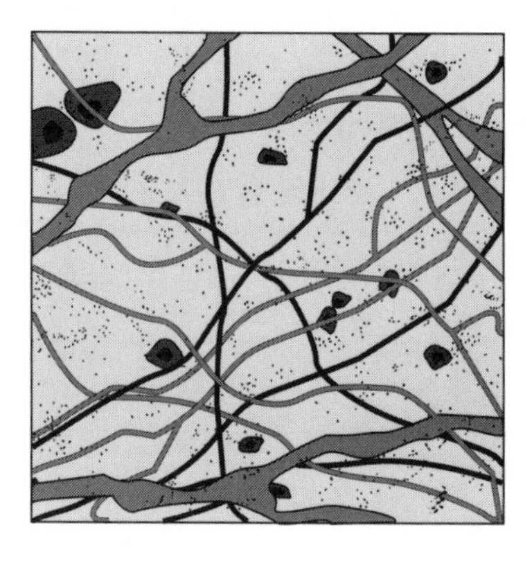

NOTES

**Table 3.3b** Adipose tissue (page 73).

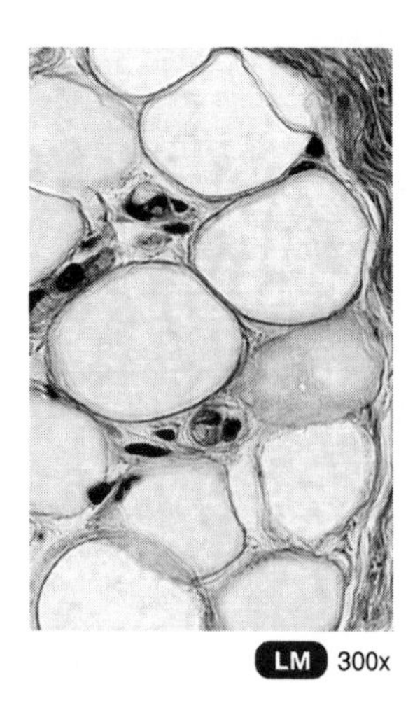


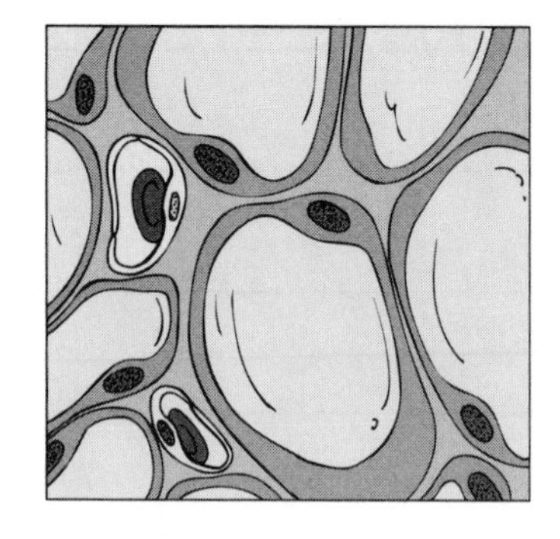

**Table 3.3c** Reticular connective tissue (page 74).

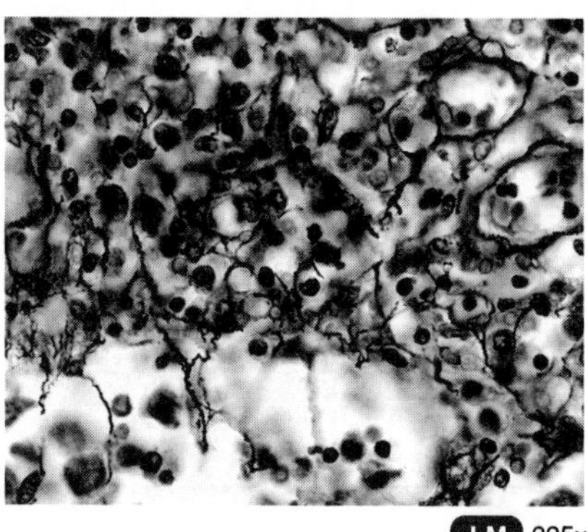


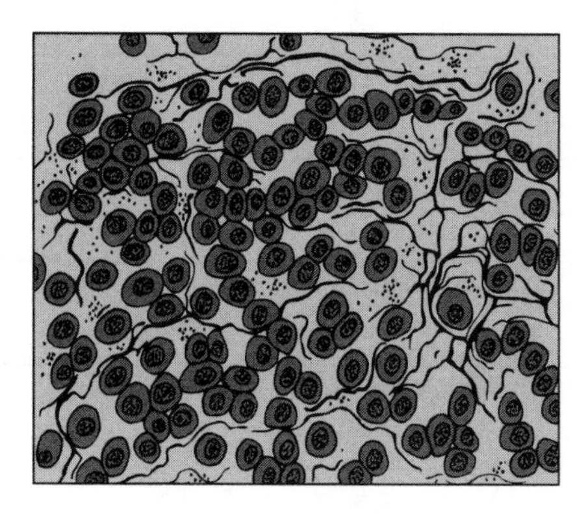

NOTES

3

**Table 3.3d** Dense regular connective tissue (page 75).

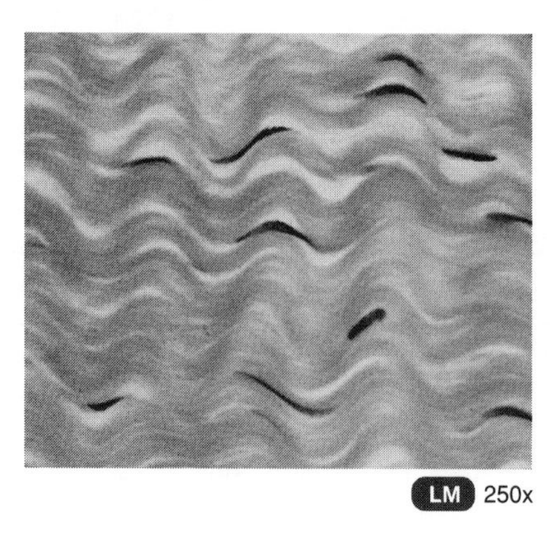

**Table 3.3e** Dense irregular connective tissue (page 76).

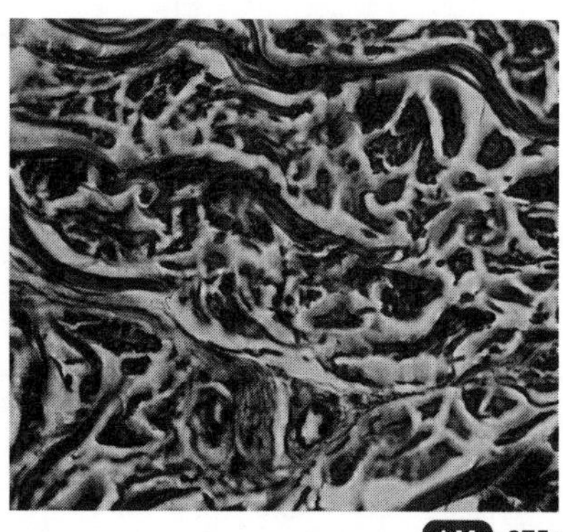


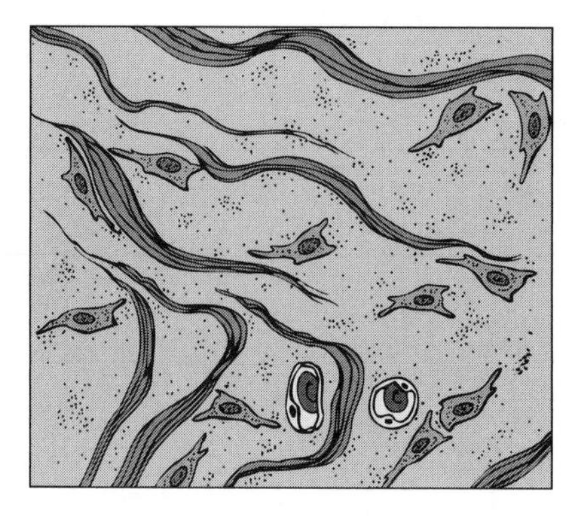

## NOTES

3

**Table 3.3f** Elastic connective tissue (page 76).

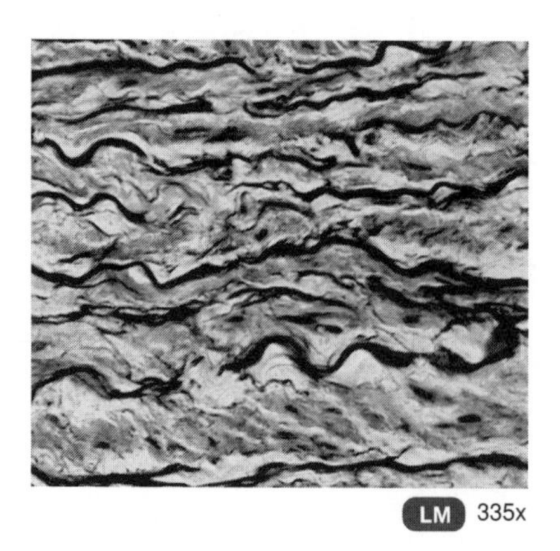


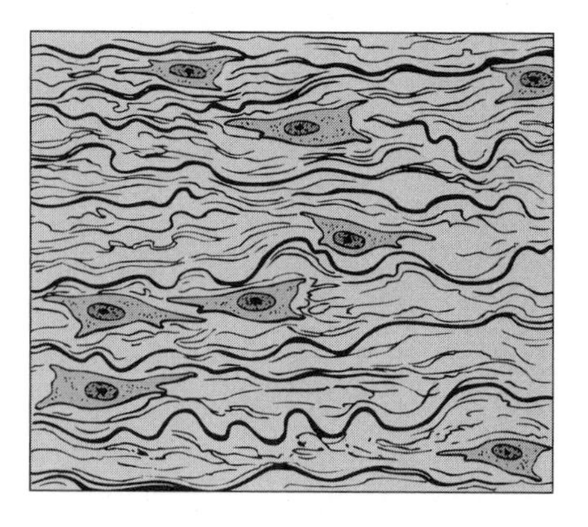

**Table 3.3g** Hyaline cartilage (page 77).

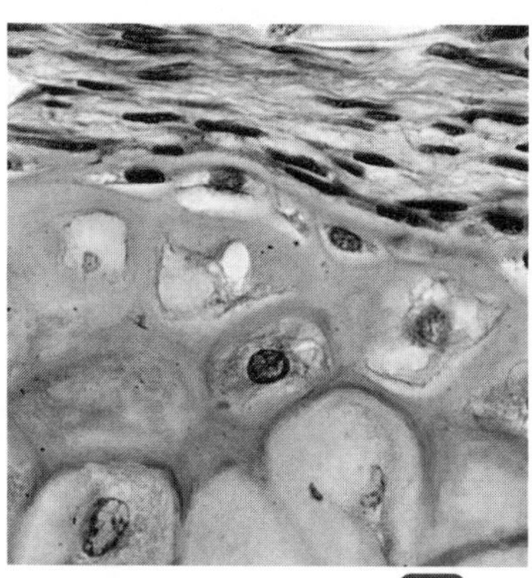


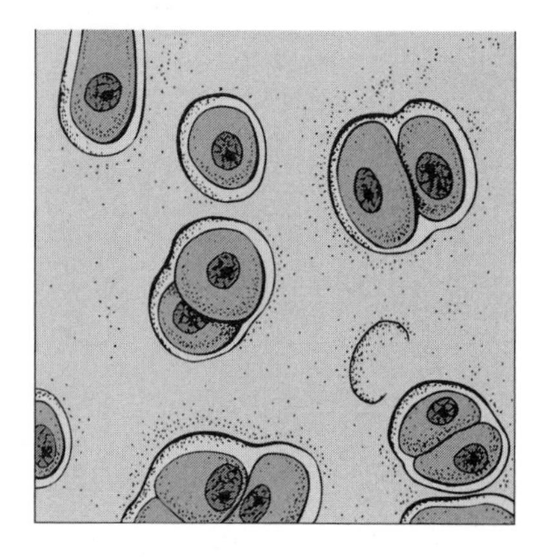

NOTES

3

**Table 3.3h** Fibrocartilage (page 77).

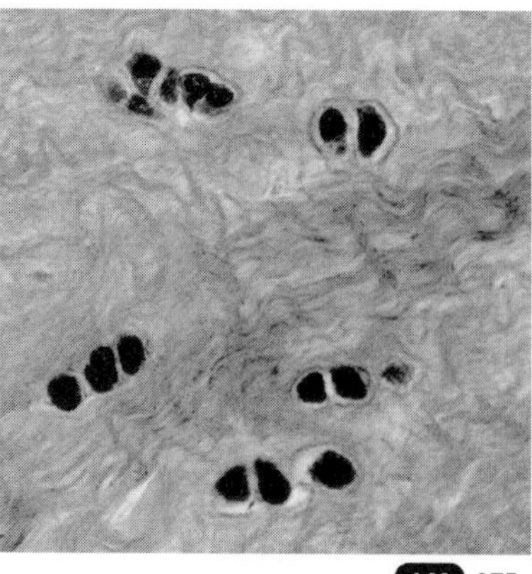


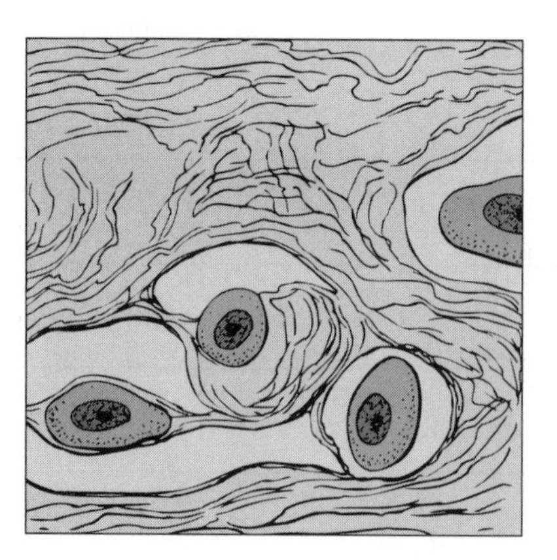

**Table 3.3i** Elastic cartilage (page 78).

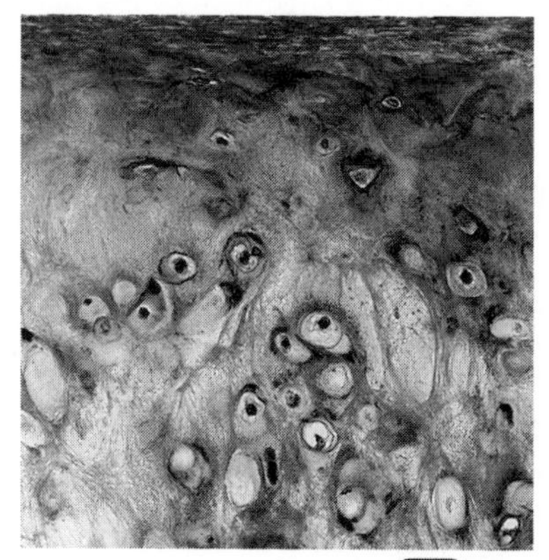

NOTES

3

**Table 3.3j** Compact bone (page 78).

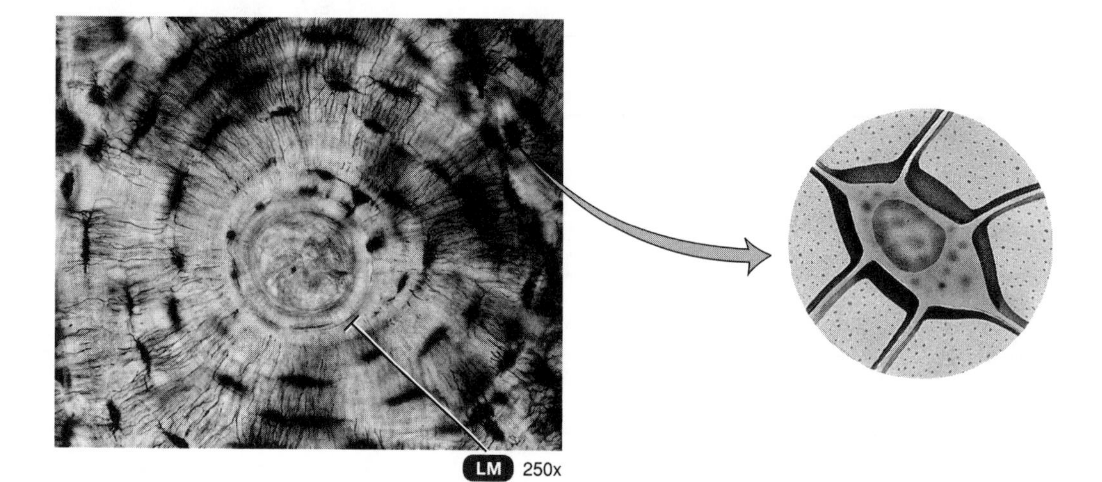


**Table 3.3k** Blood (page 79).

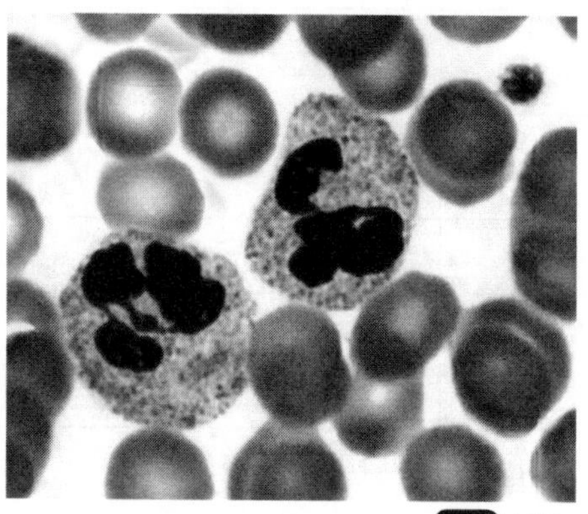


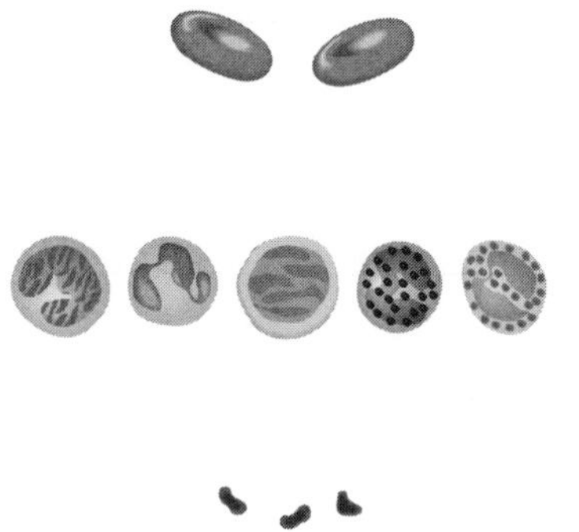

NOTES

3

**Table 3.4a** Skeletal muscle tissue (page 82).

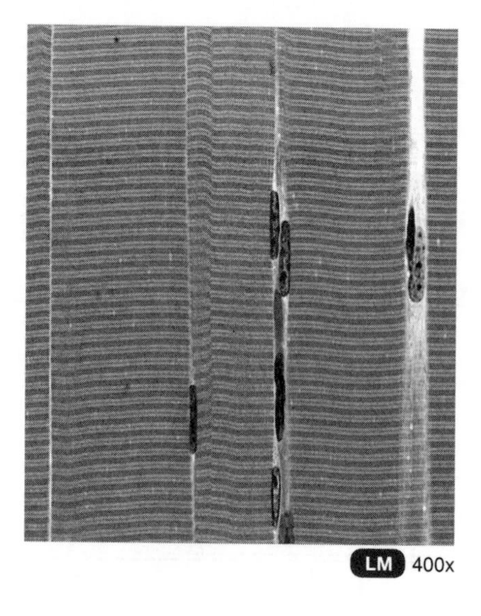


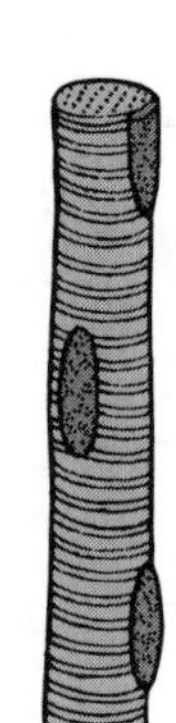

**Table 3.4b** Cardiac muscle tissue (page 82).

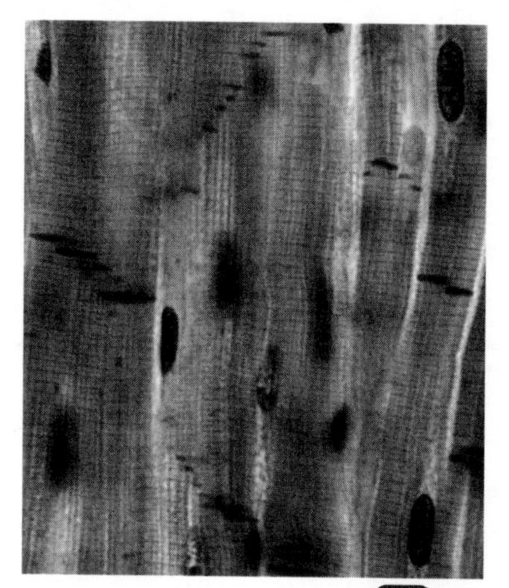


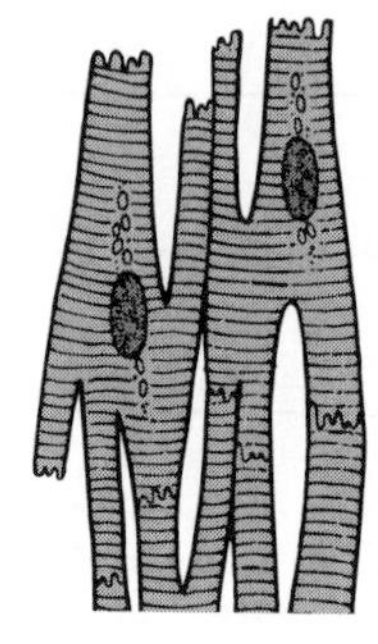

NOTES

3

**Table 3.4c** Smooth muscle tissue (page 83).

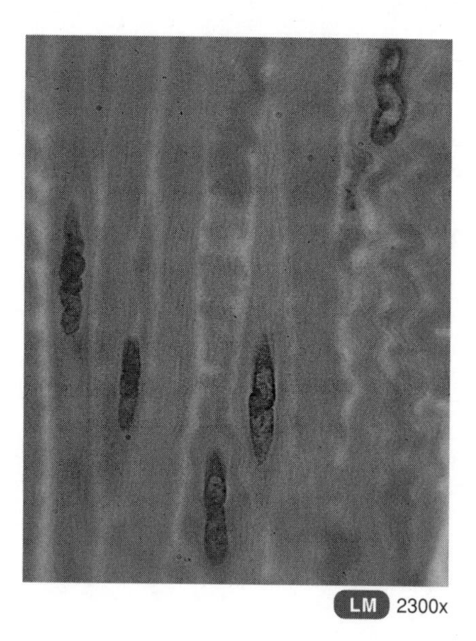


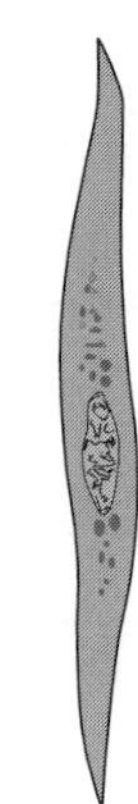

**Table 3.5** Nervous tissue (page 84).

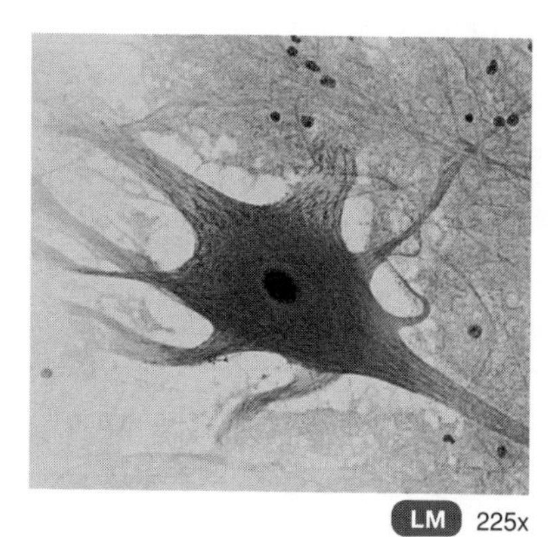

NOTES

**Figure 4.1** Components of the integumentary system (page 90).

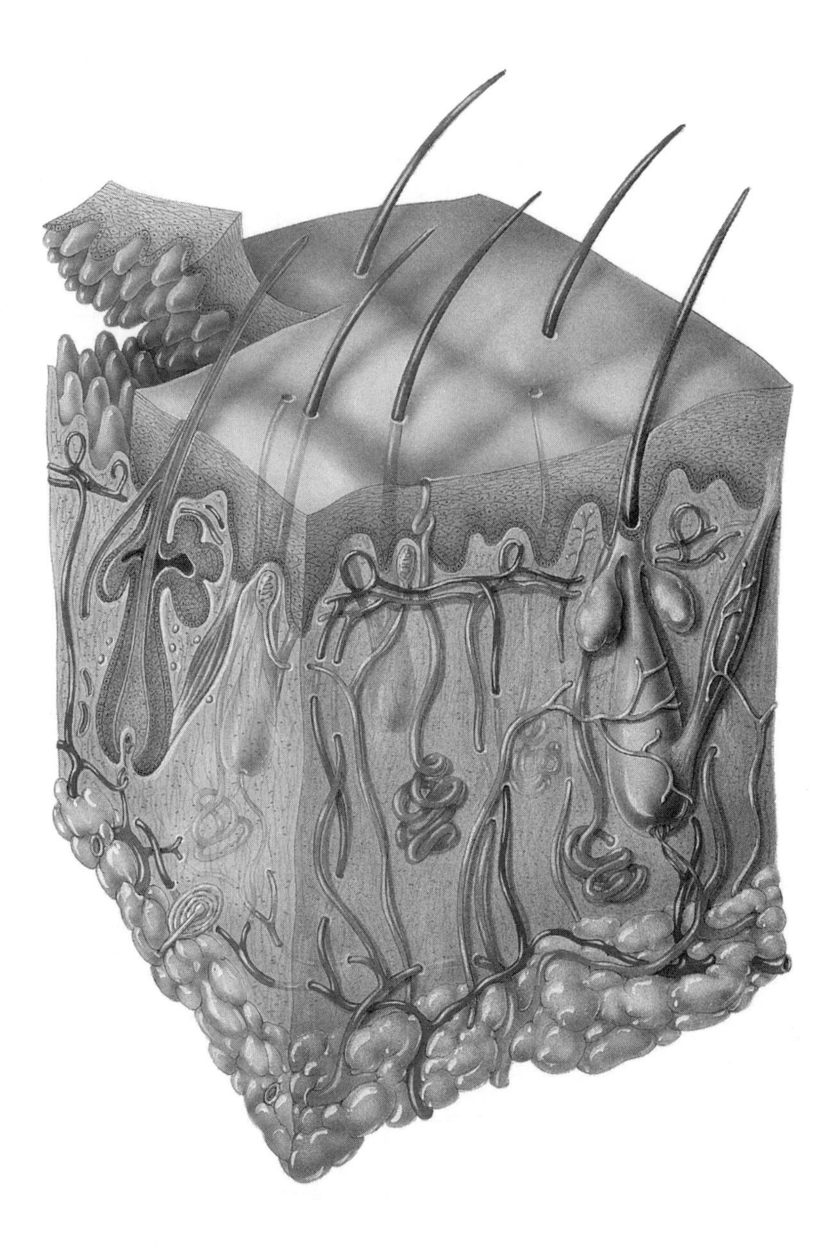

NOTES

**Figure 4.2** Types of cells in the epidermis (page 91).

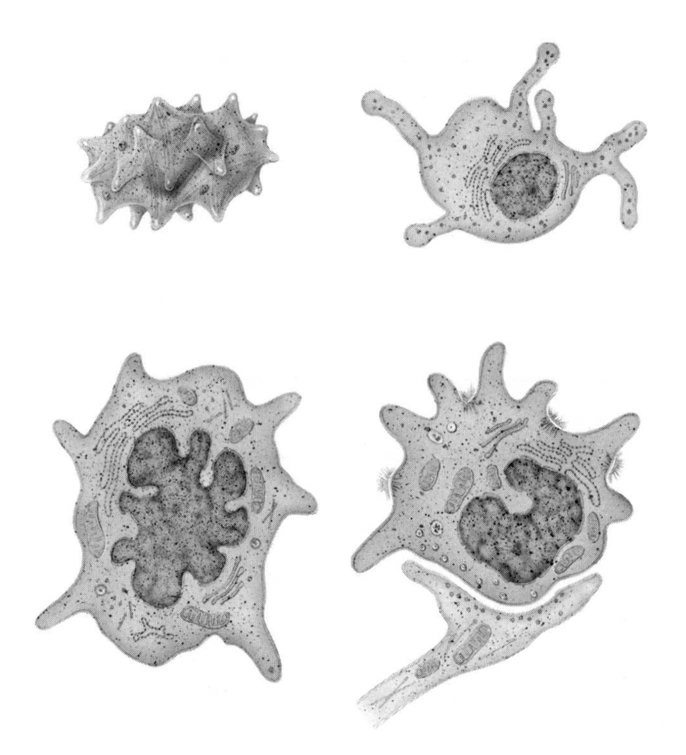

**Figure 4.3** Layers of the epidermis (page 92).

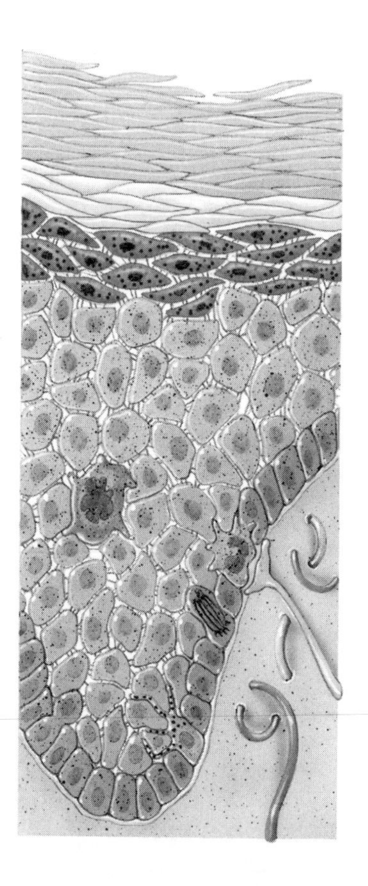

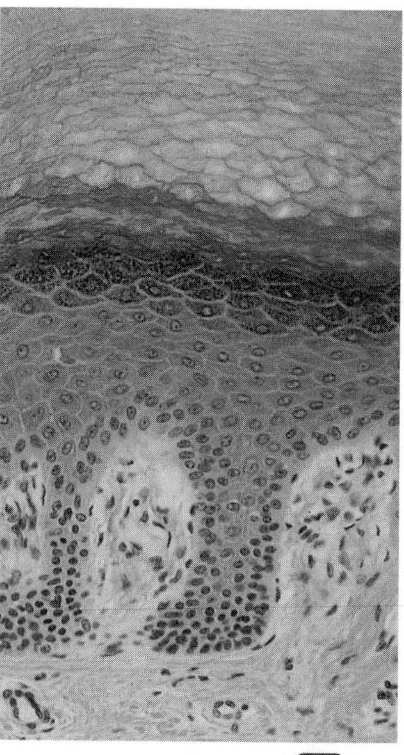

NOTES

4

**Figure 4.4** Lines of cleavage (page 94).

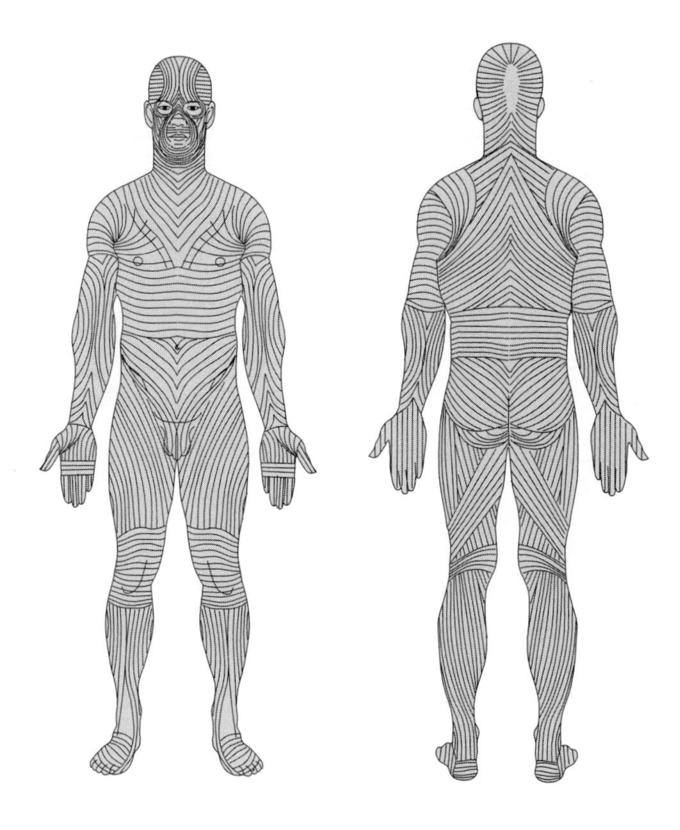

NOTES

4

**Figure 4.5** Hair (page 97).

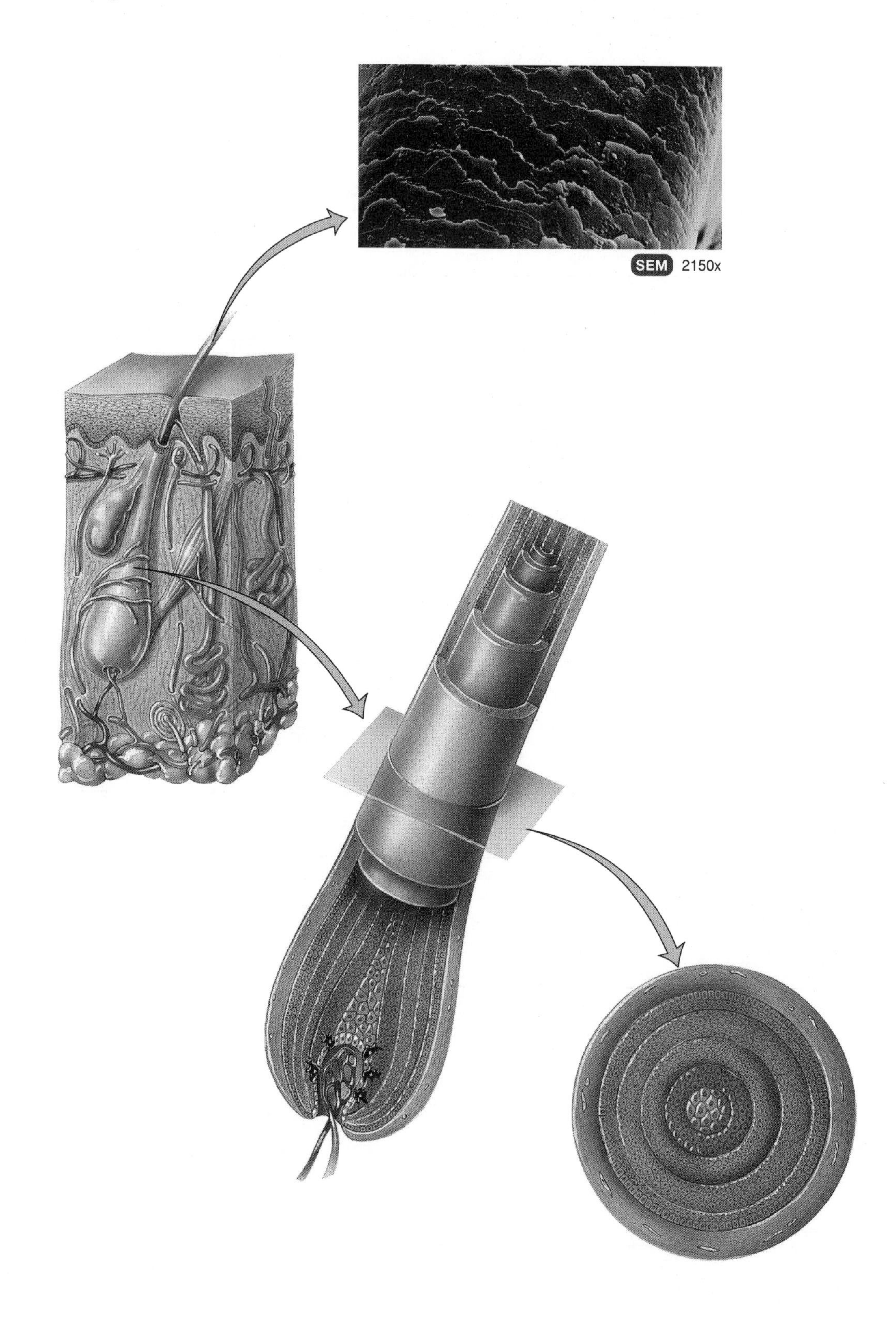

## NOTES

4

**Figure 4.6** Nails (page 100).

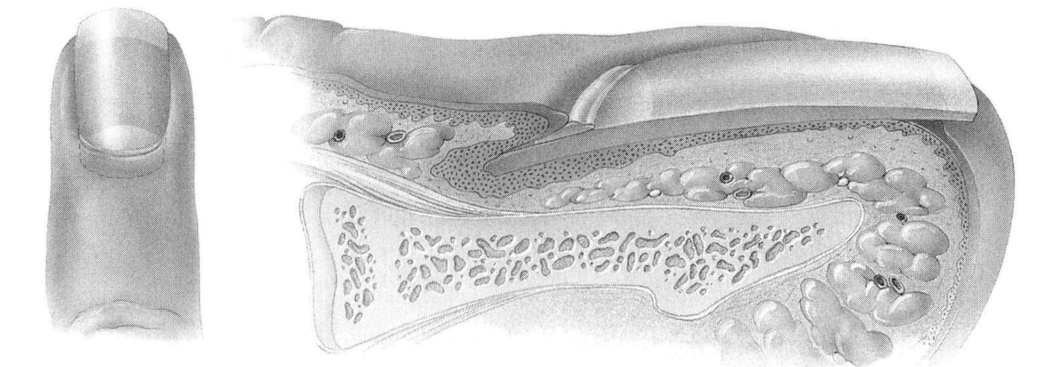

NOTES

4

**Figure 5.1** Parts of a long bone (page 111).

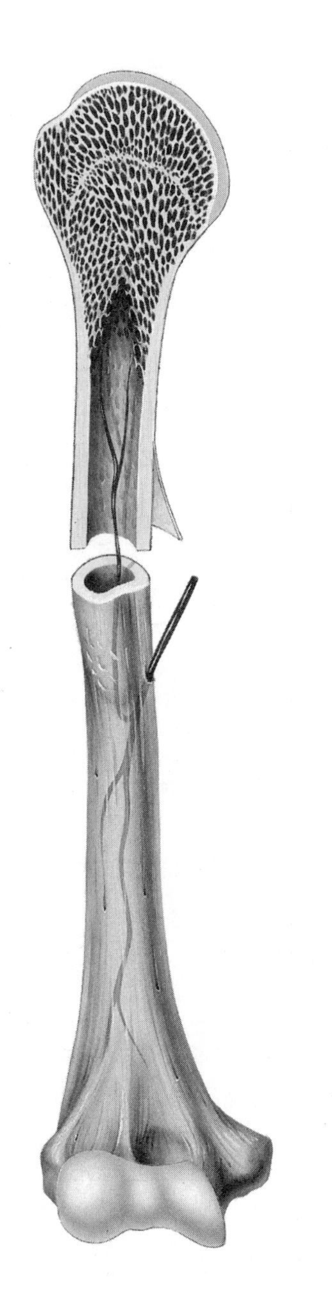

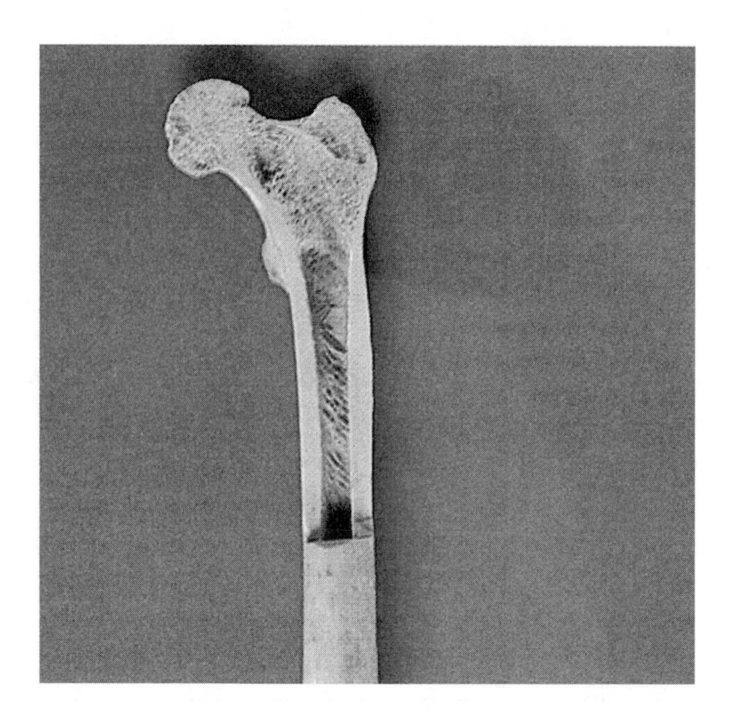

NOTES

5

**Figure 5.2** Types of cells in bone tissue (page 112).

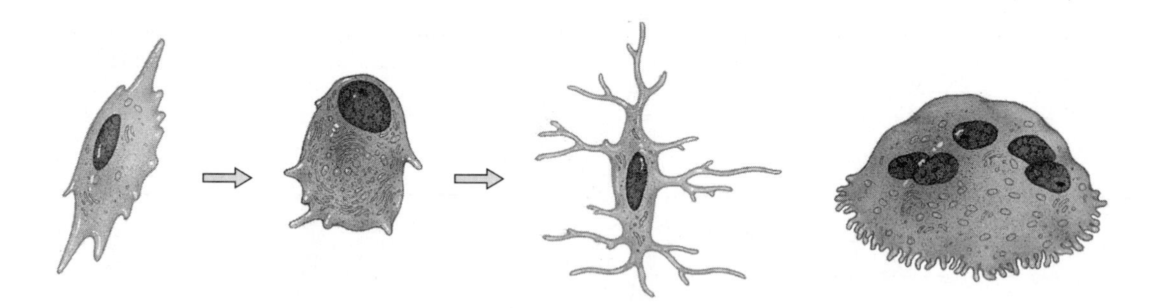

NOTES

5

**Figure 5.3** Histology of compact and spongy bone (page 113).

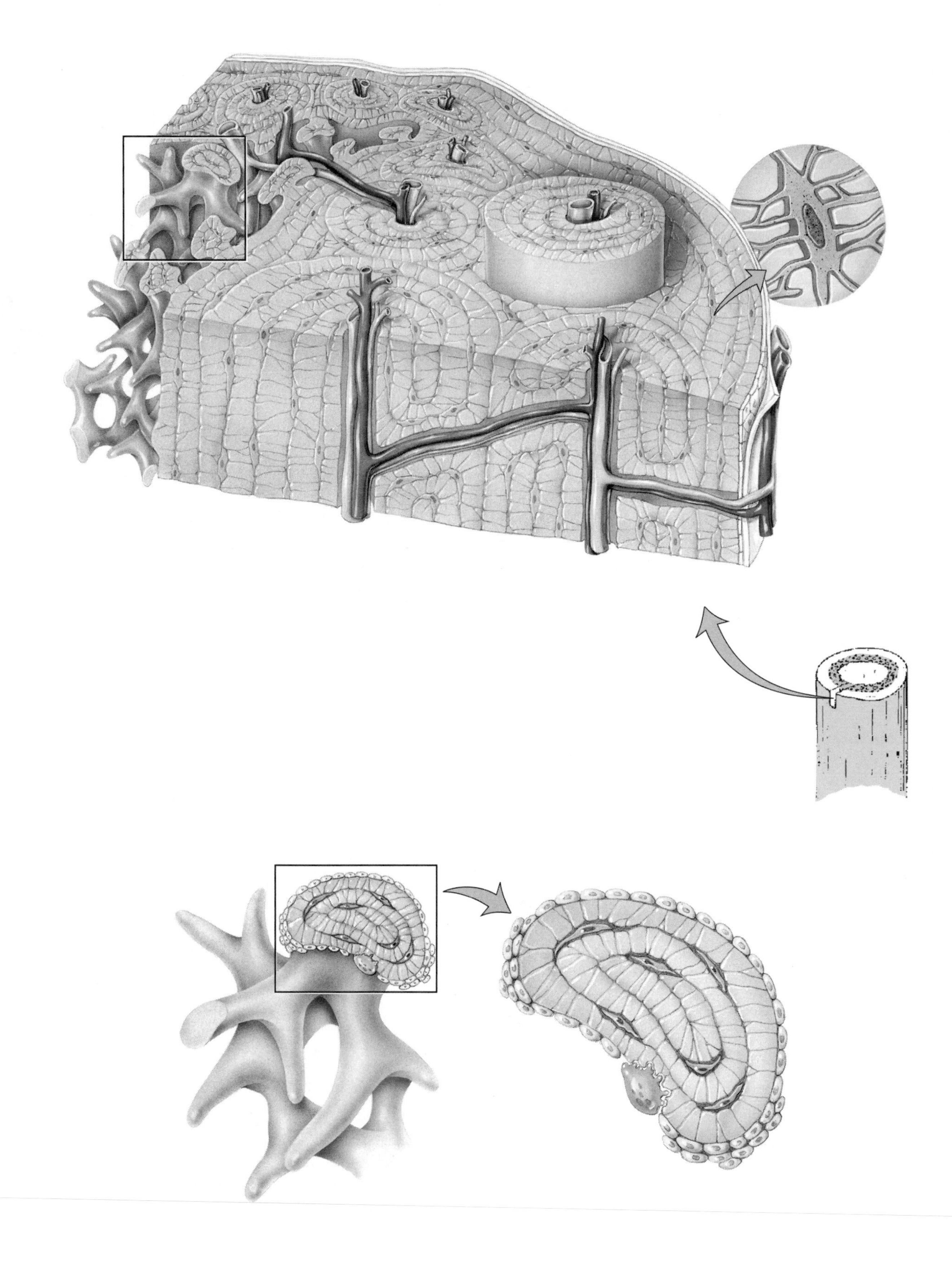

NOTES

5

**Figure 5.4** Blood supply of a mature long bone, the tibia (shinbone) (page 115).

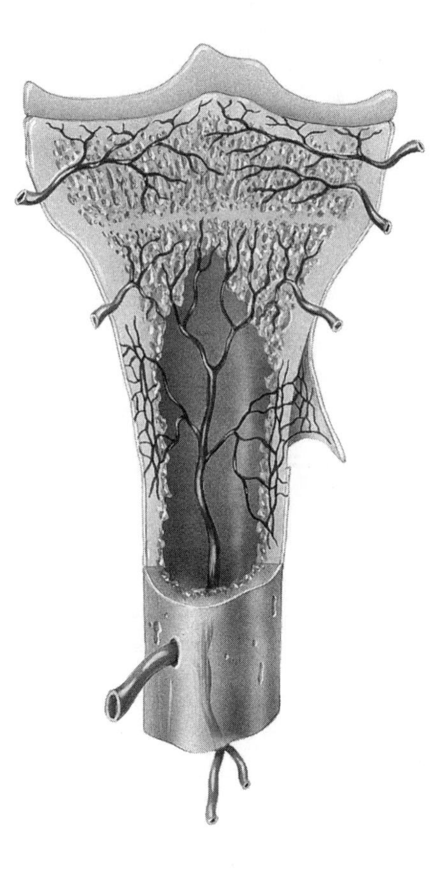

NOTES

5

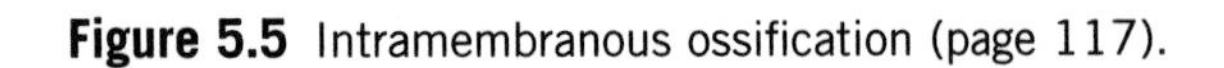

**Figure 5.5** Intramembranous ossification (page 117).

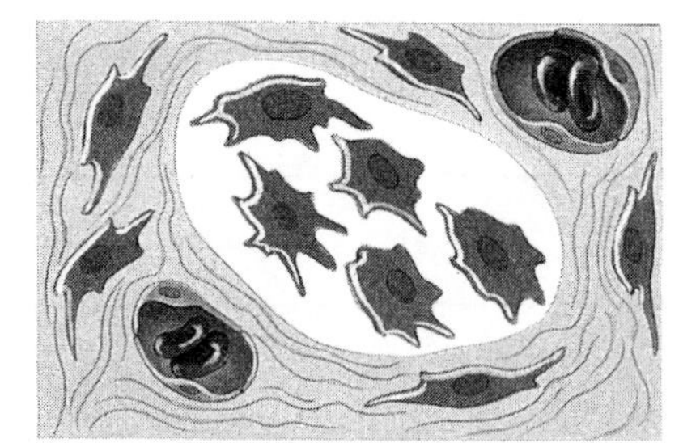

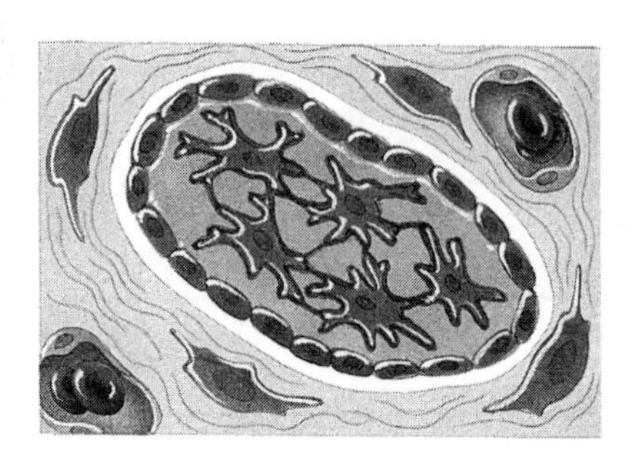

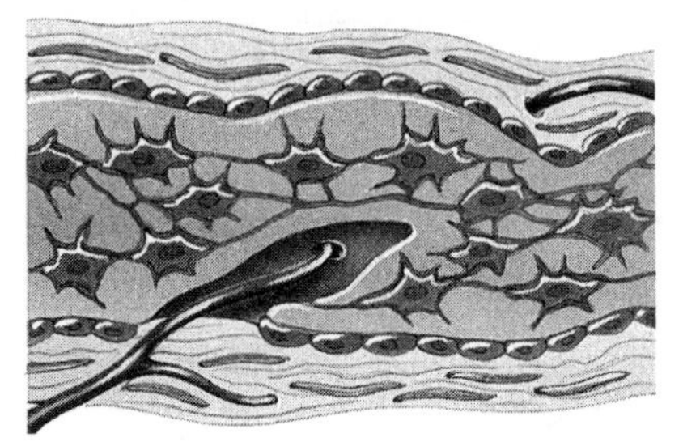

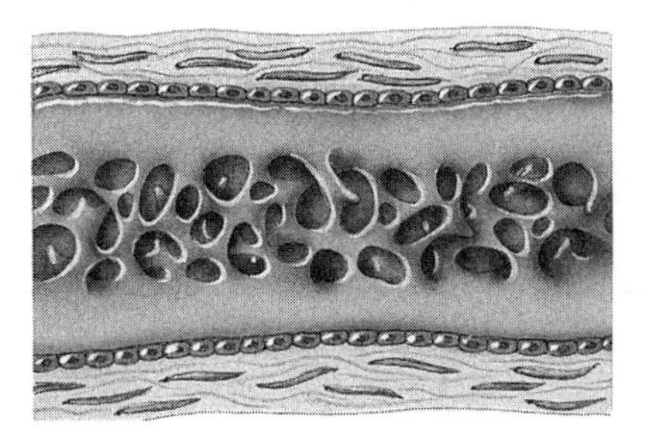

NOTES

5

**Figure 5.6** Endochondral ossification (page 118).

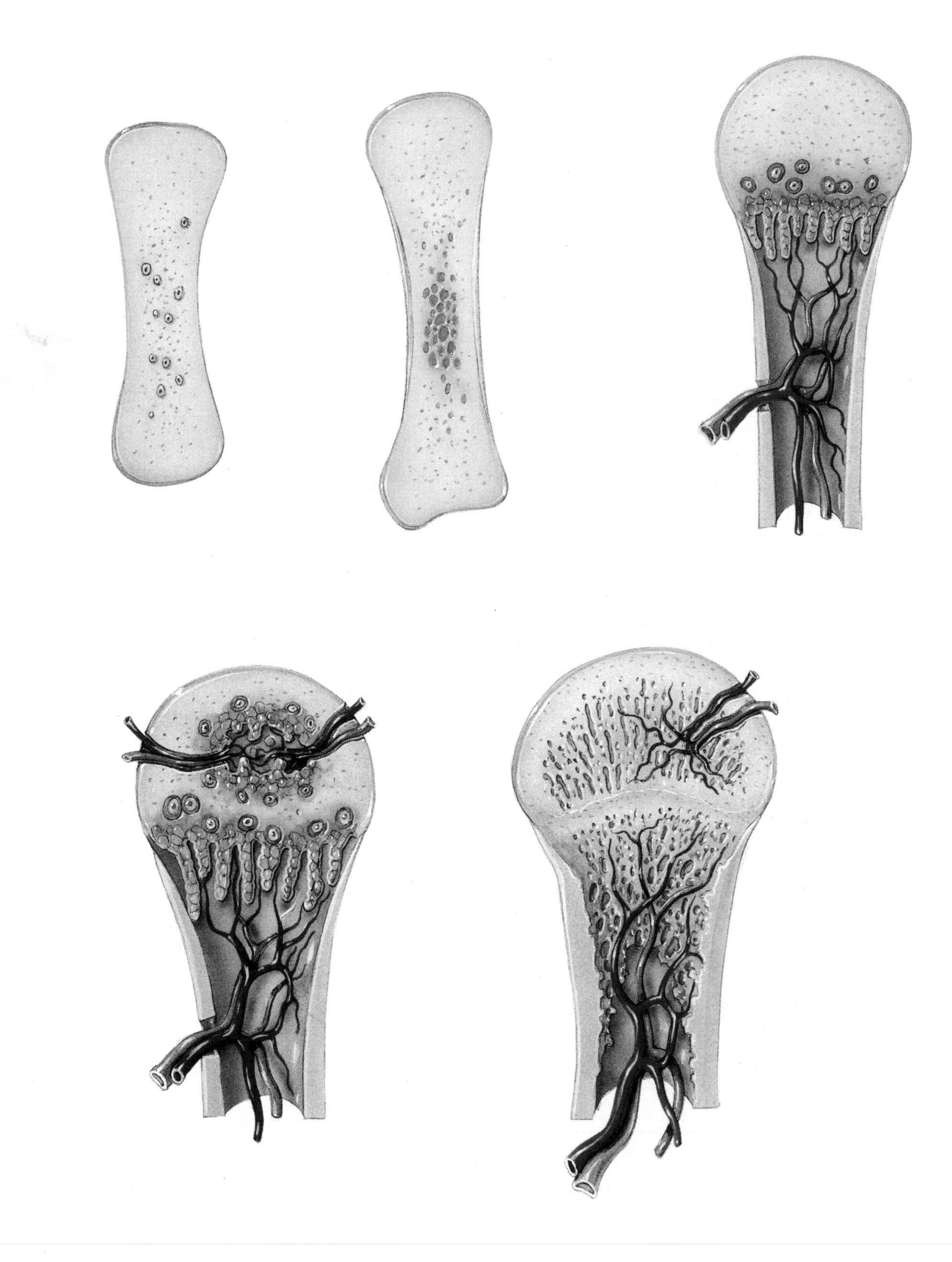

NOTES

5

**Figure 5.7** Epiphyseal plate (page 119).

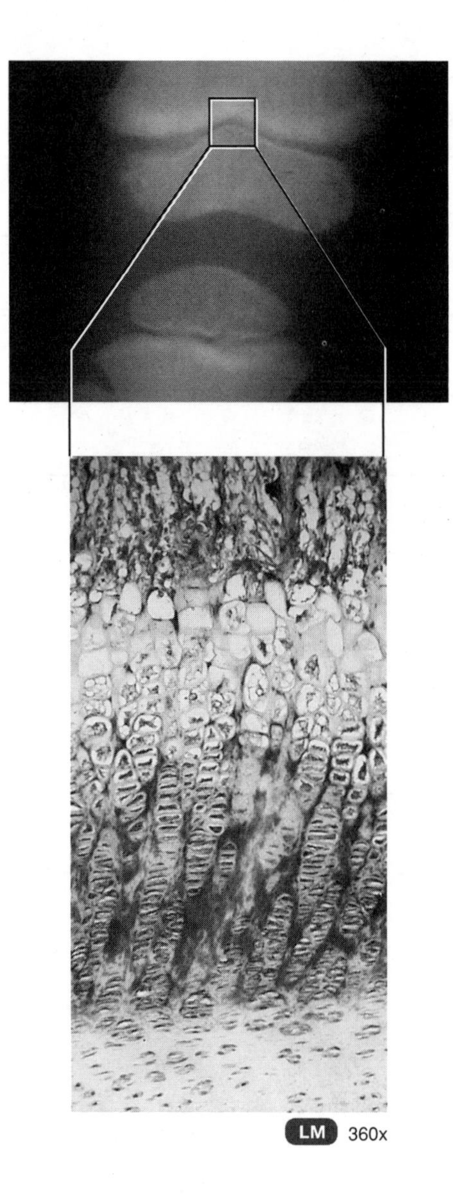

NOTES

5

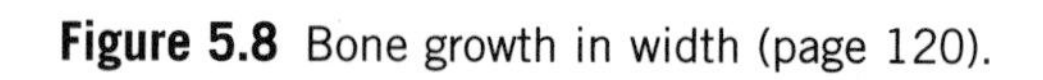

**Figure 5.8** Bone growth in width (page 120).

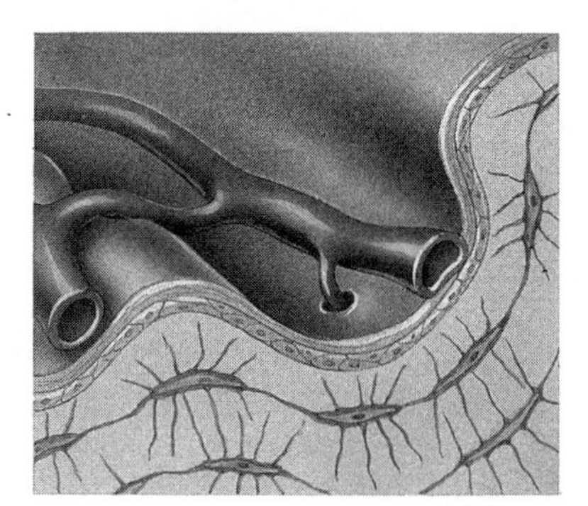

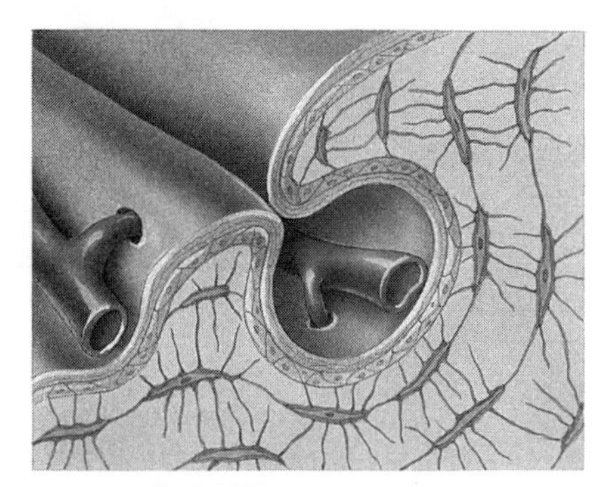

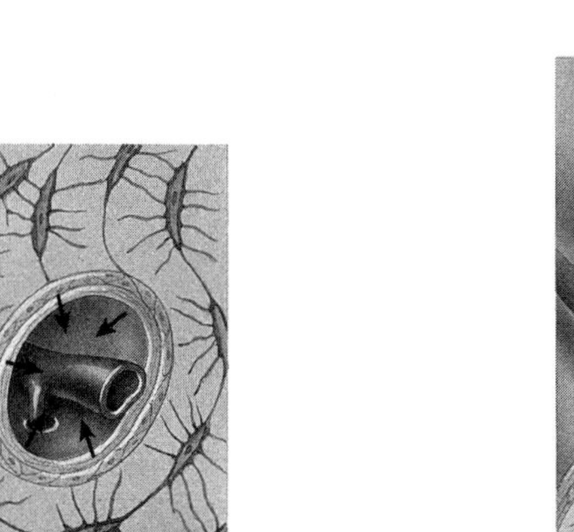

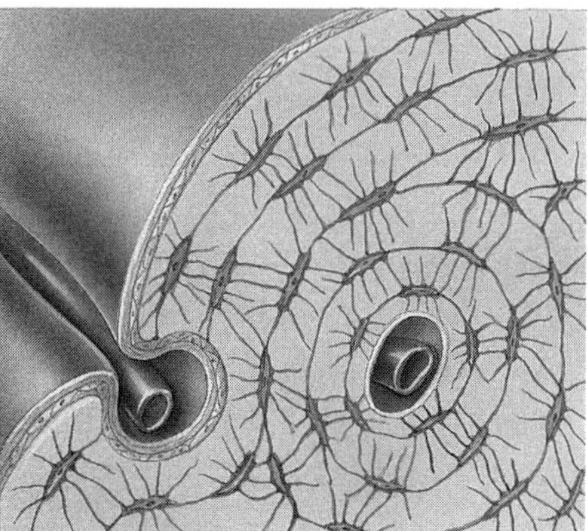

## NOTES

5

**Figure 5.9** Types of bone fractures (page 121).

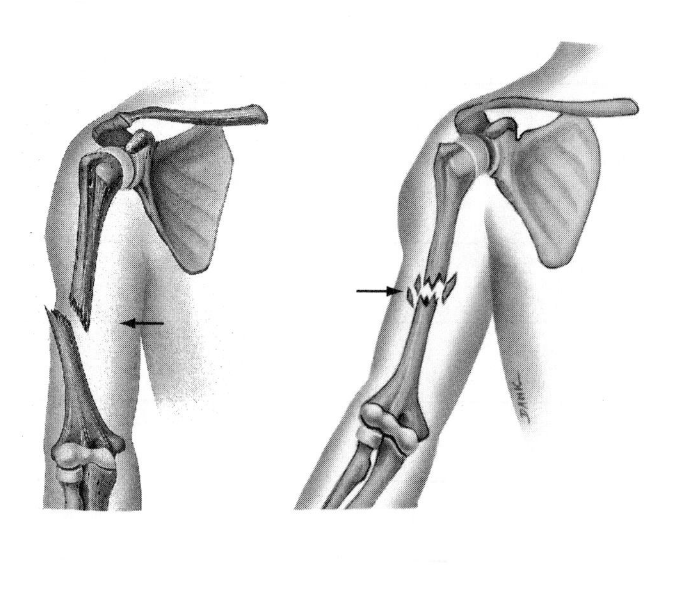

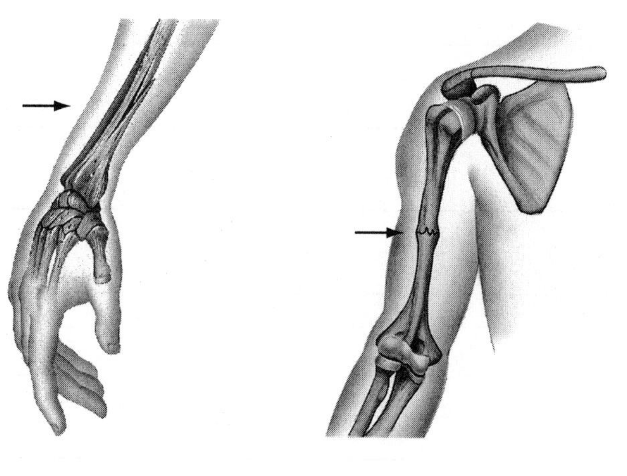

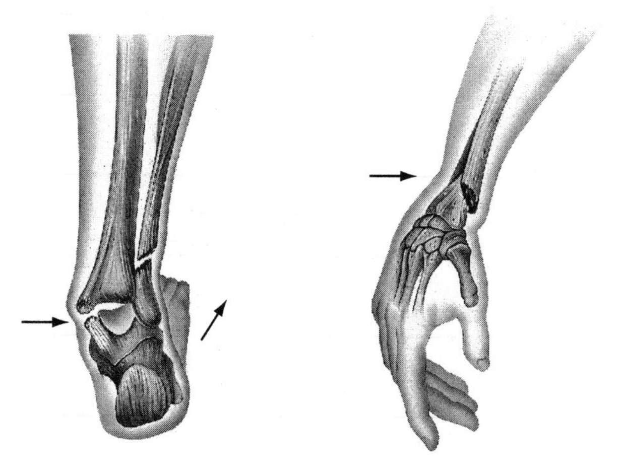

## NOTES

5

**Figure 5.10** Steps involved in repair of a bone fracture (page 123).

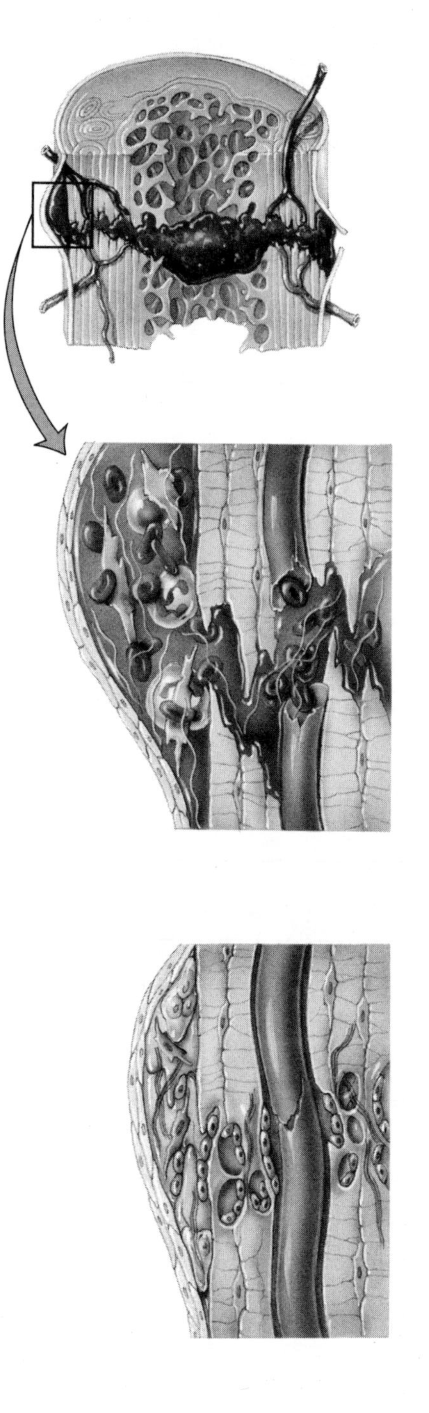

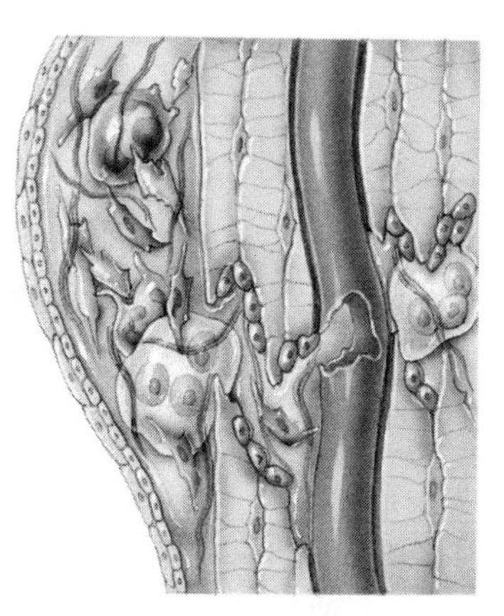

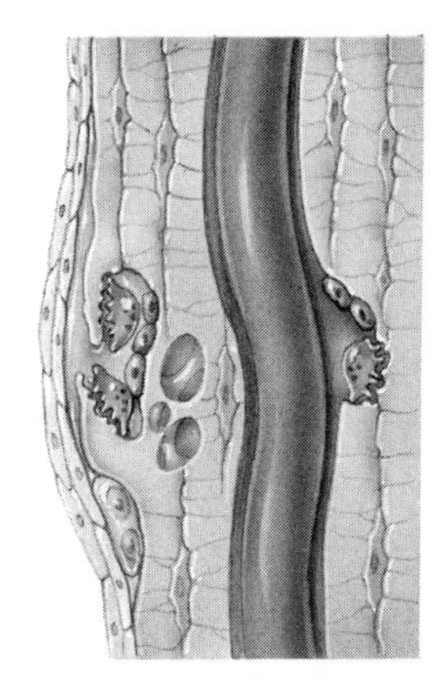

NOTES

5

**Figure 5.11** Features of a developing human embryo during weeks 5 through 8 of development (page 125).

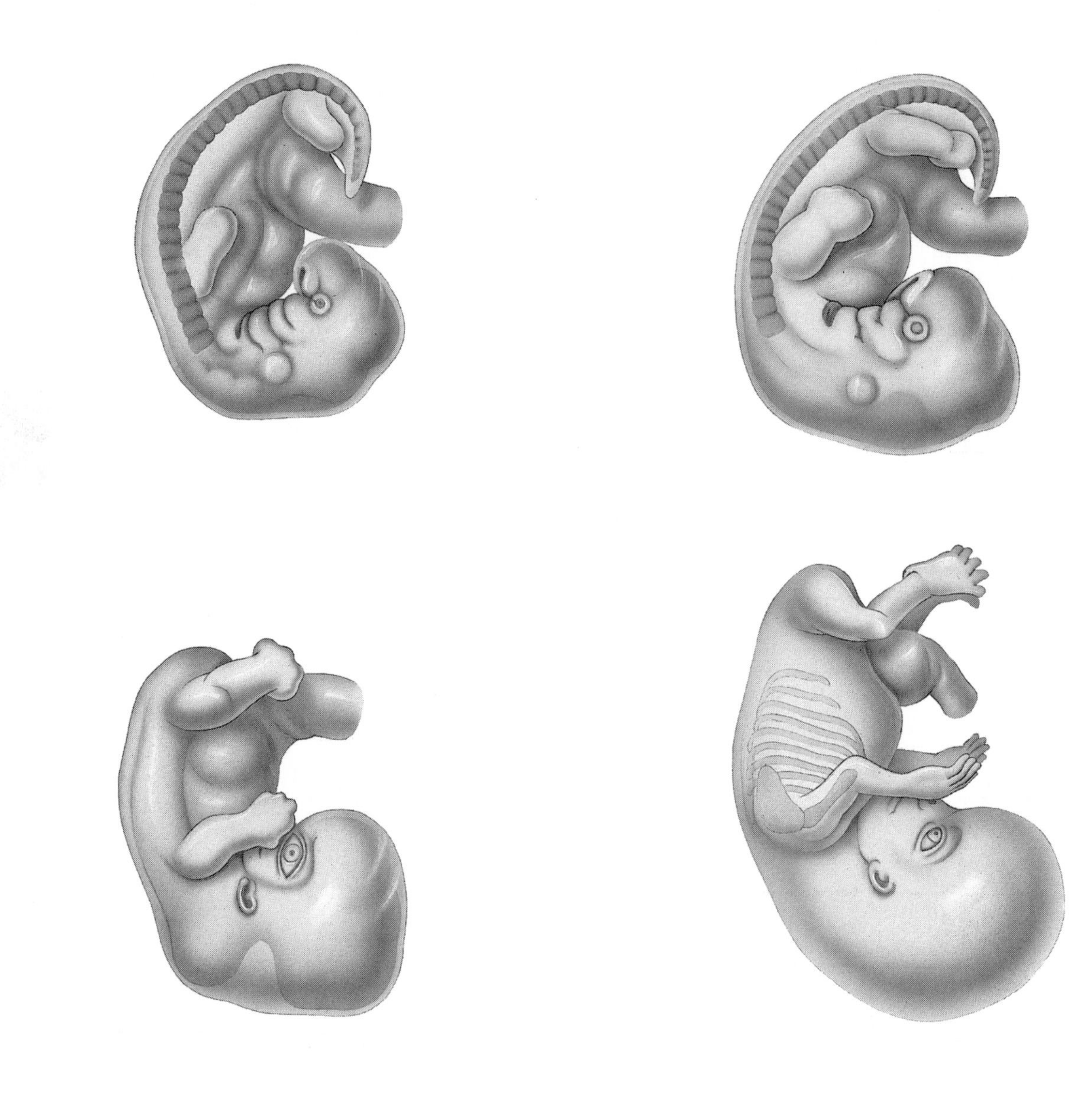

**Figure 5.12** Comparison of spongy bone tissue from (a) a normal young adult and (b) a person with osteoporosis (page 127).

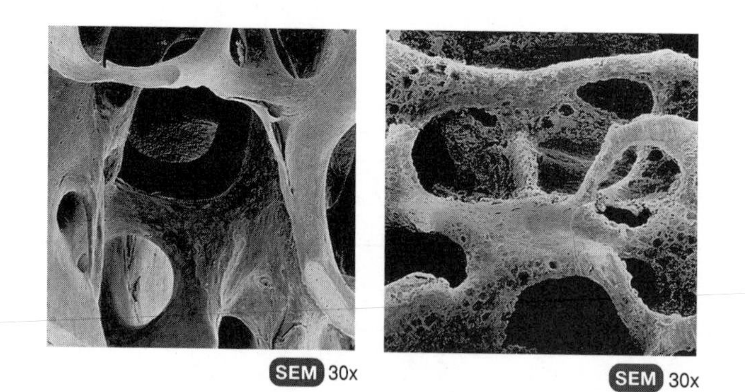

NOTES

5

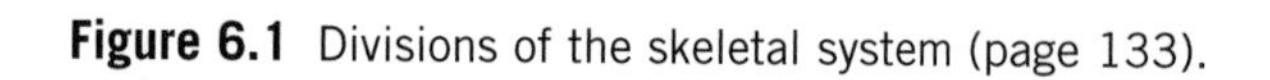

**Figure 6.1** Divisions of the skeletal system (page 133).

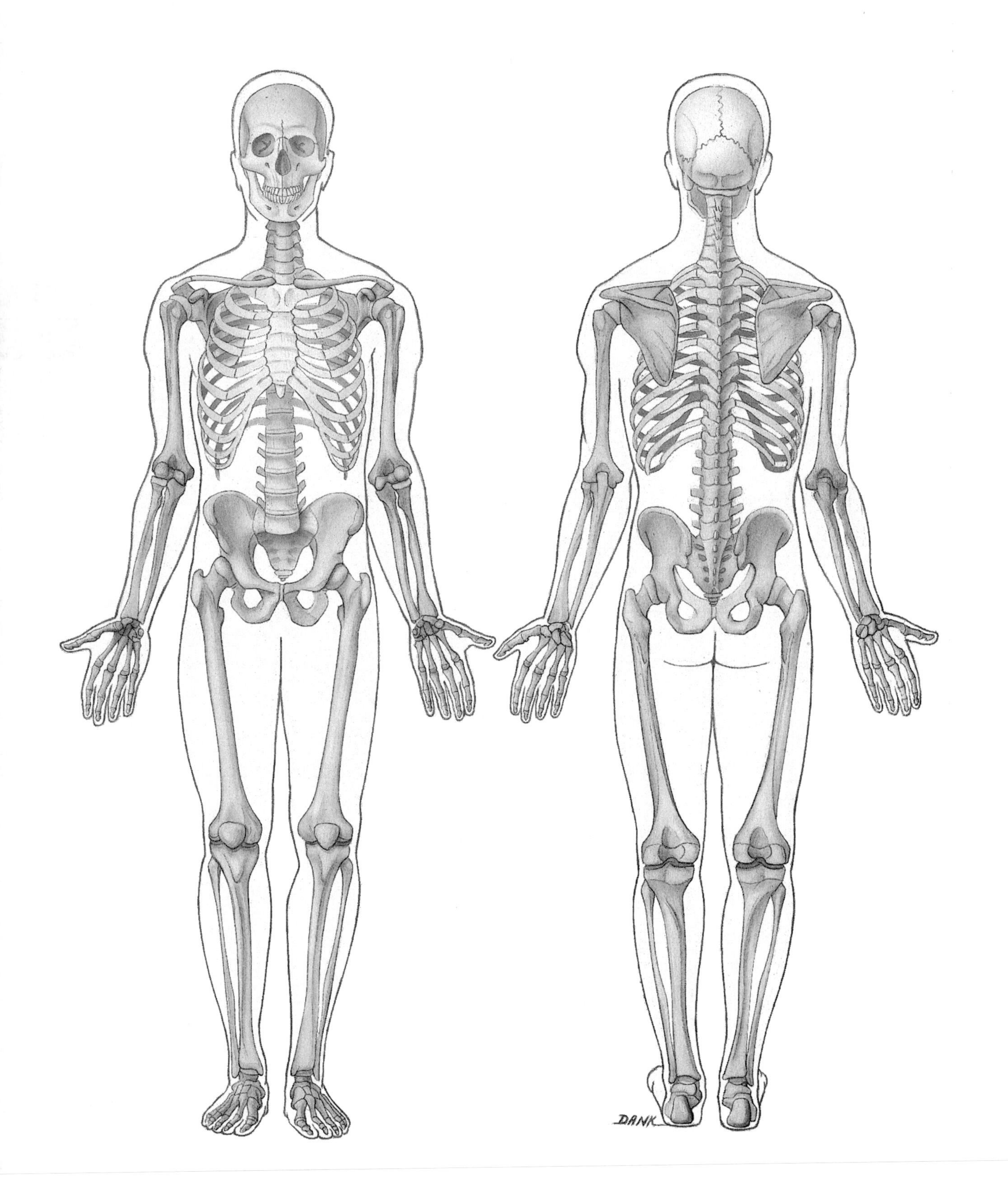

NOTES

6

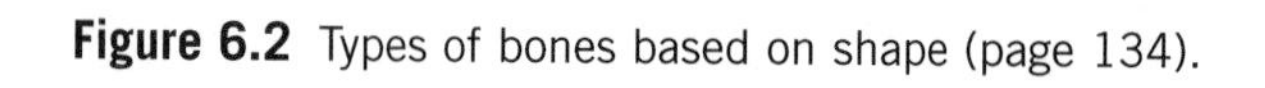

**Figure 6.2** Types of bones based on shape (page 134).

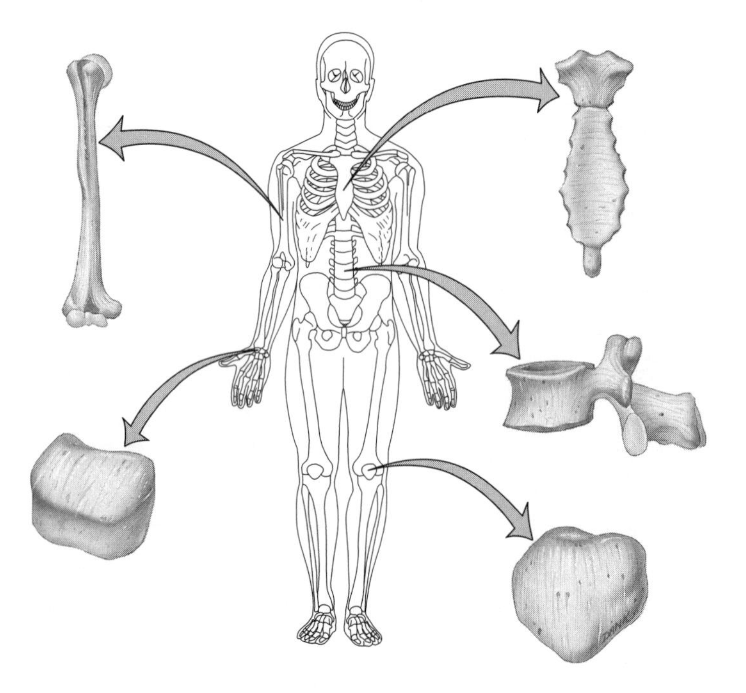

**Figure 6.3a** Skull, anterior view (page 136).

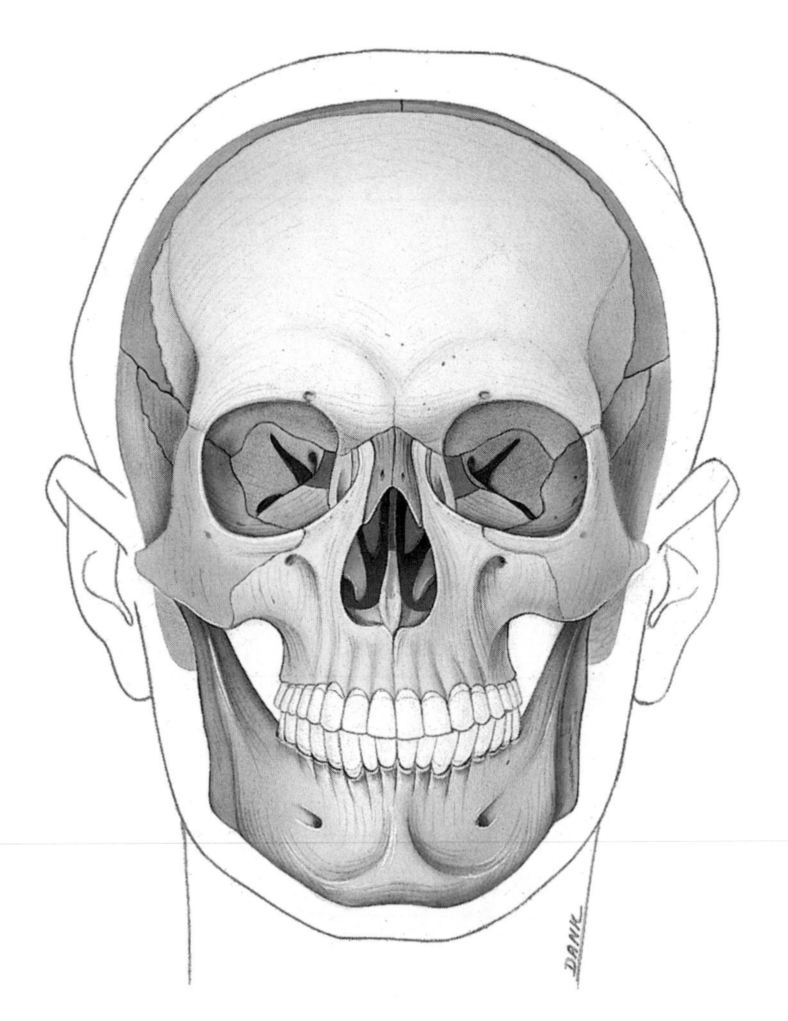

NOTES

6

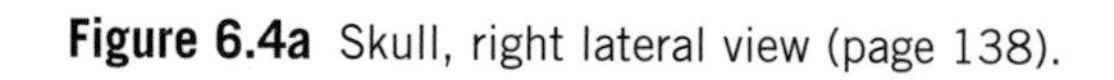

**Figure 6.4a** Skull, right lateral view (page 138).

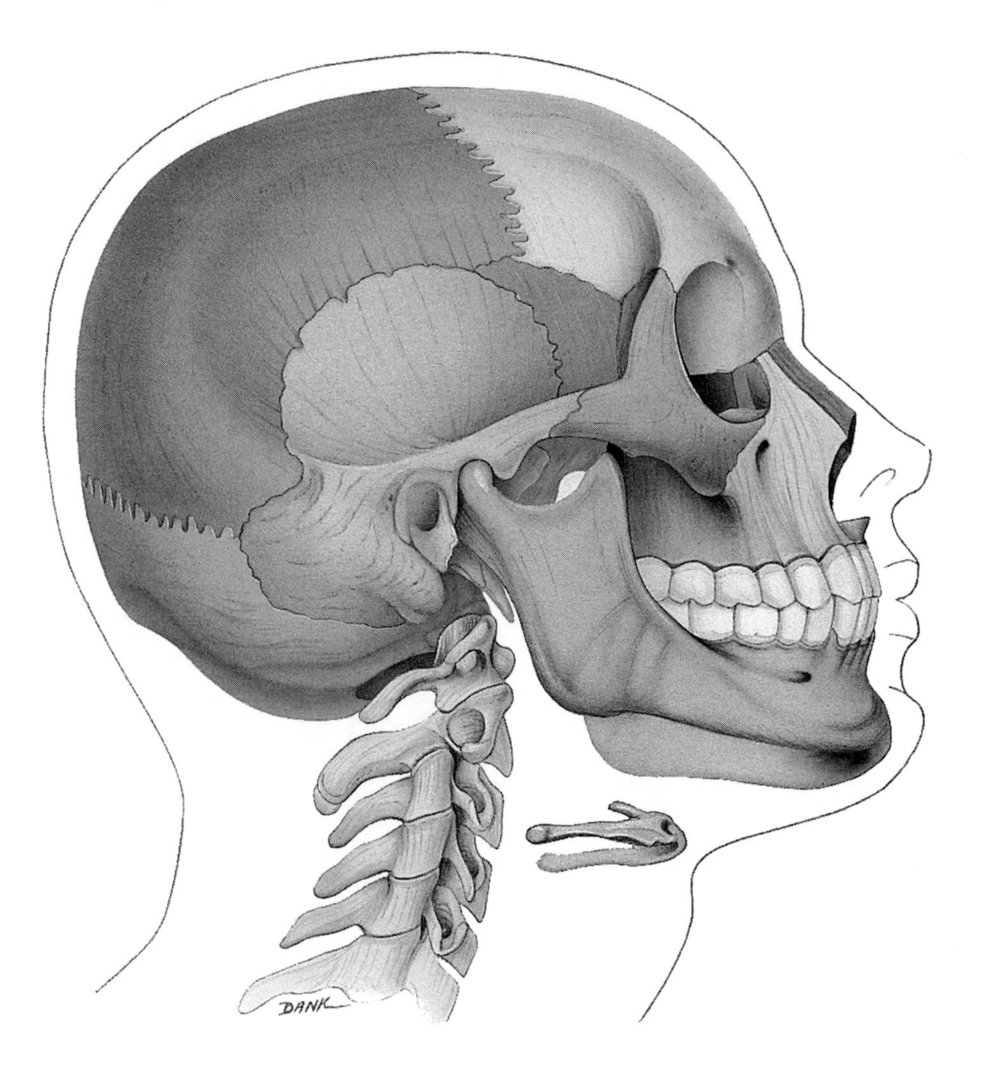

NOTES

**Figure 6.5** Skull, medial view of sagittal section (page 140).

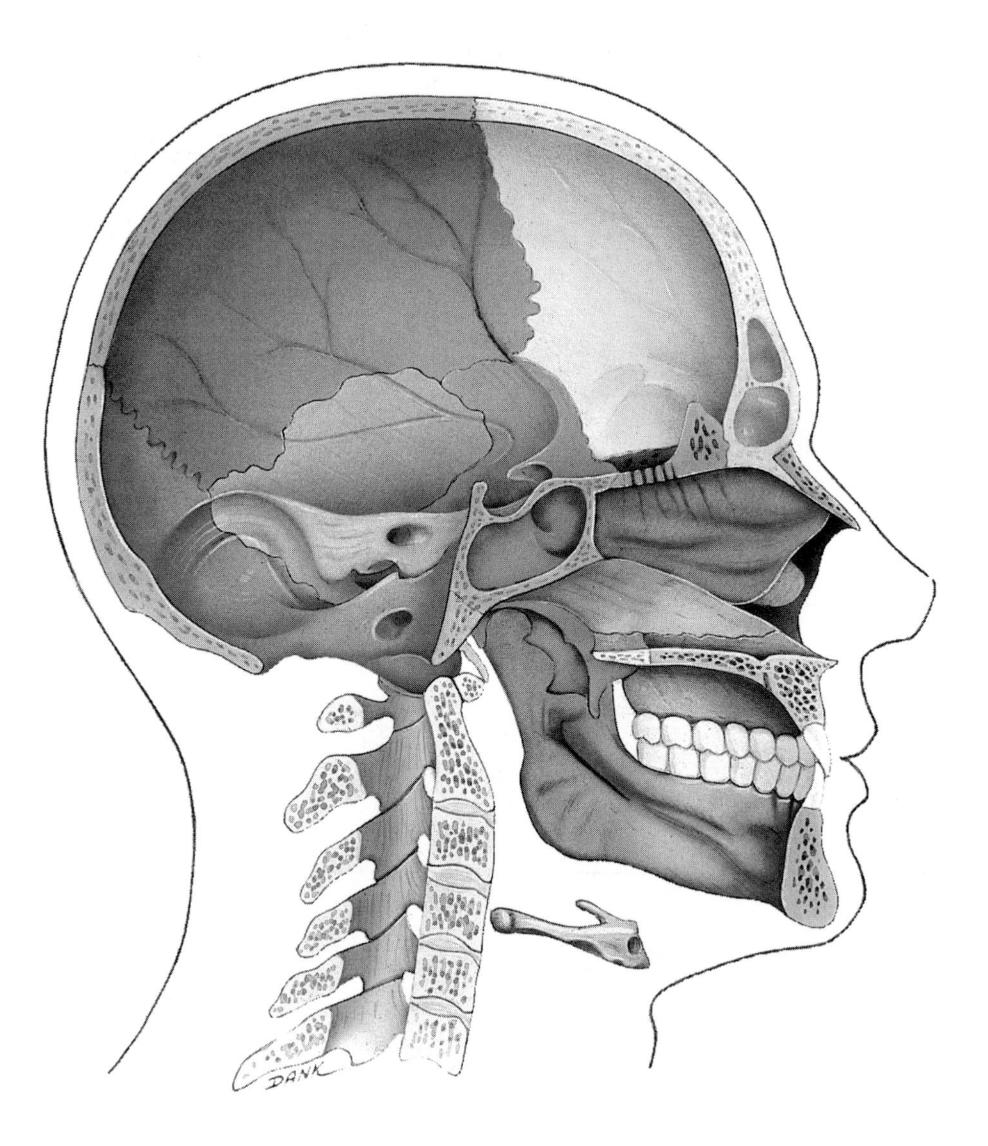

NOTES

6

**Figure 6.6** Skull, posterior view (page 141).

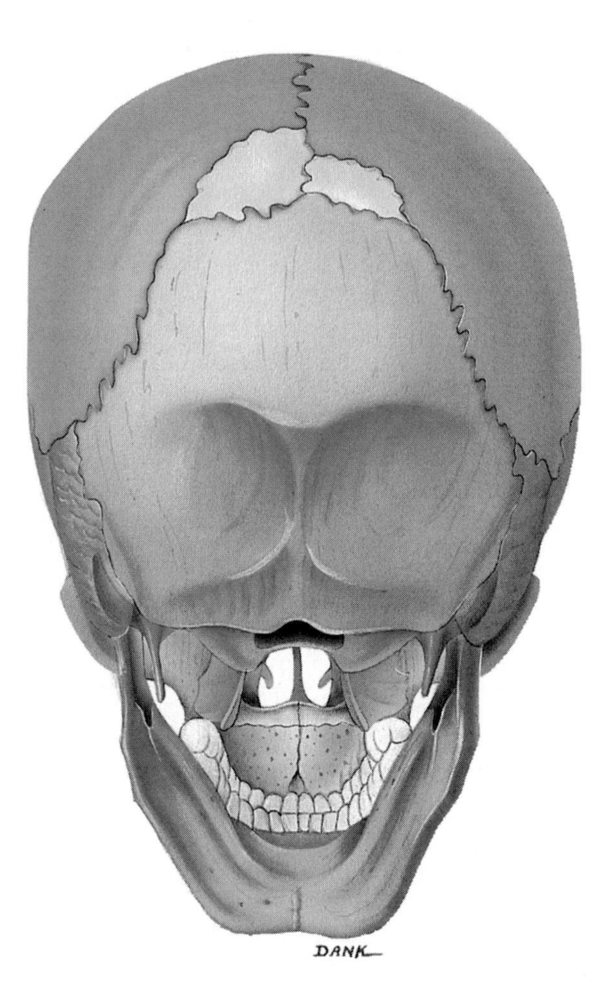

NOTES

6

**Figure 6.7a** Skull, inferior view (page 142).

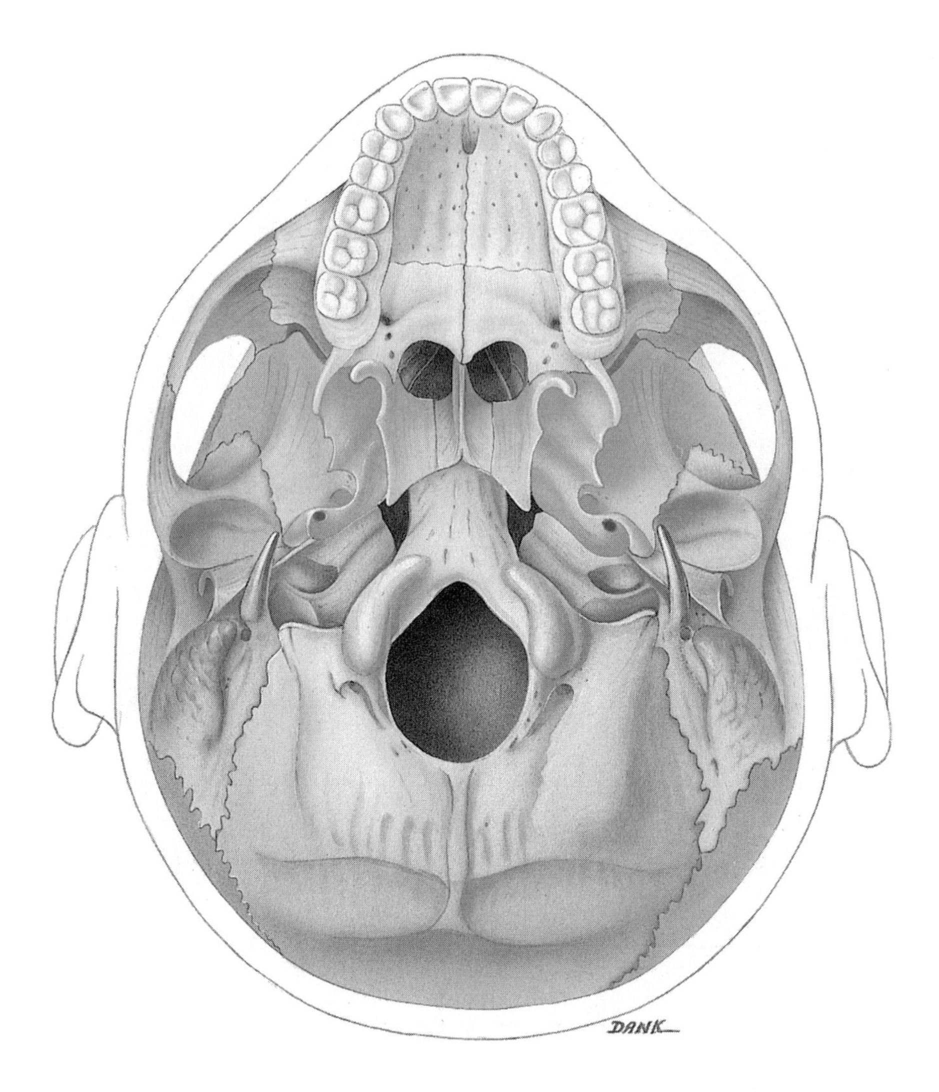

NOTES

6

**Figure 6.8a** Sphenoid bone, superior view of sphenoid bone in floor of cranium (page 144).

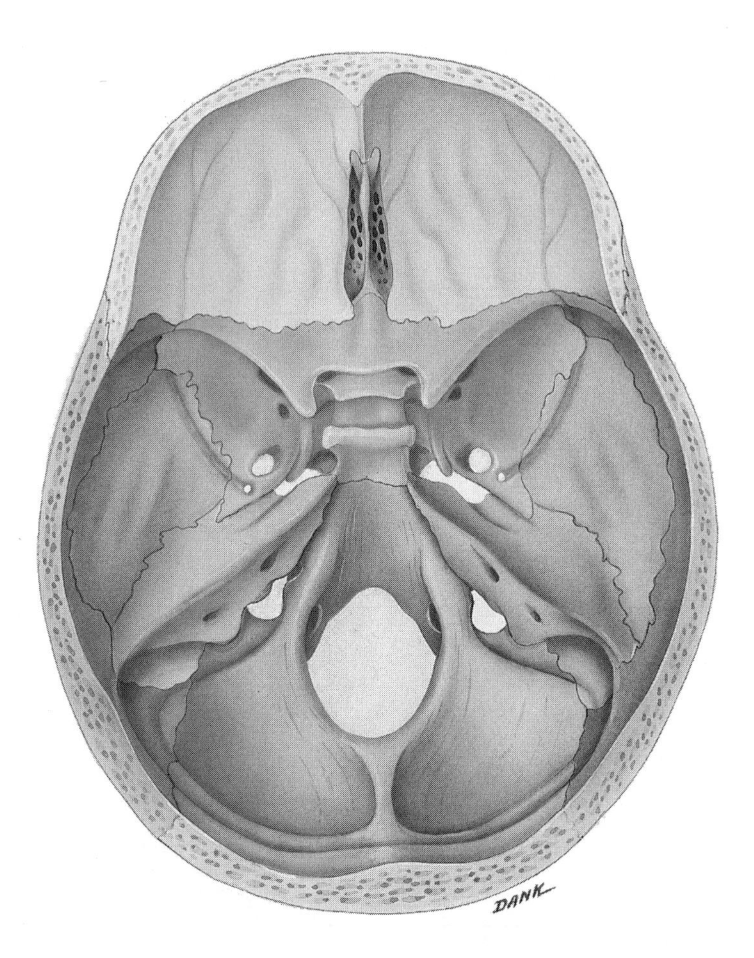


**Figure 6.8b** Sphenoid bone, anterior view of sphenoid bone (page 145).

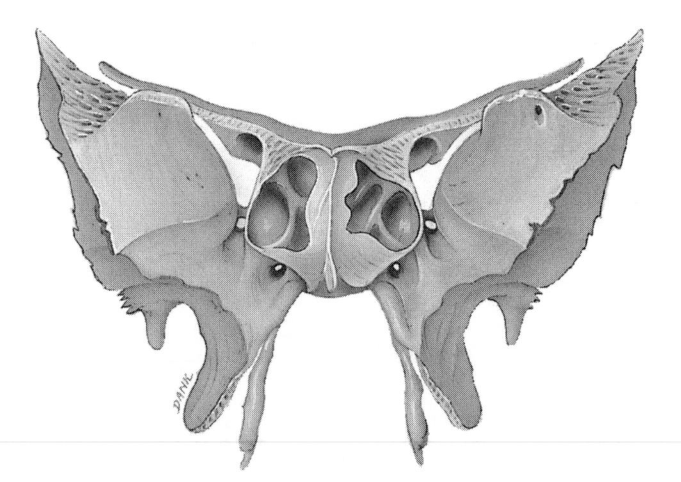

NOTES

6

**Figure 6.9a–d** Ethmoid bone (page 146).

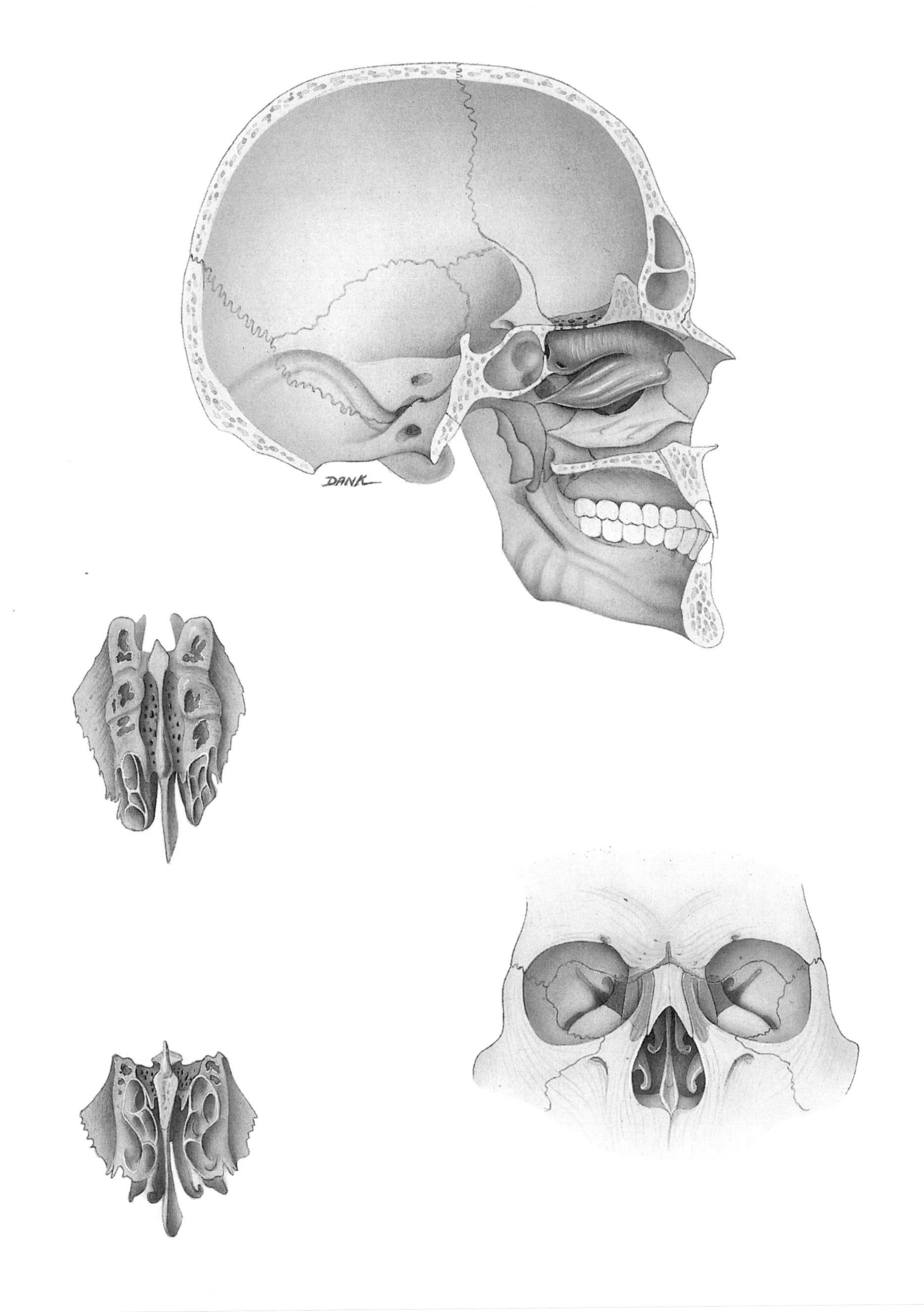

NOTES

6

**Figure 6.9e** Ethmoid bone (page 147).

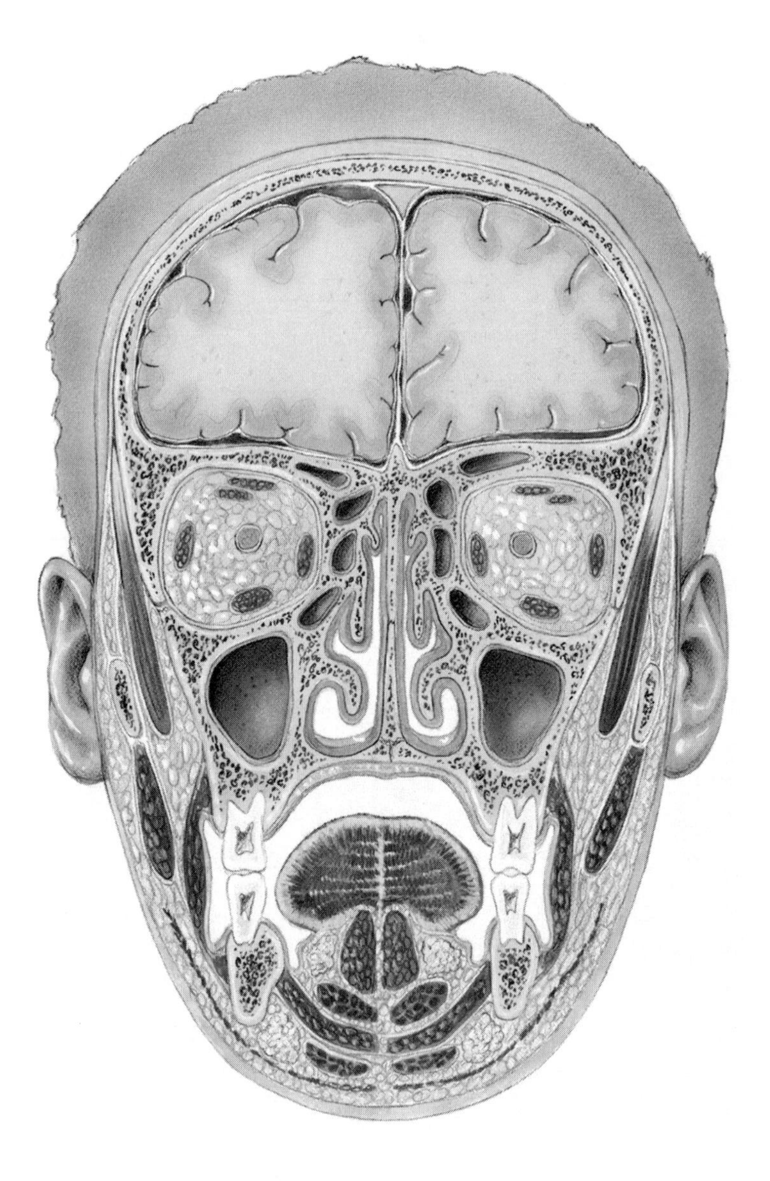

**Figure 6.10** Mandible (page 148).

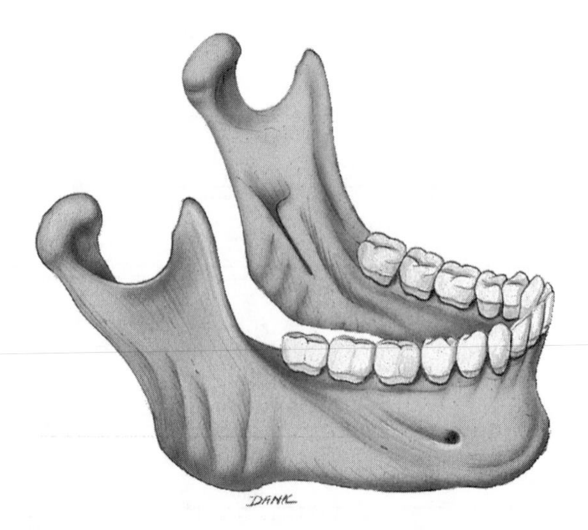

NOTES

**Figure 6.11a,c** Paranasal sinuses (page 150).

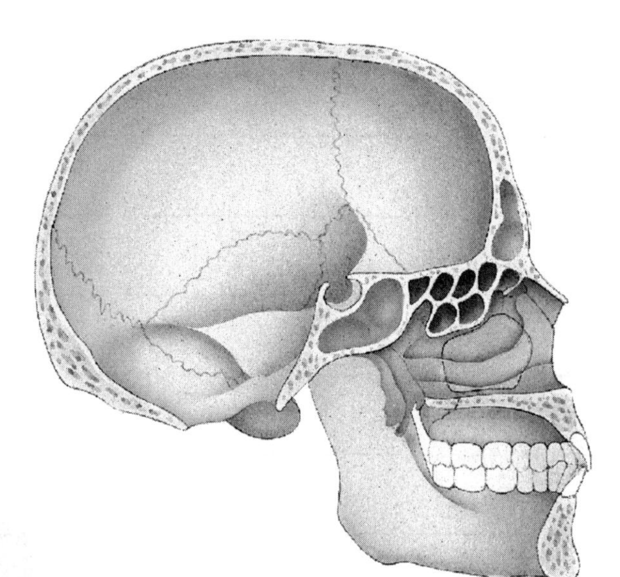

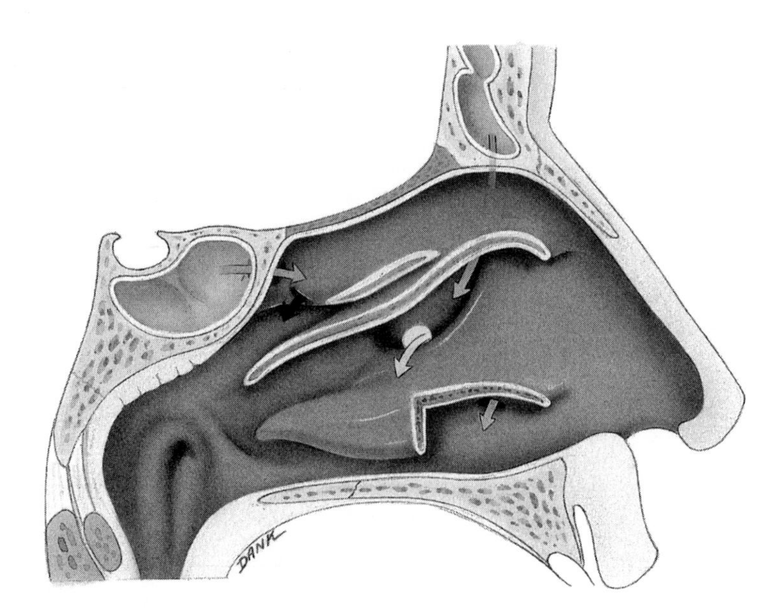


**Figure 6.12** Fontanels at birth (page 151).

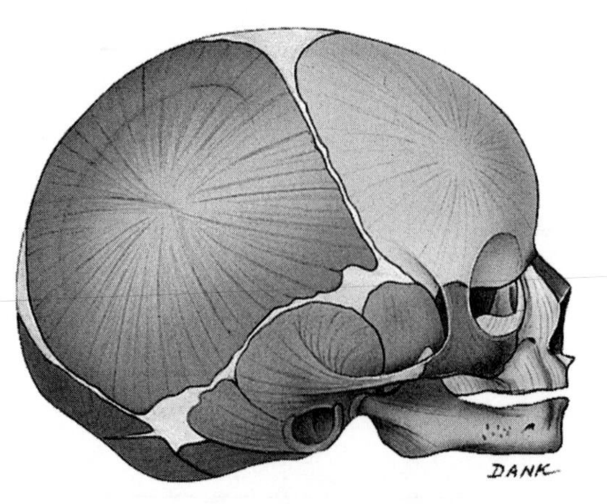

NOTES

6

**Figure 6.13** Details of the orbit (eye socket) (page 153).

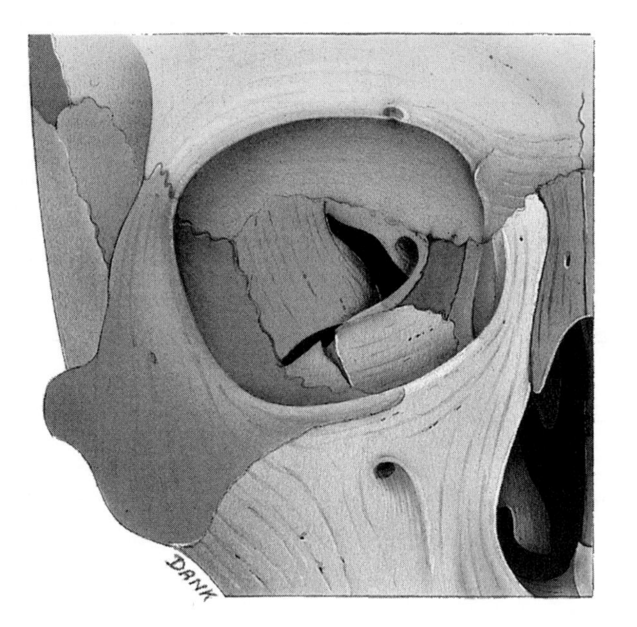


**Figure 6.14** Nasal septum (page 153).

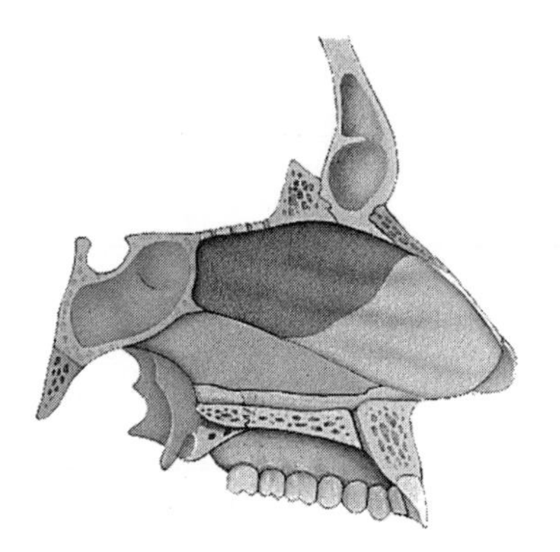

NOTES

6

**Figure 6.15** Cranial fossae (page 154).

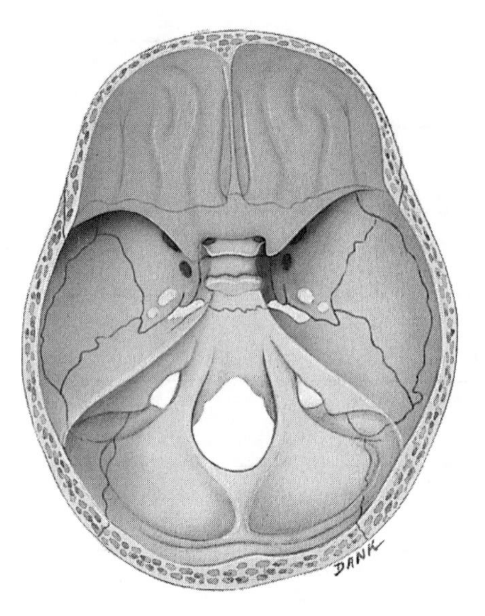

**Figure 6.16** Hyoid bone (page 155).

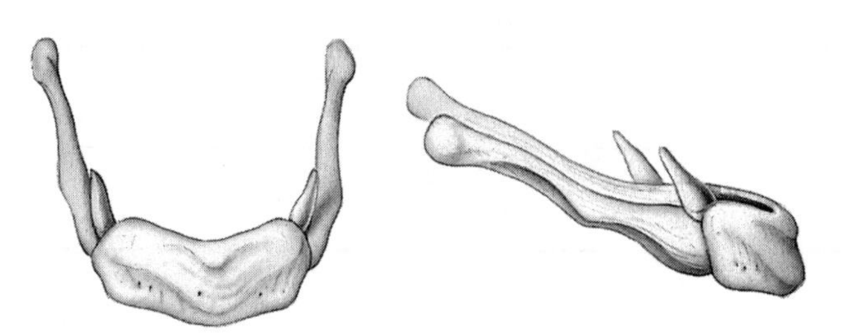

NOTES

6

**Figure 6.17** Vertebral column (page 157).

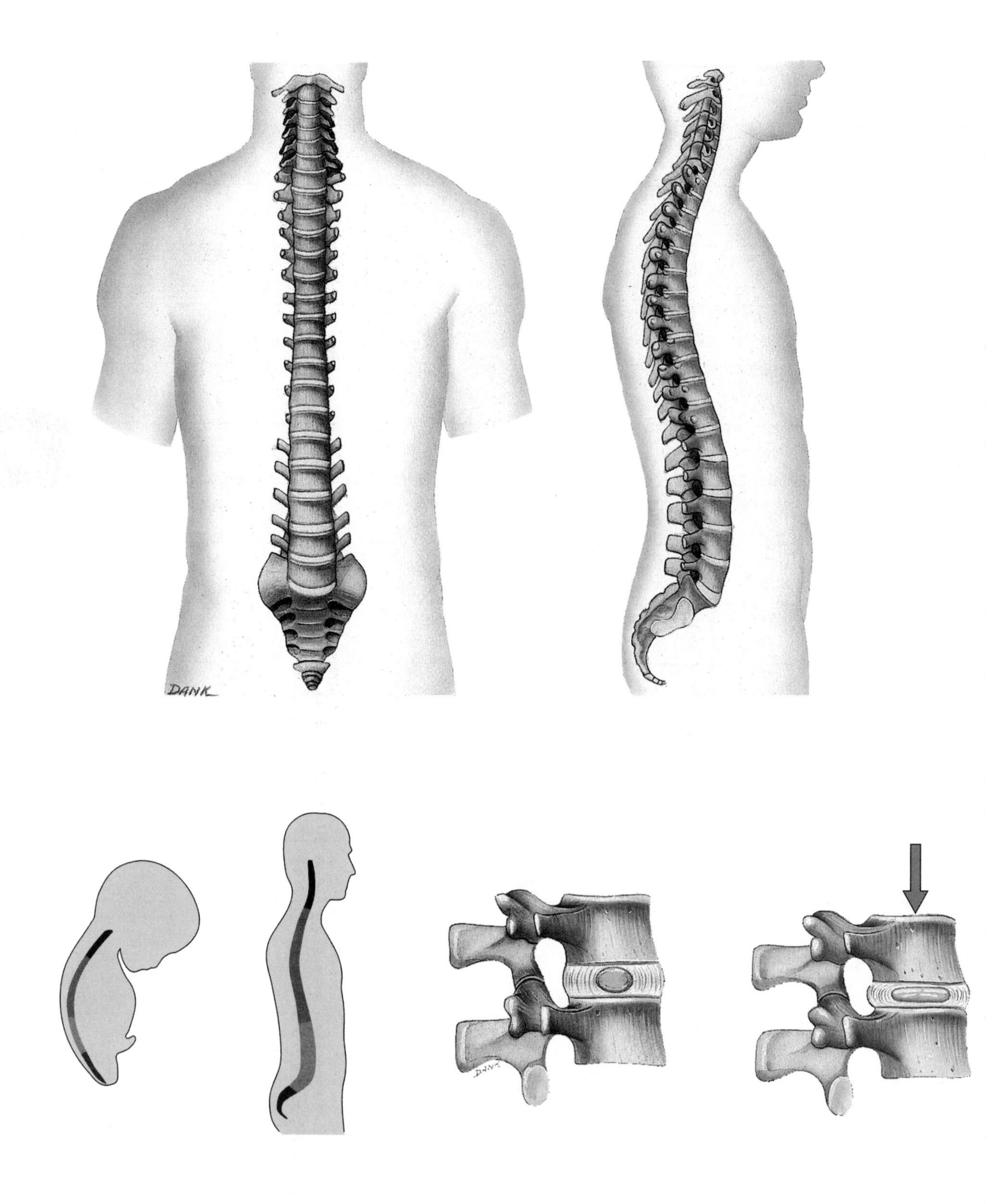

NOTES

6

**Figure 6.18a,b** Structure of a typical vertebra, as illustrated by a thoracic vertebra (page 158).

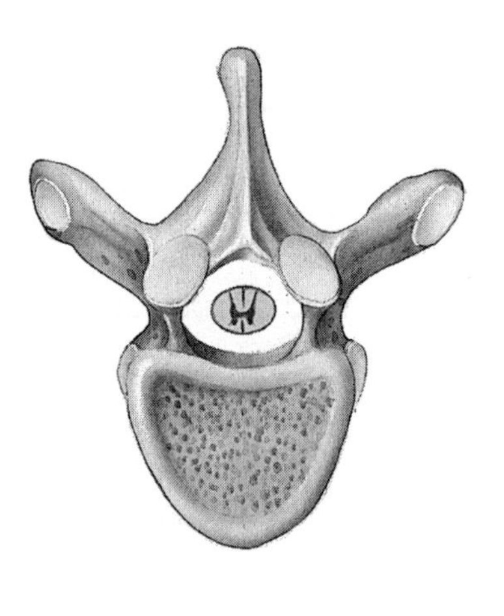

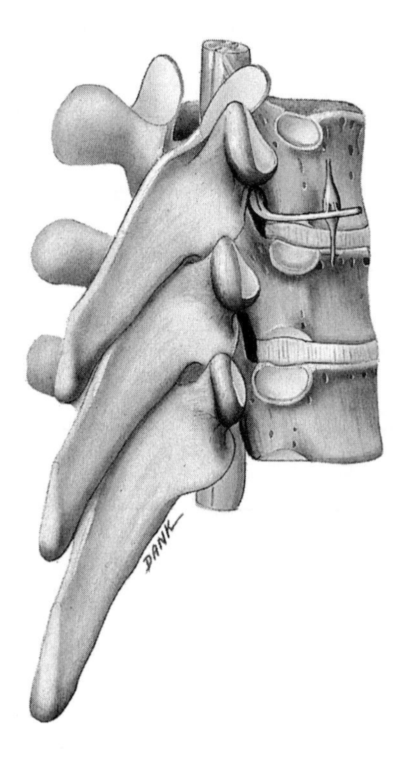


**Figure 6.19a** Cervical vertebrae (page 160).

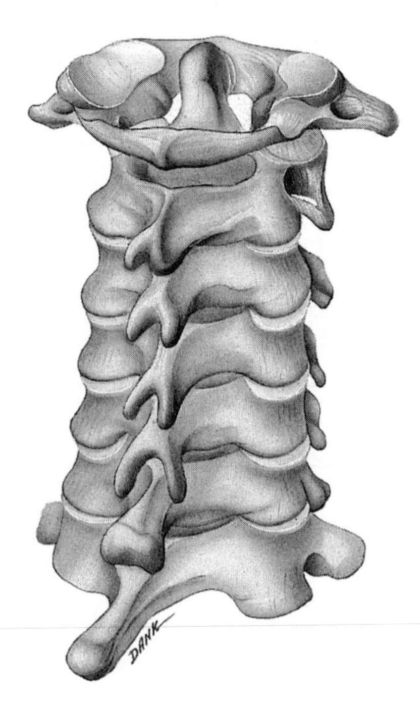

NOTES

6

**Figure 6.20a** Thoracic vertebrae (page 162).

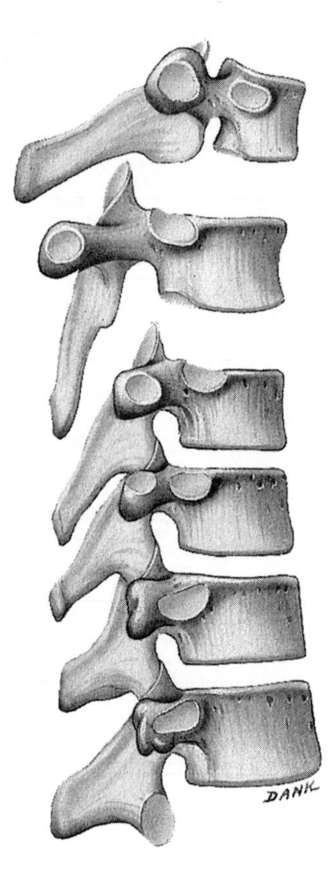


**Figure 6.21a** Lumbar vertebrae (page 163).

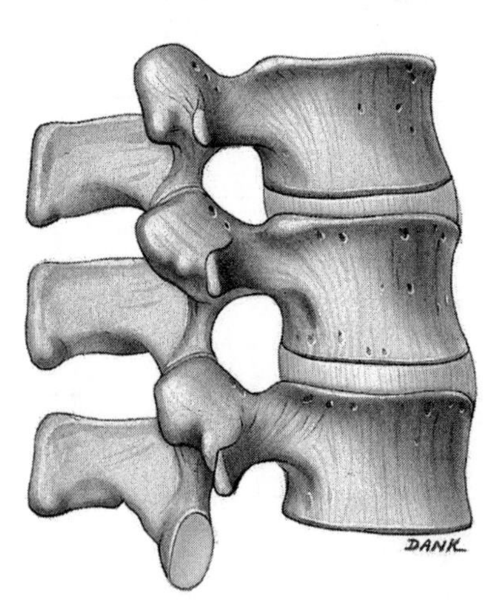

NOTES

6

**Figure 6.22** Sacrum and coccyx (page 165).

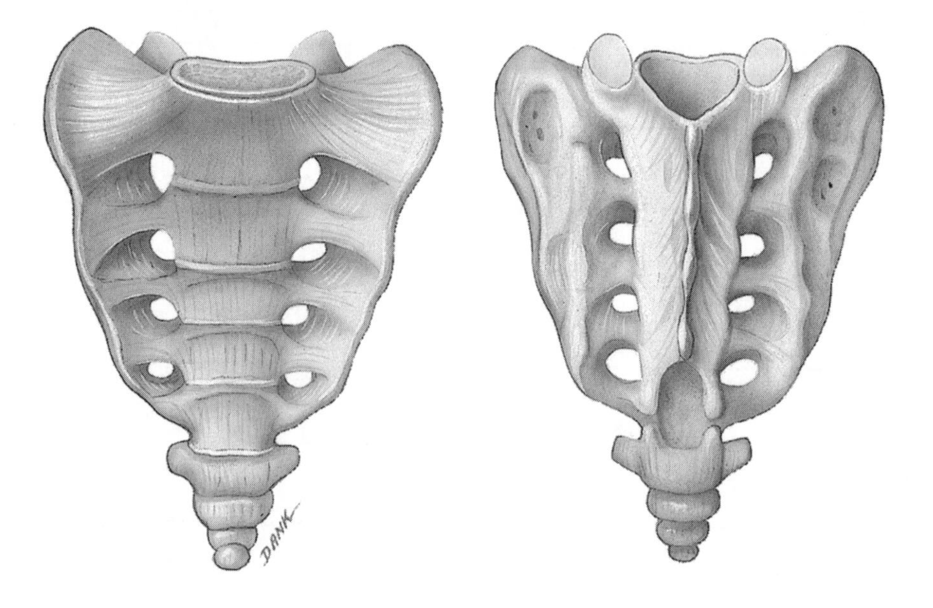

**Figure 6.23** Skeleton of the thorax (page 166).

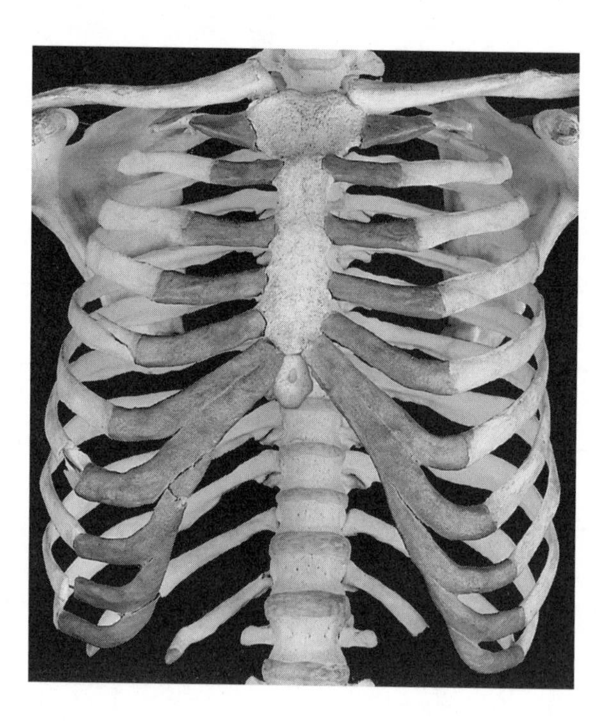

NOTES

6

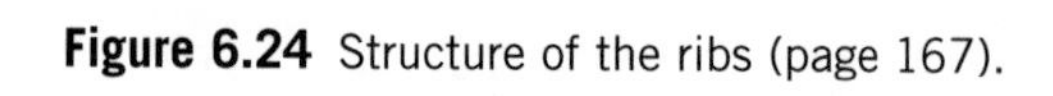

**Figure 6.24** Structure of the ribs (page 167).

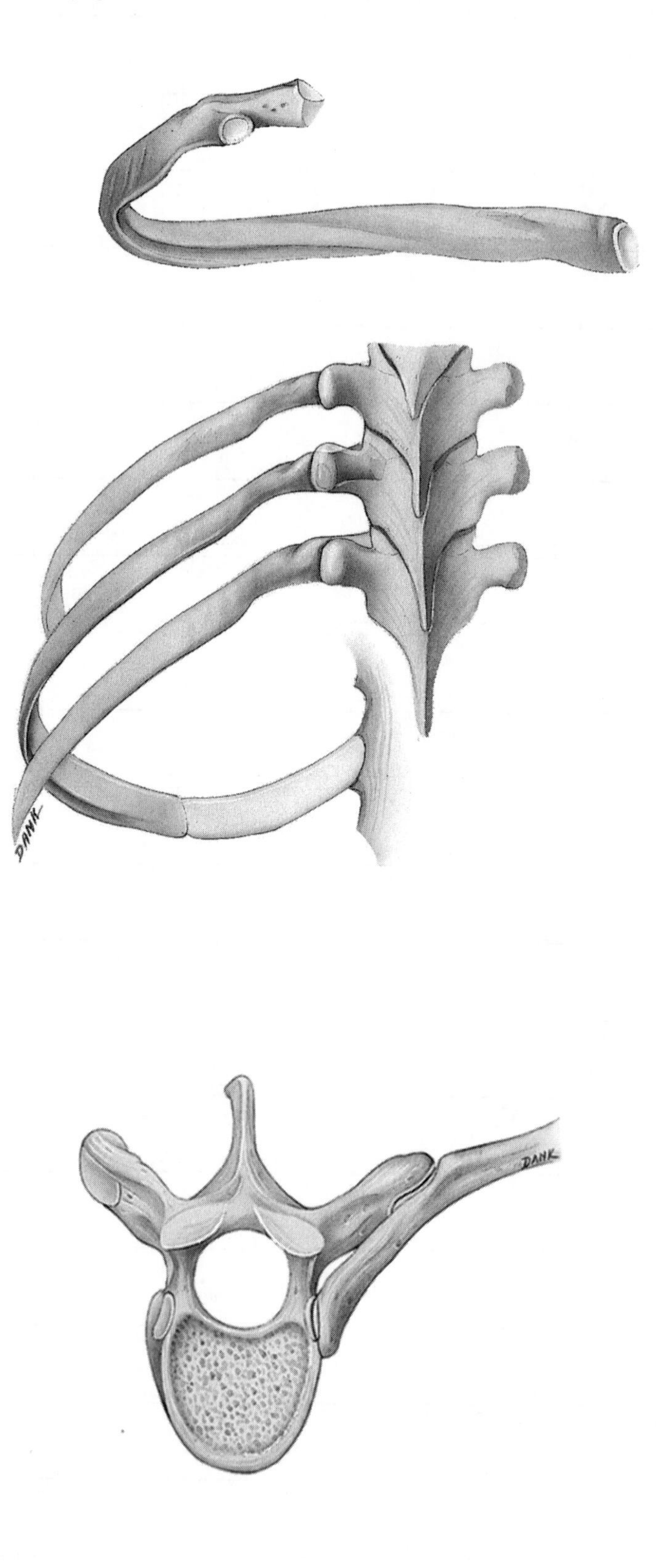

## NOTES

6

**Figure 6.25** Herniated (slipped) disc (page 168).

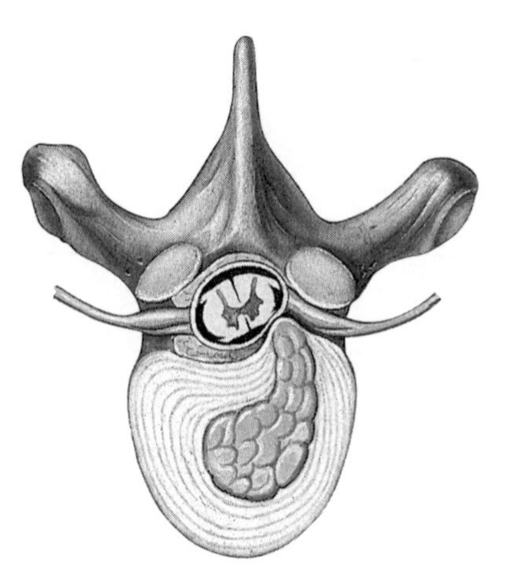

## NOTES

6

**Figure 7.1** Right pectoral (shoulder) girdle (page 173).

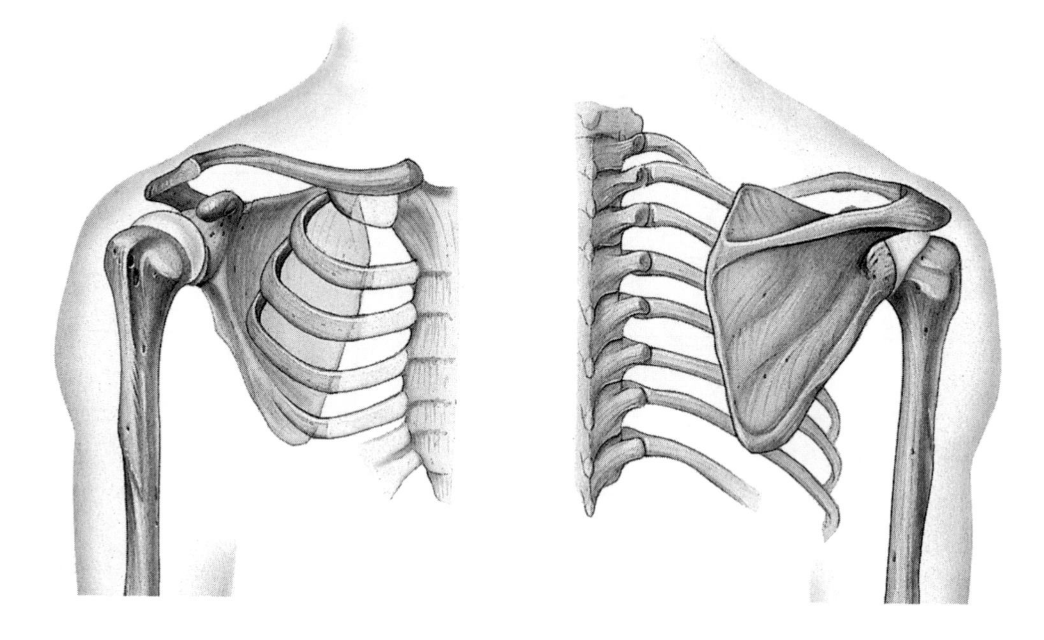

**Figure 7.2** Right clavicle (page 174).

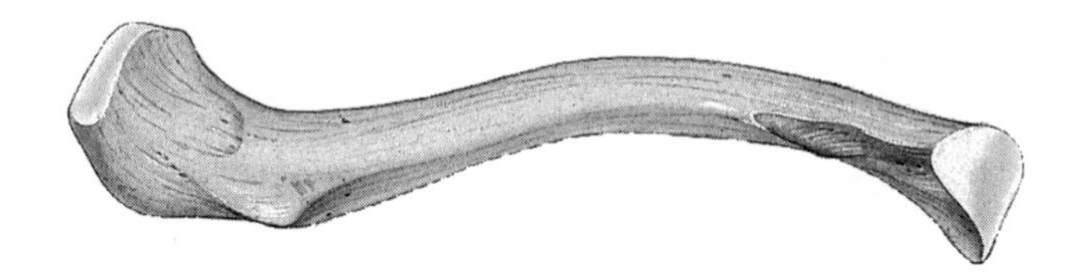

NOTES

**Figure 7.3** Right scapula (shoulder blade) (page 175).

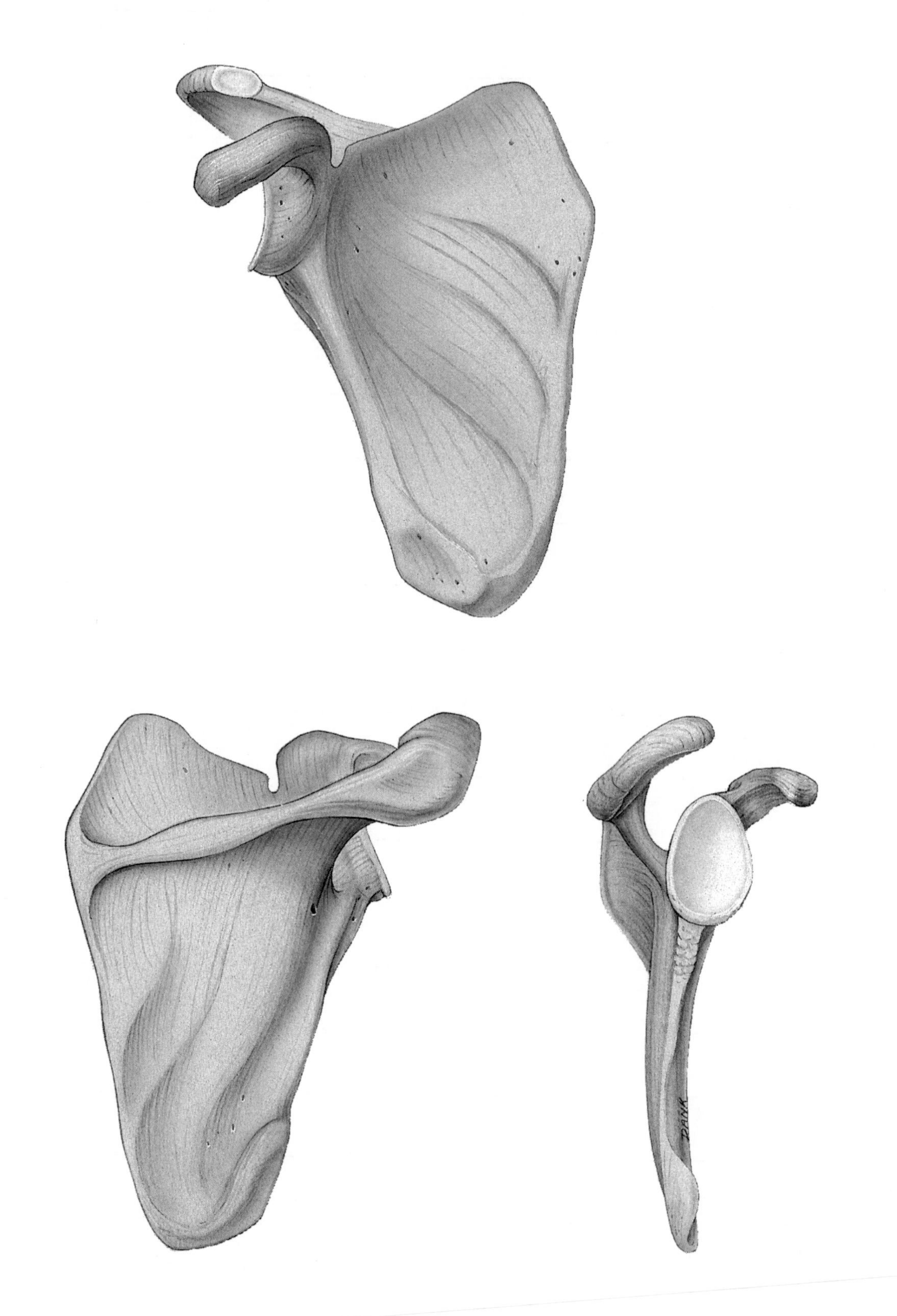

NOTES

7

**Figure 7.4** Right upper limb (page 176).

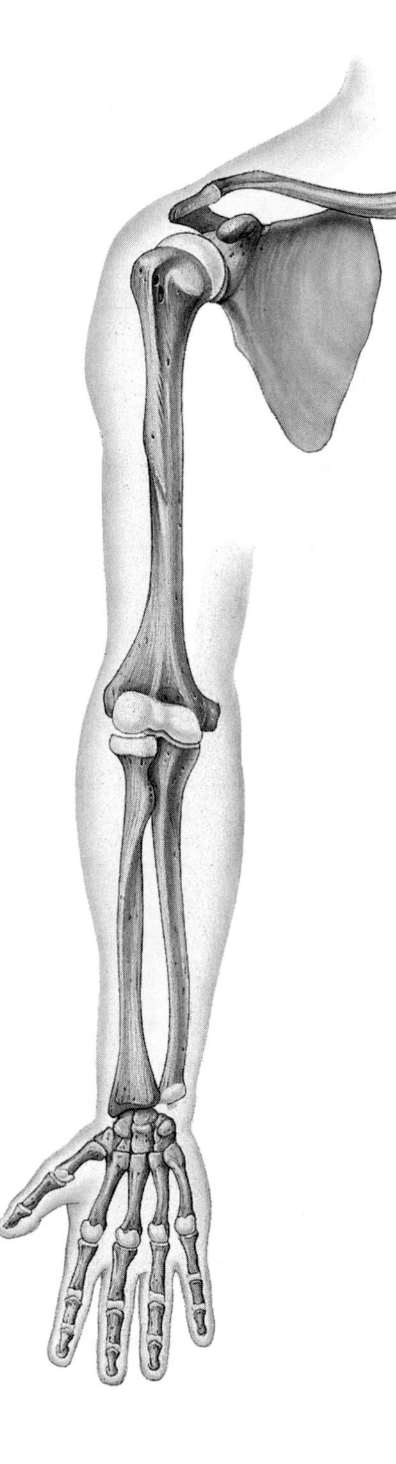

NOTES

7

**Figure 7.5** Right humerus in relation to the scapula, ulna, and radius (page 177).

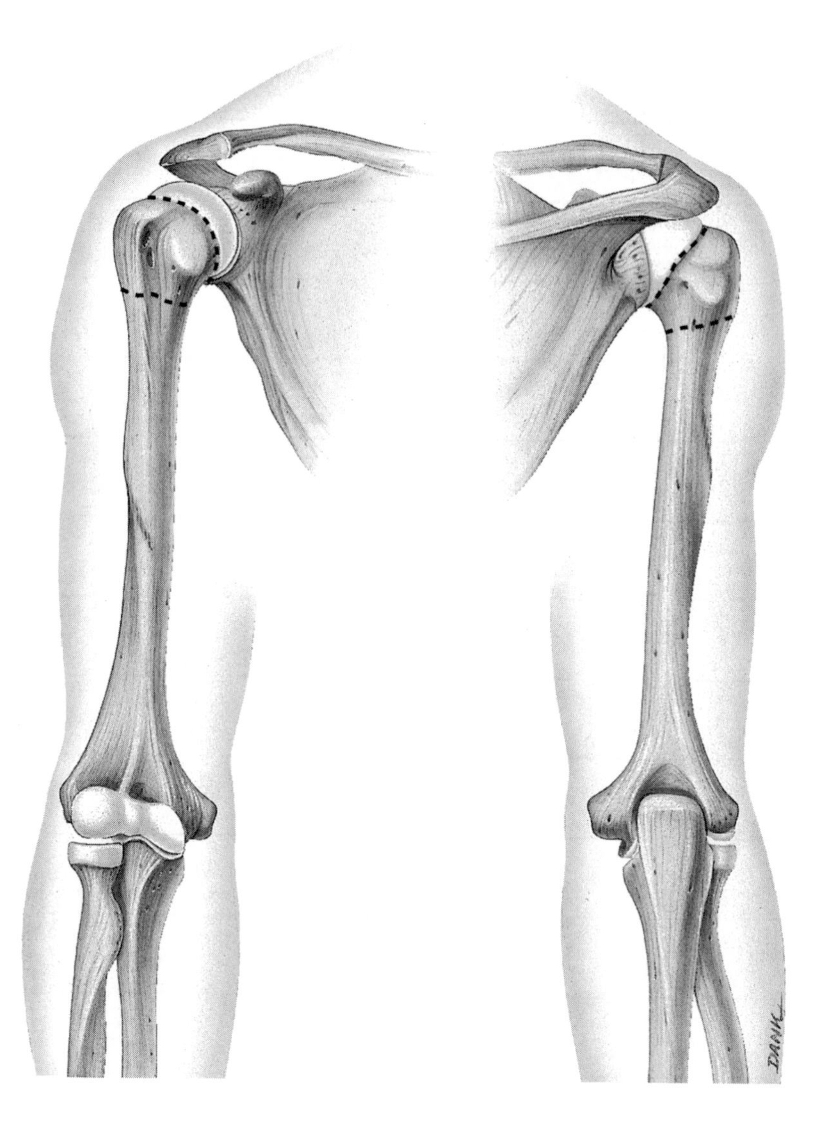

NOTES

7

**Figure 7.6** Right ulna and radius in relation to the humerus and carpals (page 179).

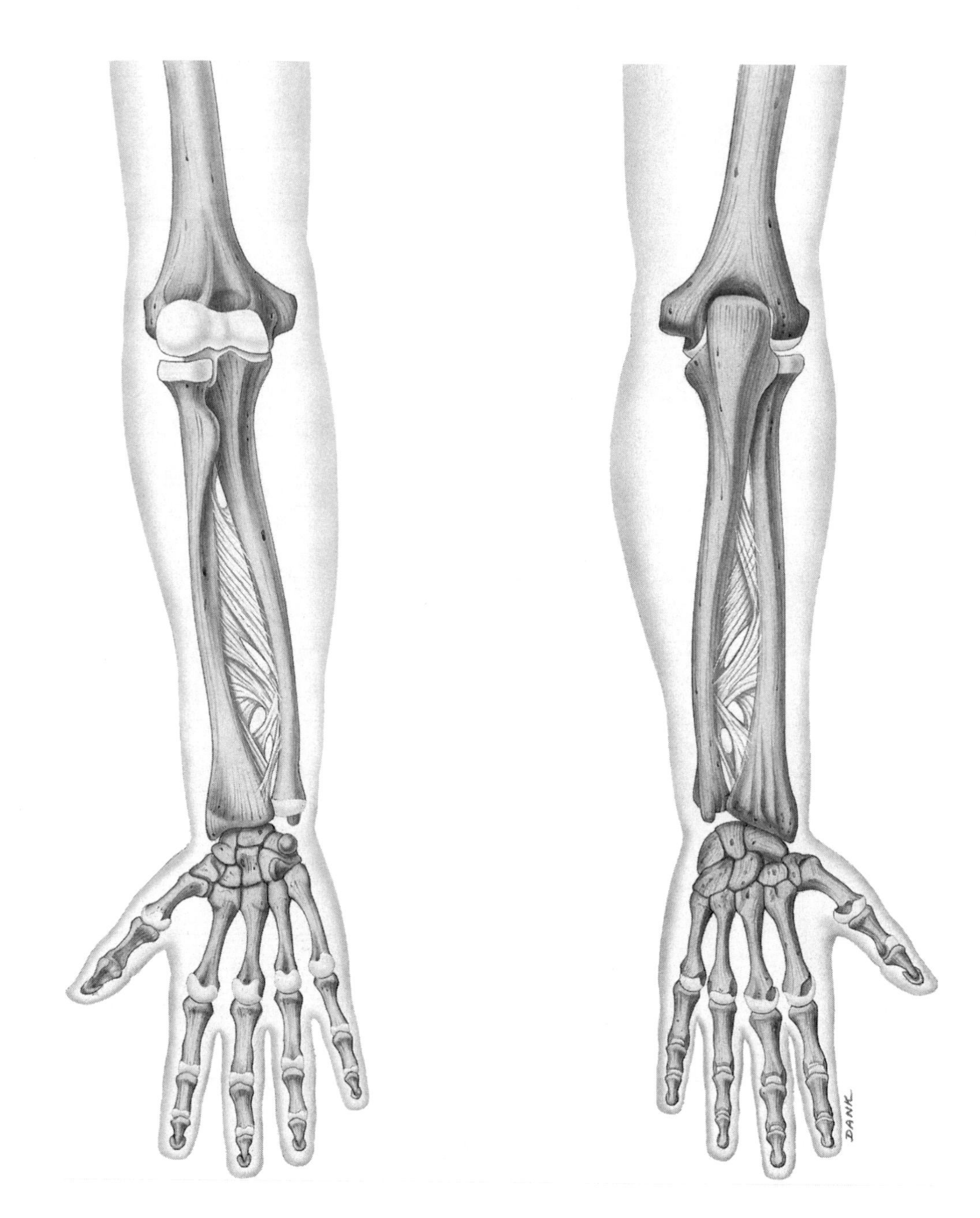

NOTES

7

**Figure 7.7** Articulations formed by the ulna and radius (page 180).

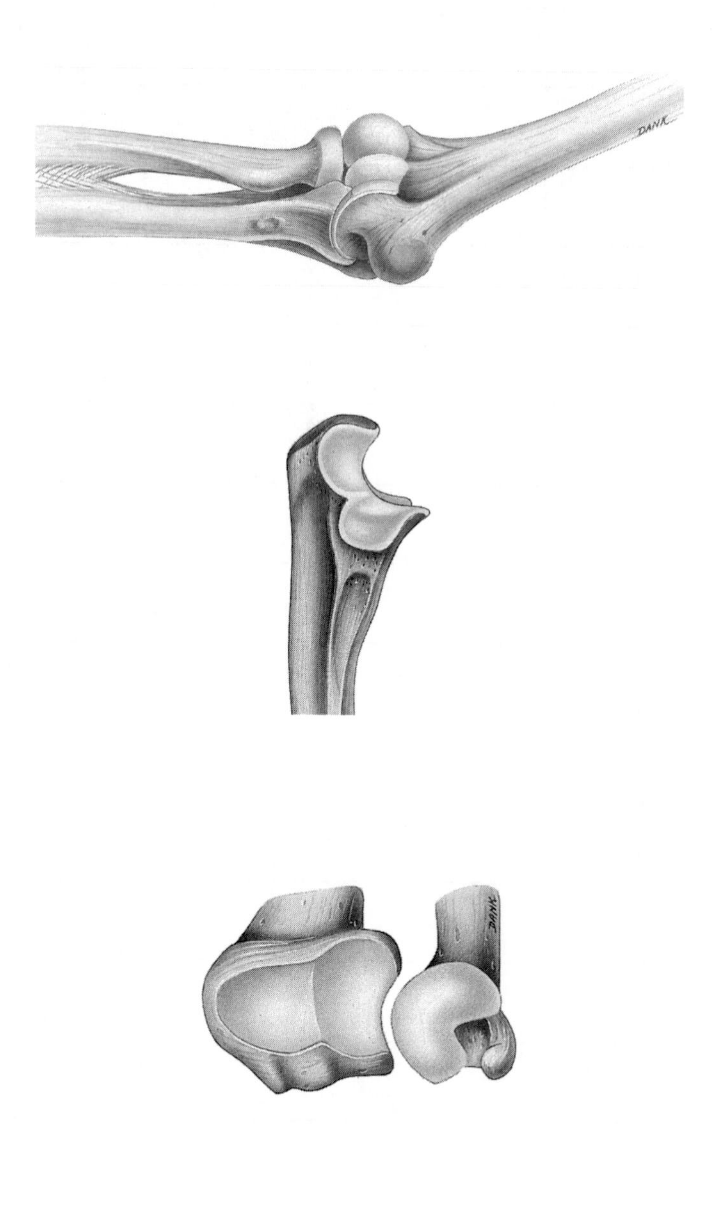

NOTES

7

**Figure 7.8** Right wrist and hand in relation to the ulna and radius (page 181).

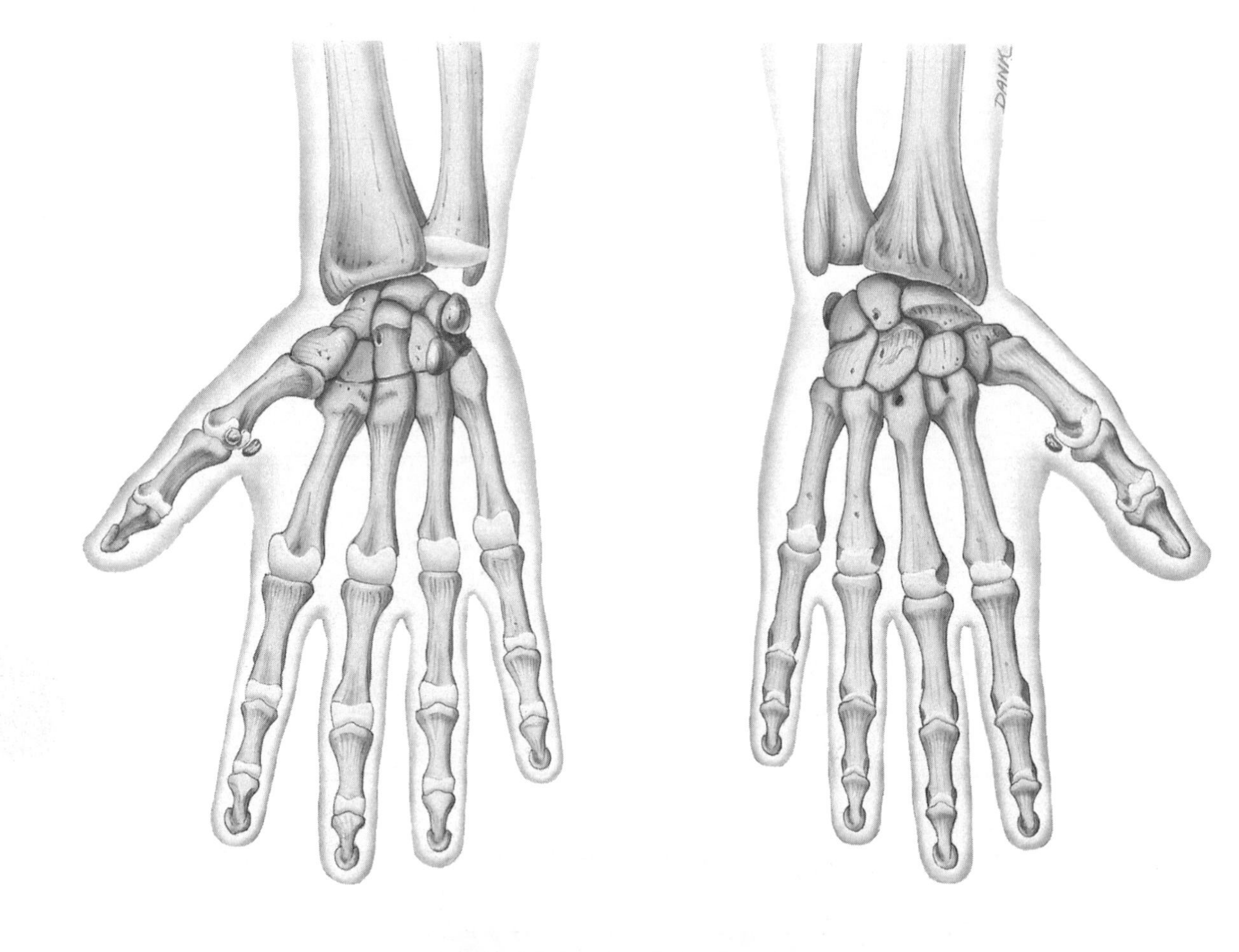


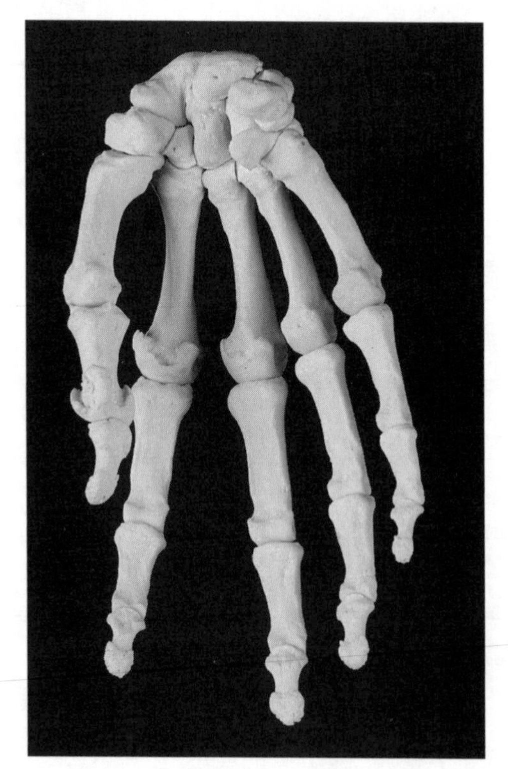

NOTES

7

**Figure 7.9** Bony pelvis (page 182).

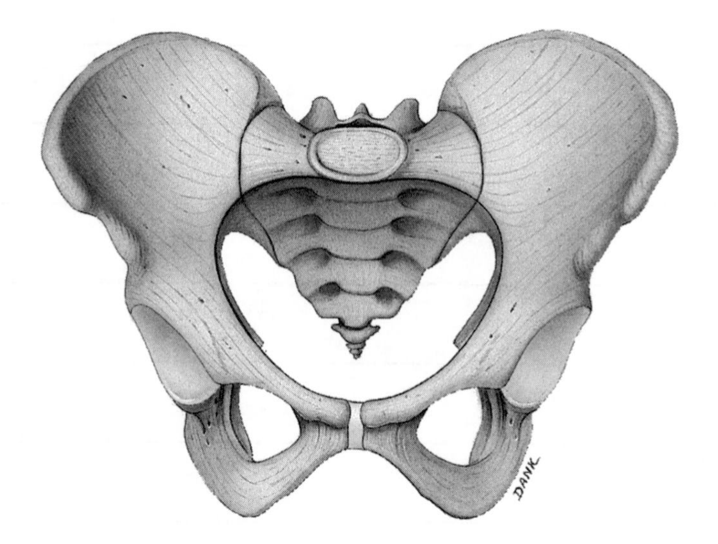

NOTES

7

**Figure 7.10** Right hip bone (page 183).

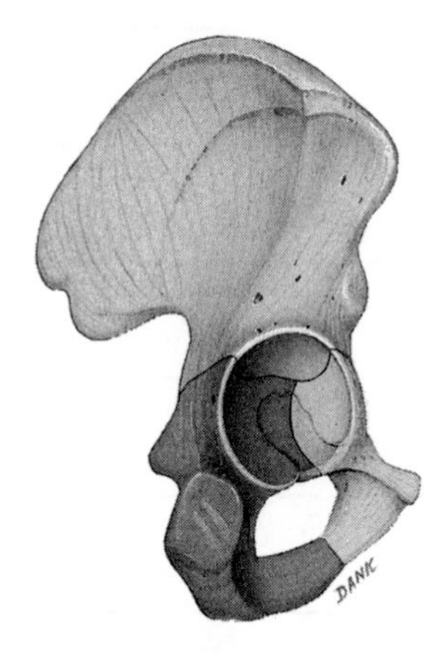

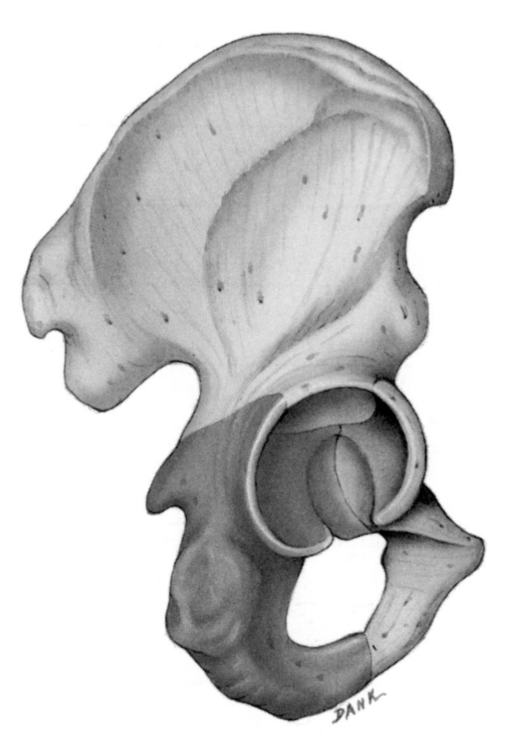

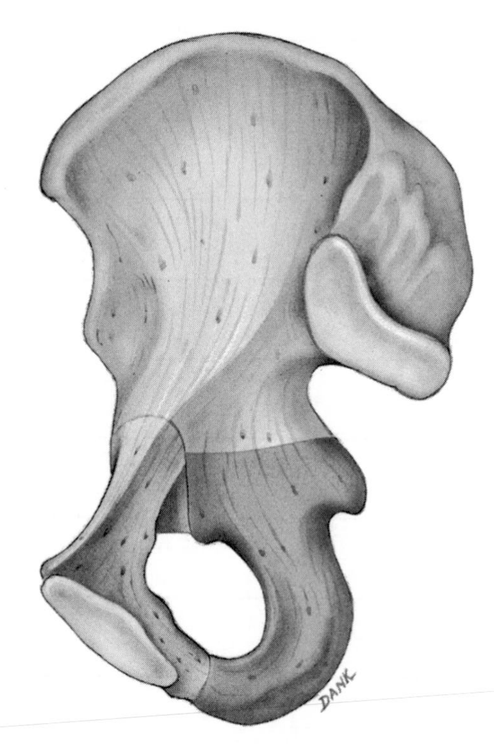

NOTES

7

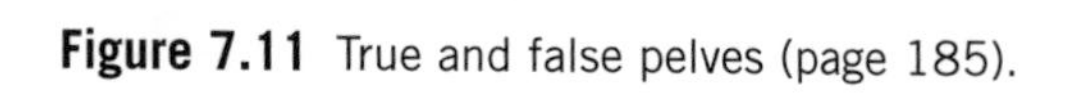

**Figure 7.11** True and false pelves (page 185).

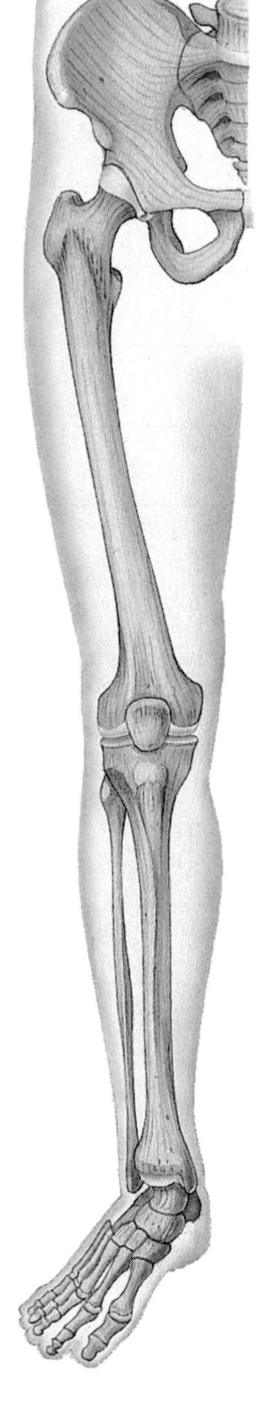

**Figure 7.12** Right lower limb (page 187).

NOTES

7

**Figure 7.13a,b** Right femur in relation to the hip bone, patella, tibia, and fibula (page 188).

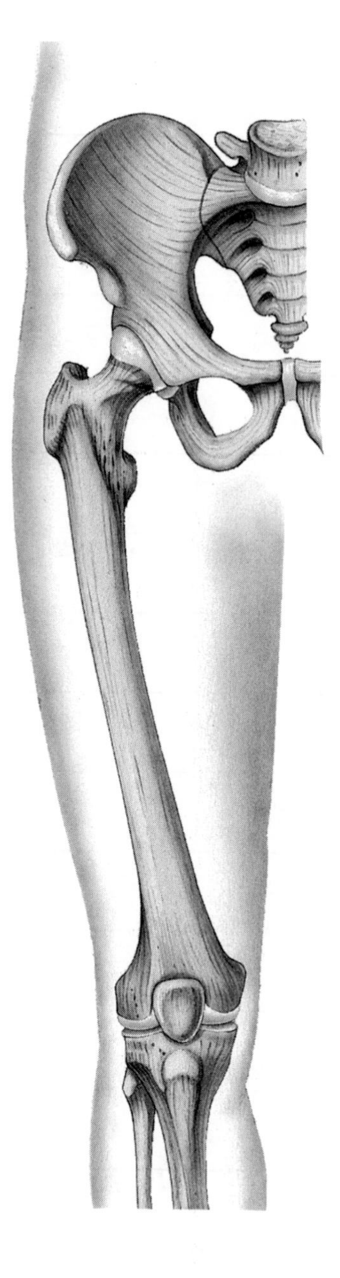

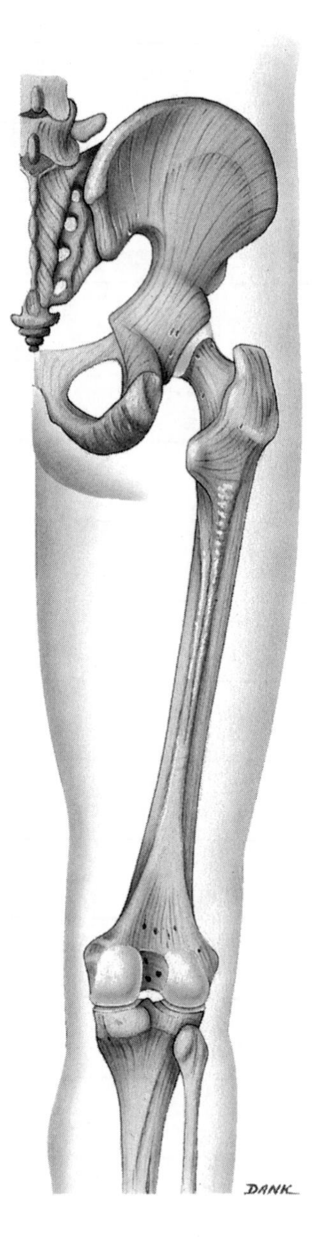

NOTES

7

**Figure 7.13c** Right femur in relation to the hip bone, patella, tibia, and fibula (page 189).

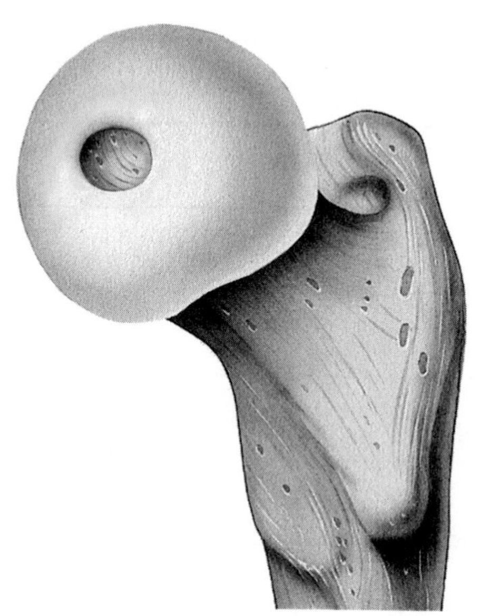

**Figure 7.14** Right patella (page 189).

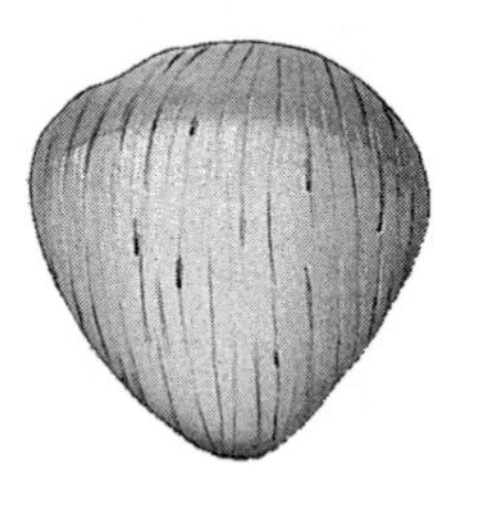

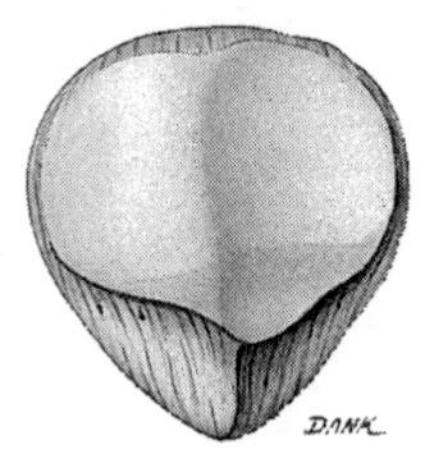

## NOTES

7

**Figure 7.15a,b** Right tibia and fibula in relation to the femur, patella, and talus (page 190).

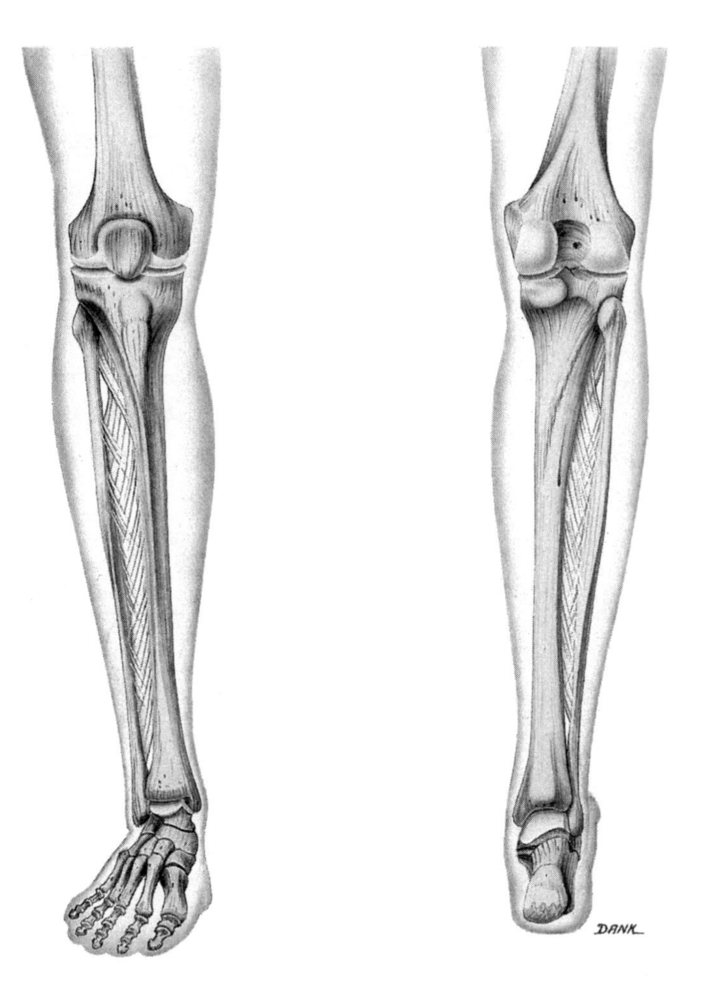

**Figure 7.15c** Right tibia and fibula in relation to the femur, patella, and talus (page 191).

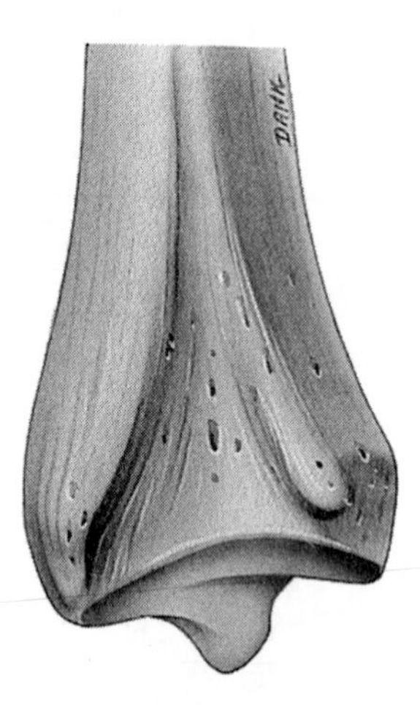

NOTES

7

**Figure 7.16a,b** Right foot (page 192).

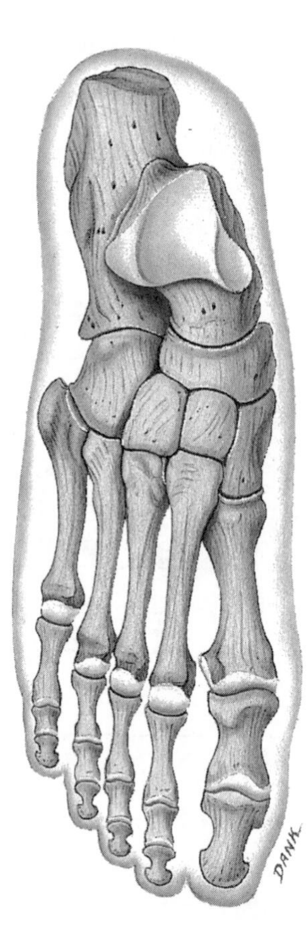


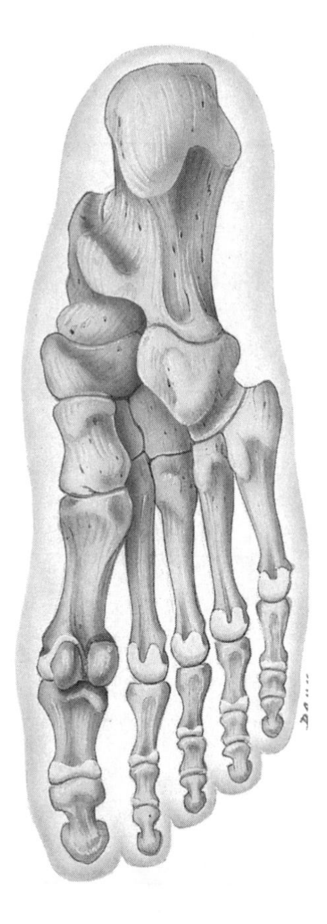

**Figure 7.17** Arches of the right foot (page 193).

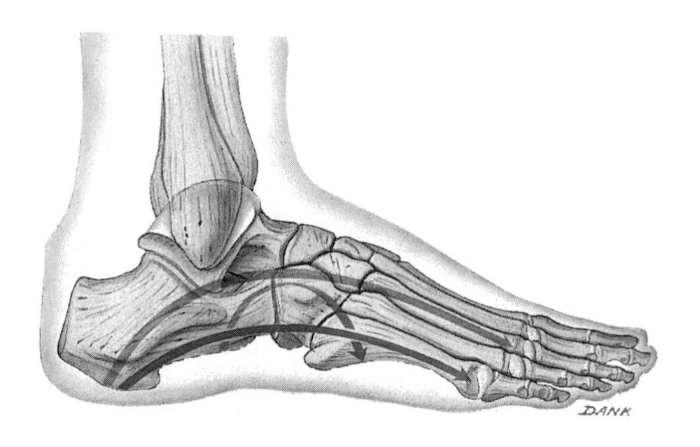

NOTES

7

**Figure 8.1** Fibrous joints (page 198).

**Figure 8.2** Cartilaginous joints (page 199).

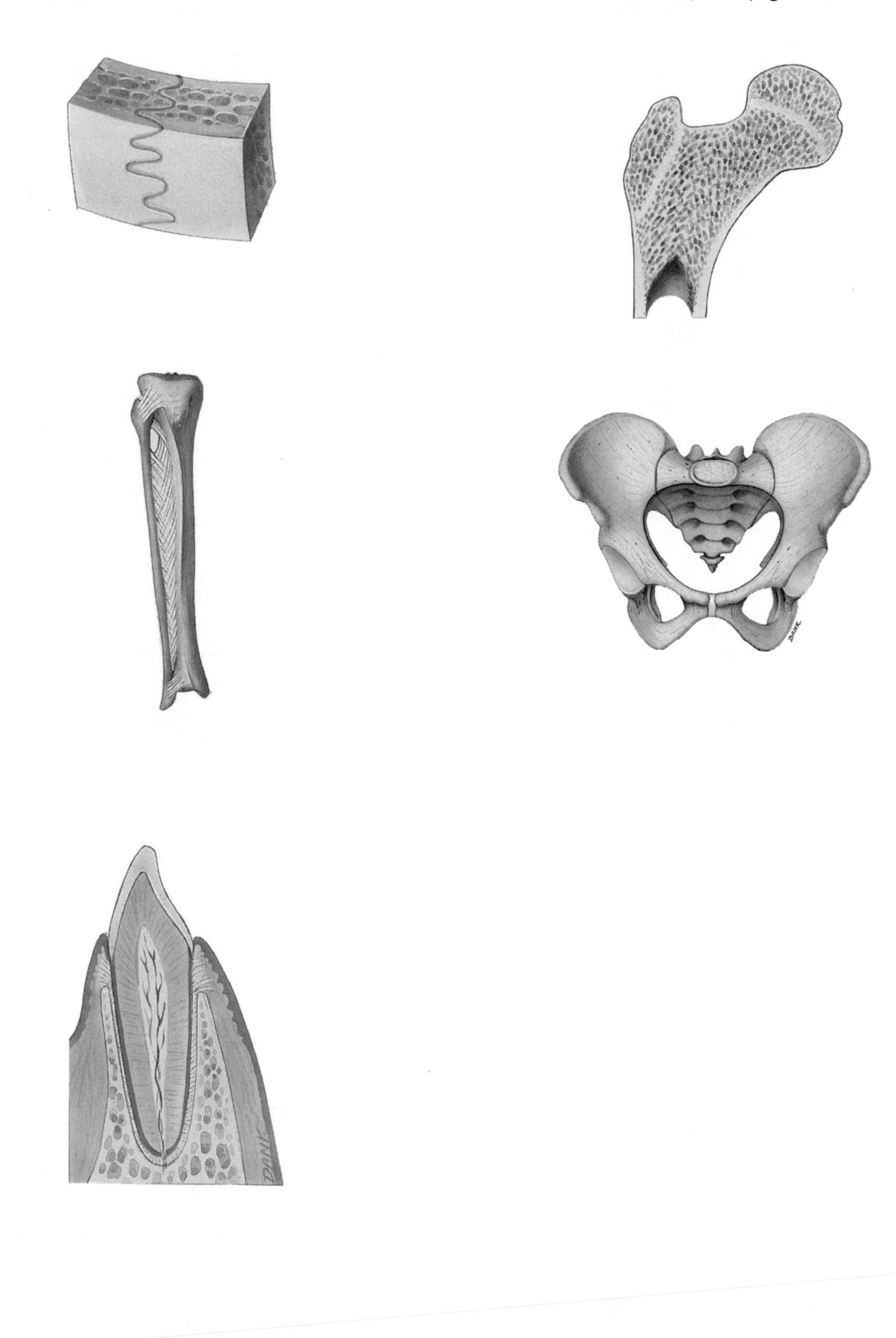

NOTES

8

**Figure 8.3a** Structure of a typical synovial joint (page 200).

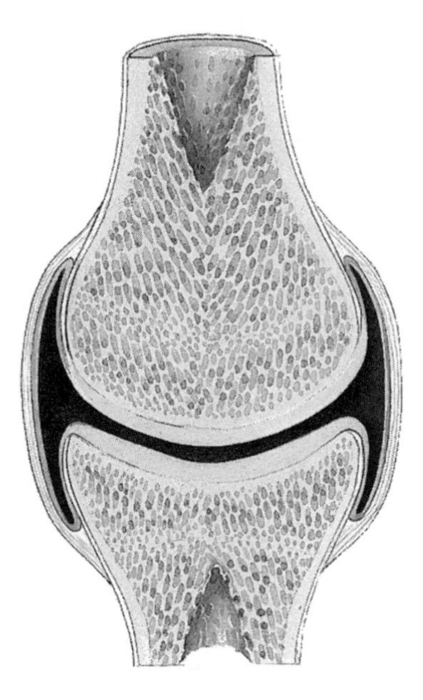

NOTES

8

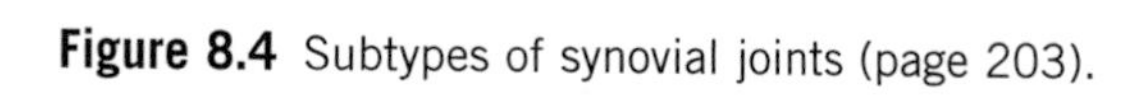
**Figure 8.4** Subtypes of synovial joints (page 203).

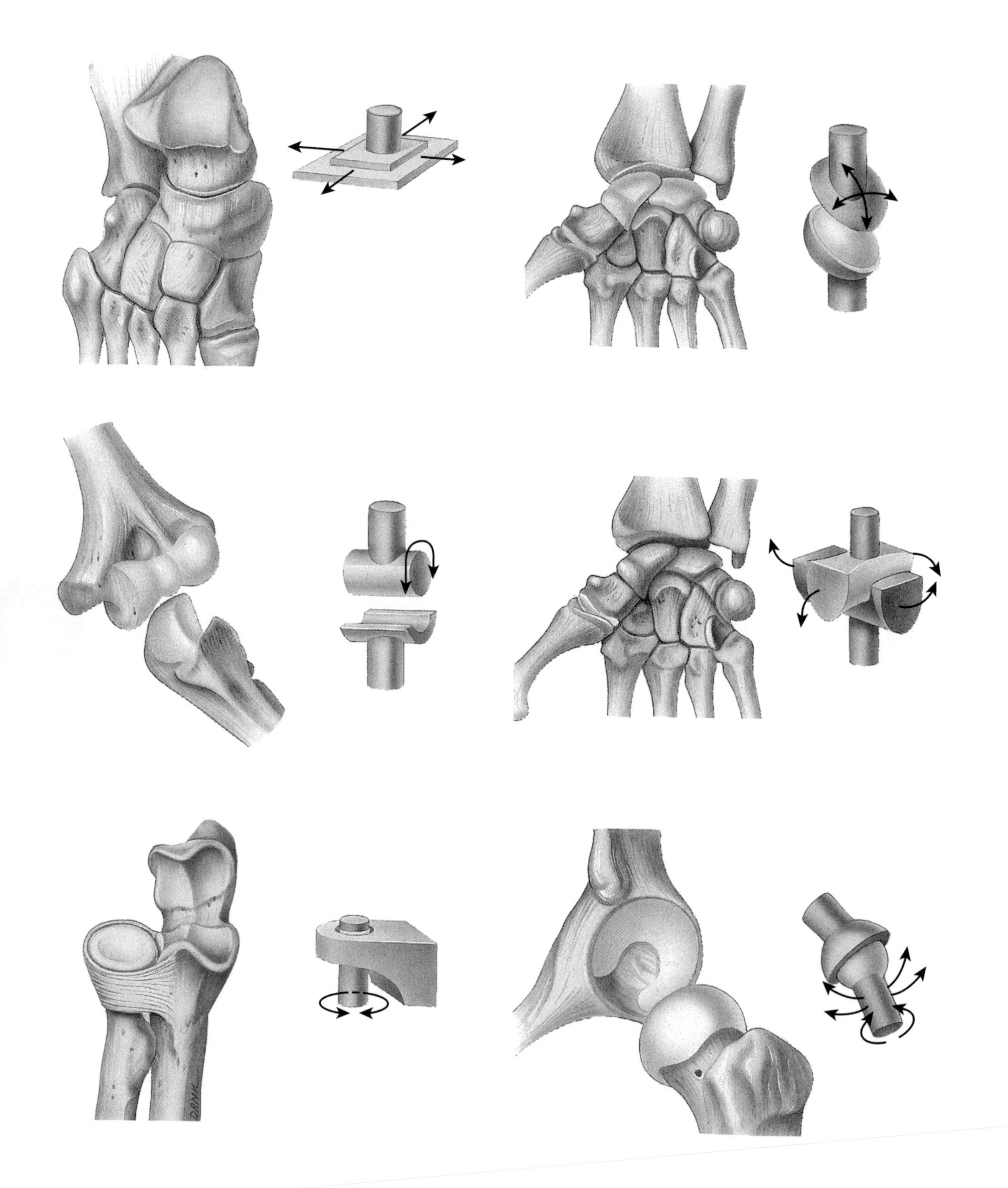

NOTES

8

**Figure 8.11** Right temporomandibular joint (TMJ) (page 213).

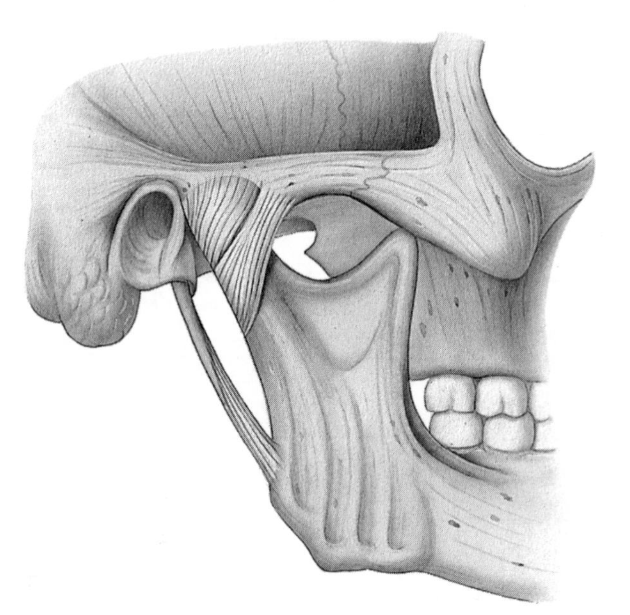

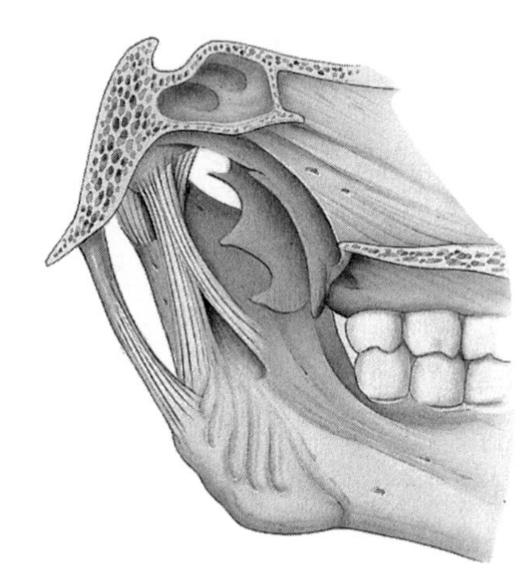

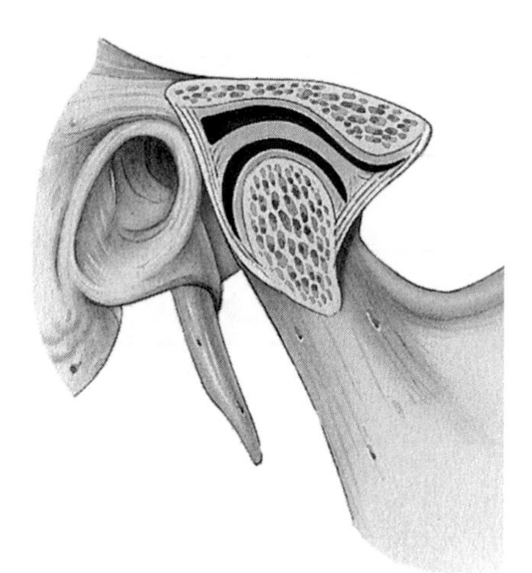

NOTES

8

**Figure 8.12a** Right shoulder (humeroscapular or glenohumeral) joint (page 214).

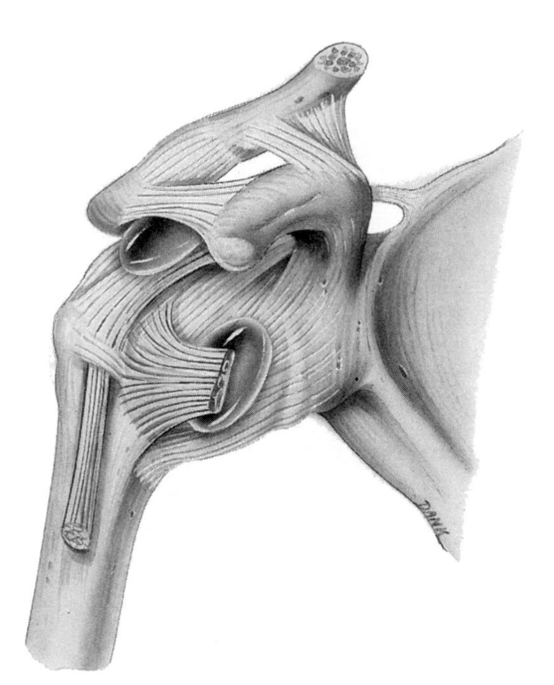

NOTES

8

**Figure 8.12b,c** Right shoulder (humeroscapular or glenohumeral) joint (page 215).

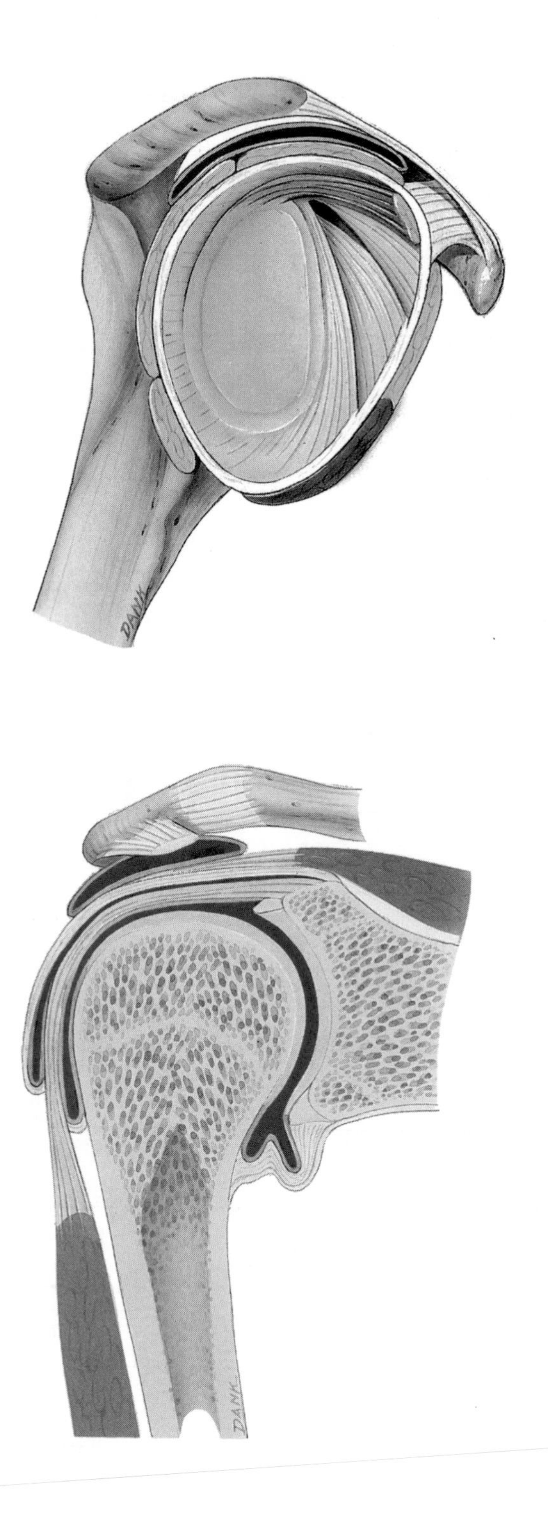

NOTES

8

**Figure 8.13** Right elbow joint (page 217).

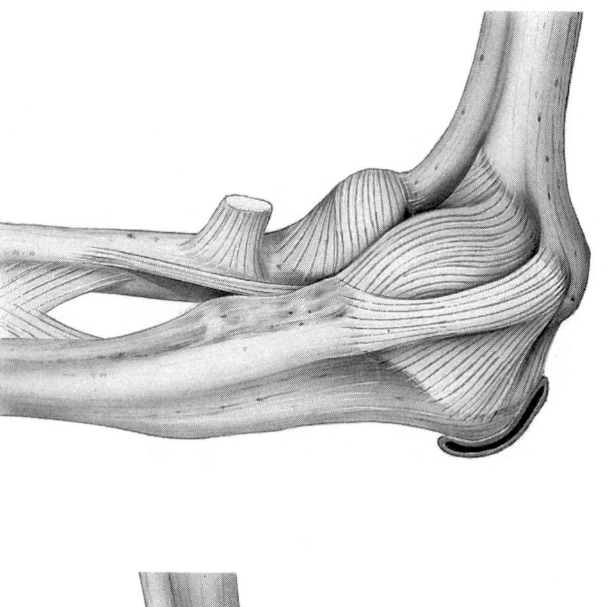

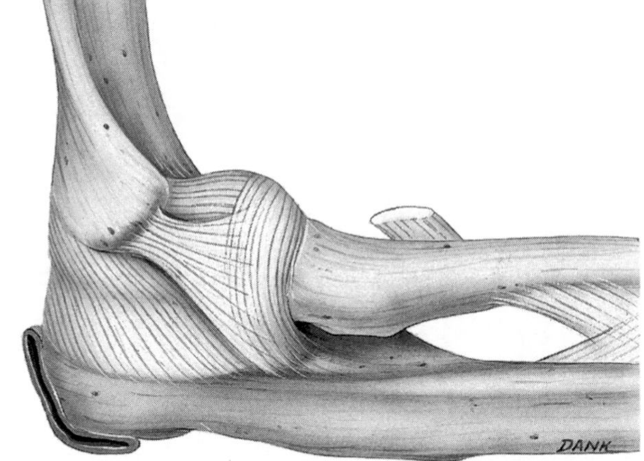


**Figure 8.14a** Right hip (coxal) joint (page 218).

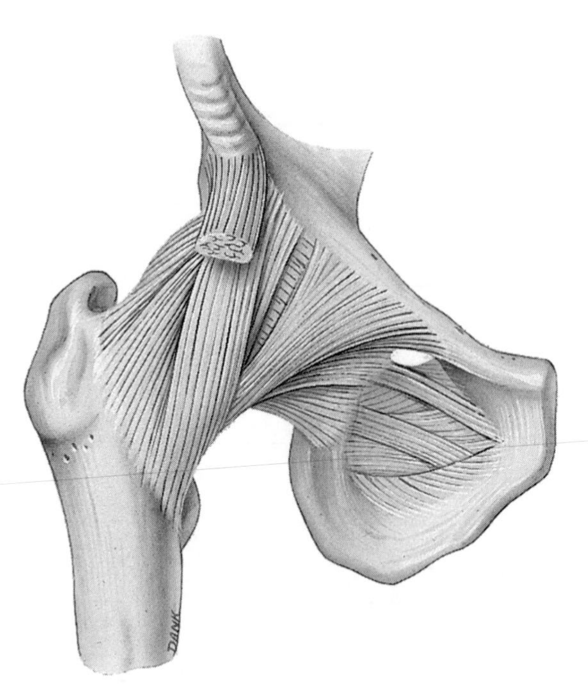

NOTES

8

**Figure 8.14b,c** Right hip (coaxal) joint (page 219).

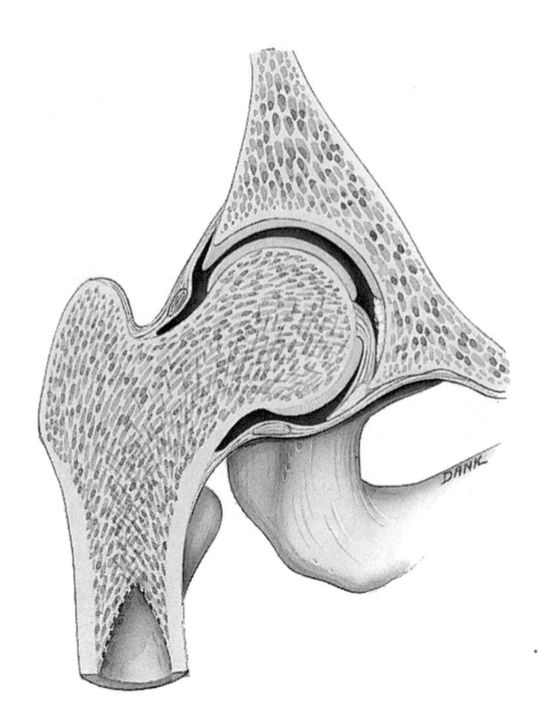


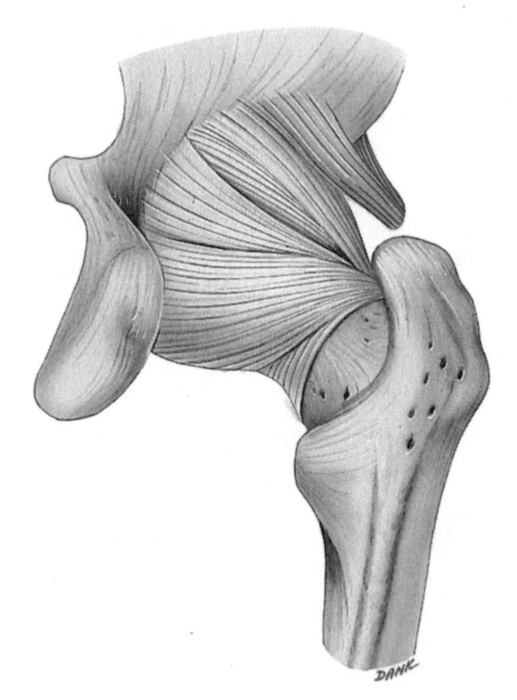

NOTES

8

**Figure 8.15** Right knee (tibiofemoral) joint (page 222).

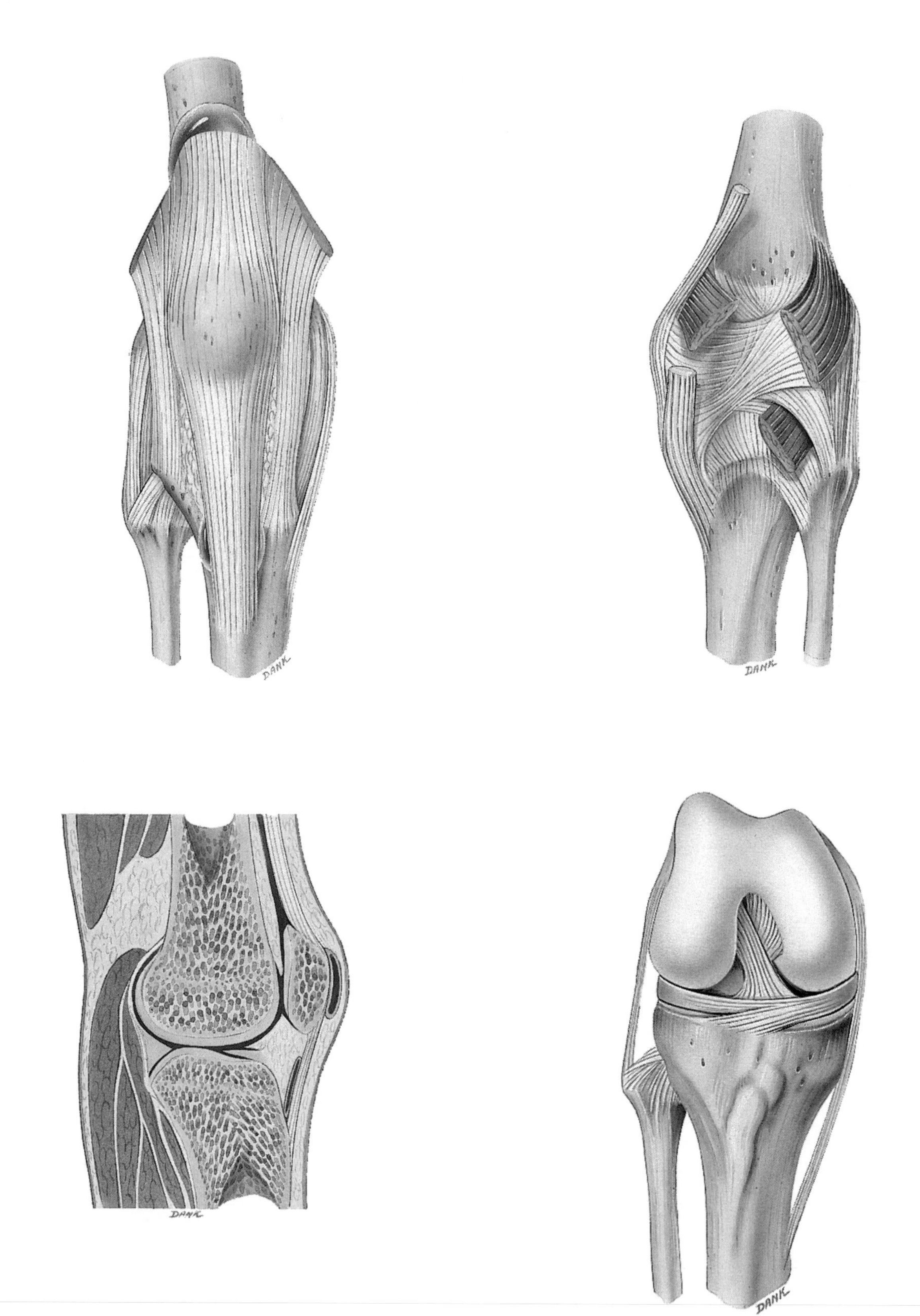

NOTES

8

**Figure 9.1** Organization of skeletal muscle and its connective tissue coverings (page 232).

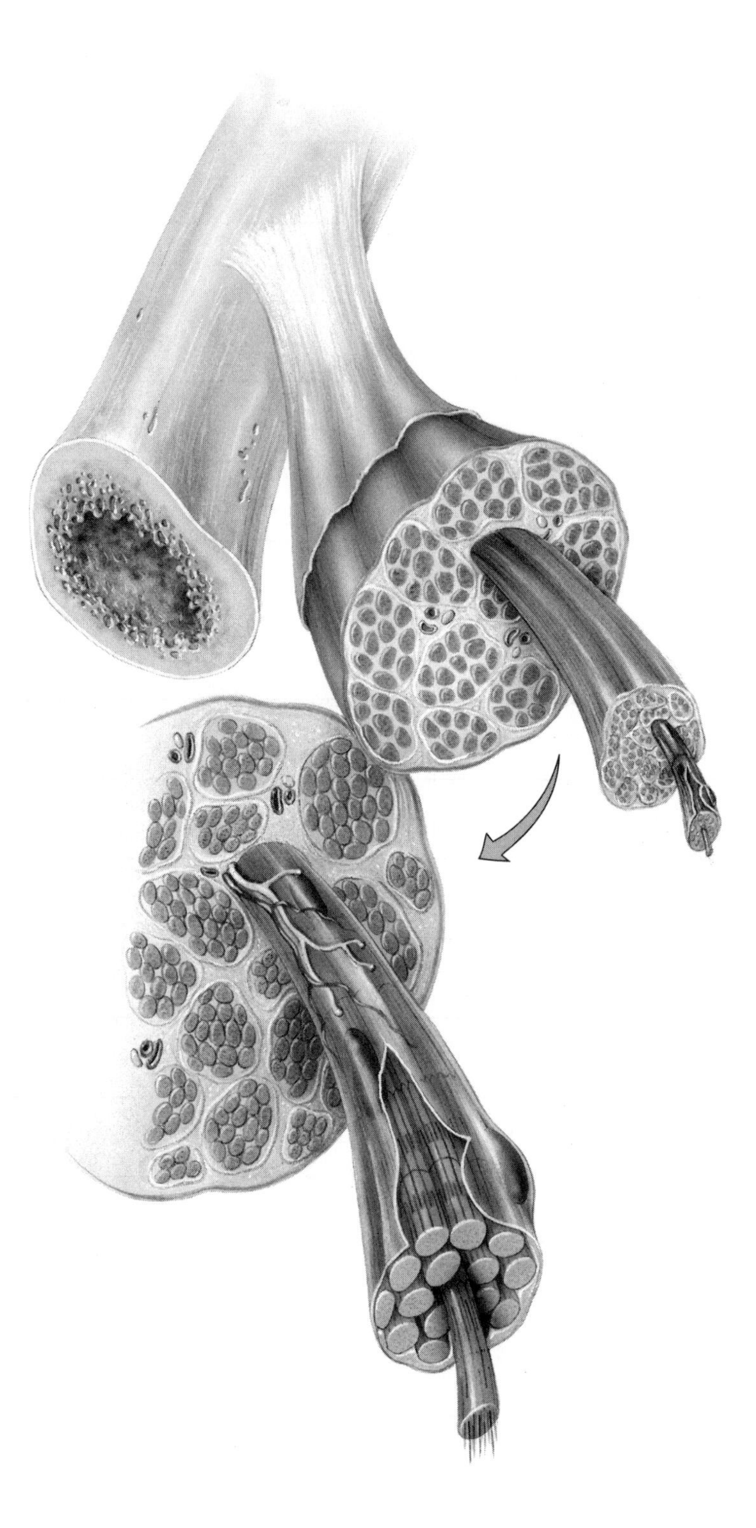

NOTES

9

**Figure 9.2** Microscopic organization of skeletal muscle (page 233).

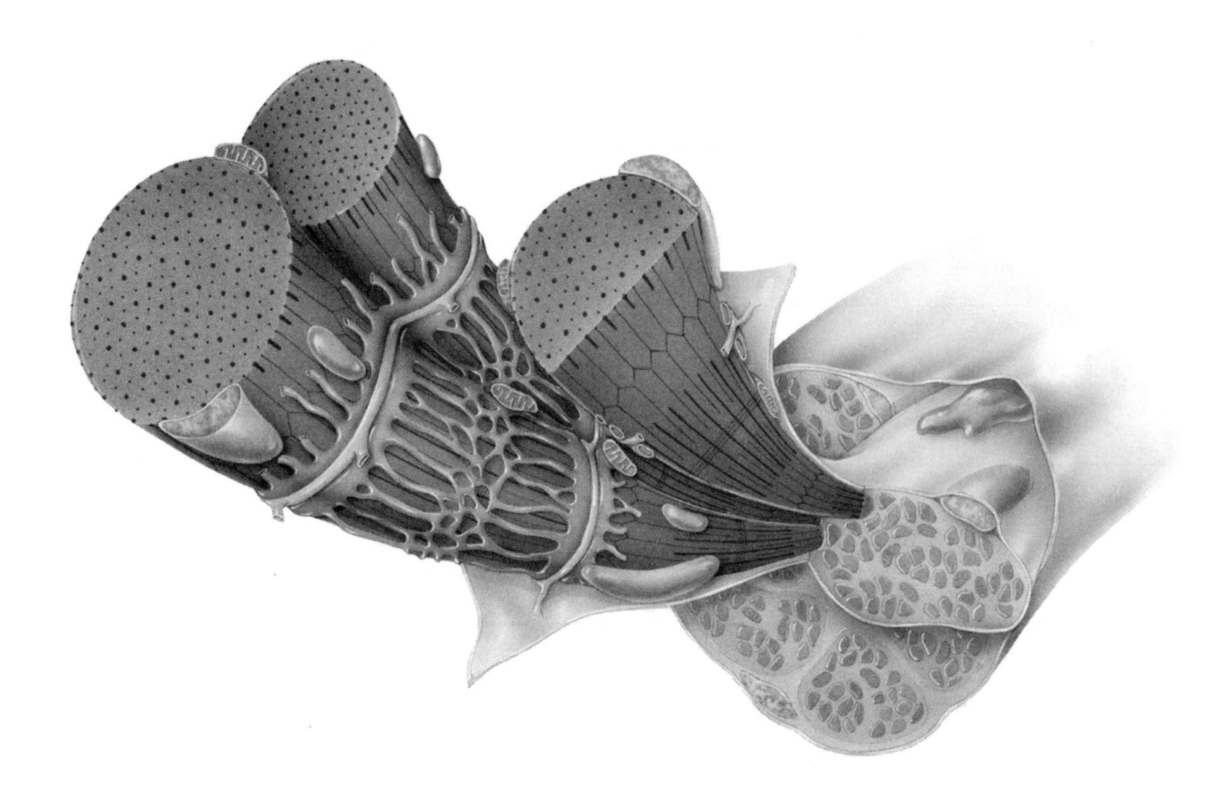

NOTES

9

**Figure 9.3** Arrangement of filaments within a sarcomere (page 235).

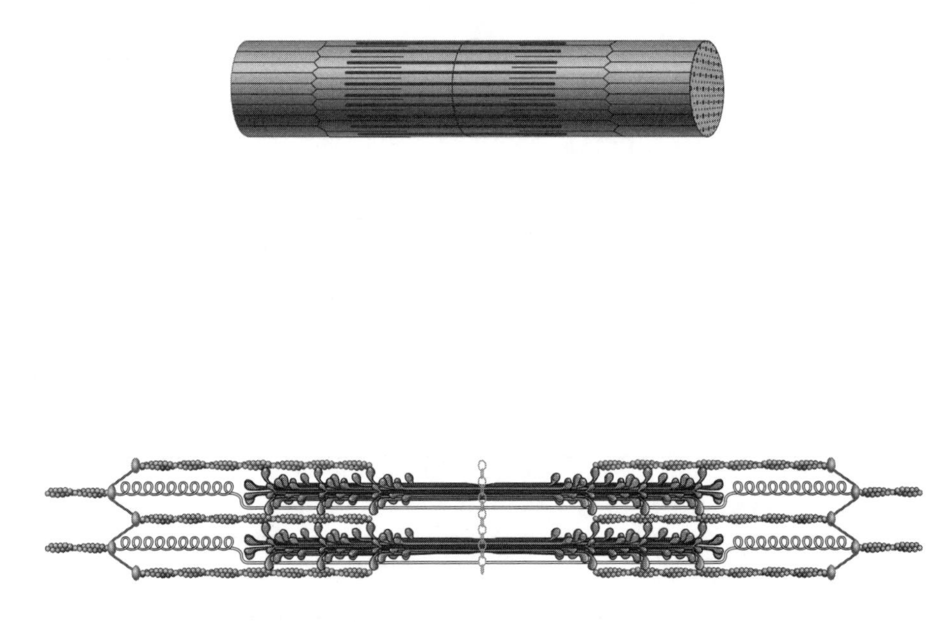

**Figure 9.4** Transmission electron micrograph showing the characteristic zones and bands of a sarcomere (page 235).

TEM 21,600x

## NOTES

9

**Figure 9.5** Structure of thick and thin filaments (page 236).

**Figure 9.6a–c** Structure of the neuromuscular junction (NMJ) (page 238).

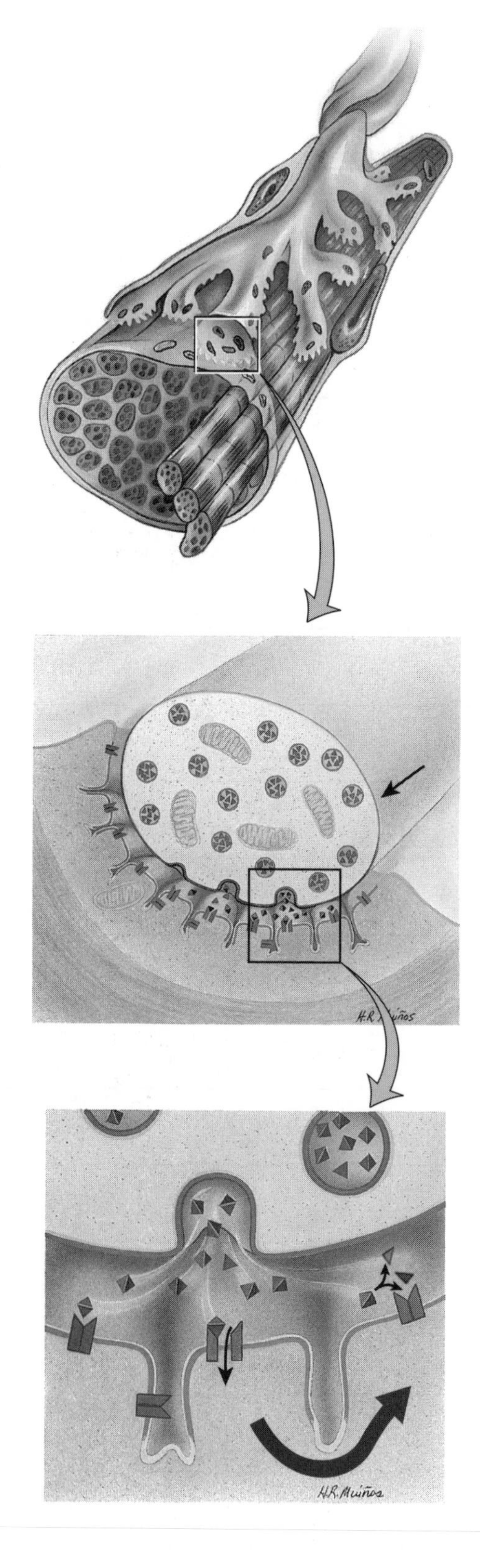

NOTES

9

**Figure 9.7** Sliding filament mechanism of muscle contraction (page 240).

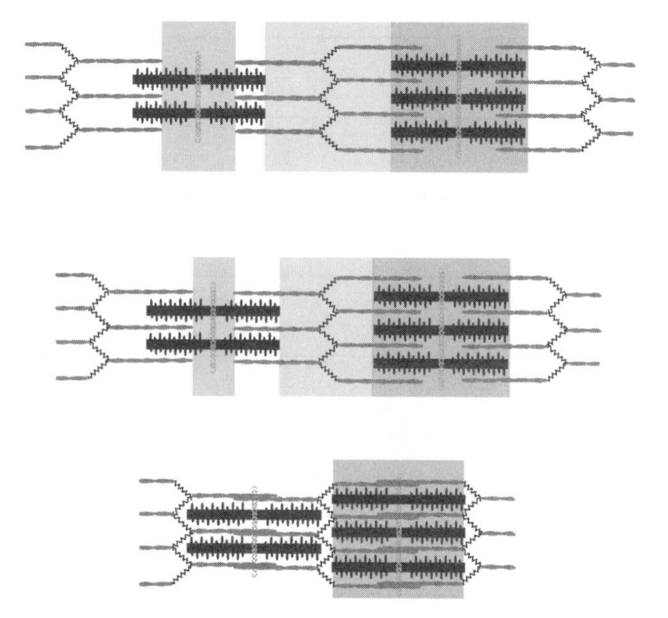

NOTES

9

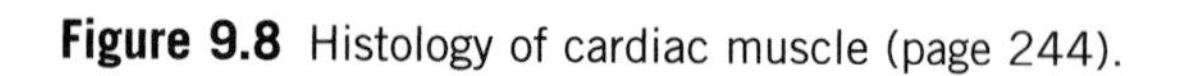

**Figure 9.8** Histology of cardiac muscle (page 244).

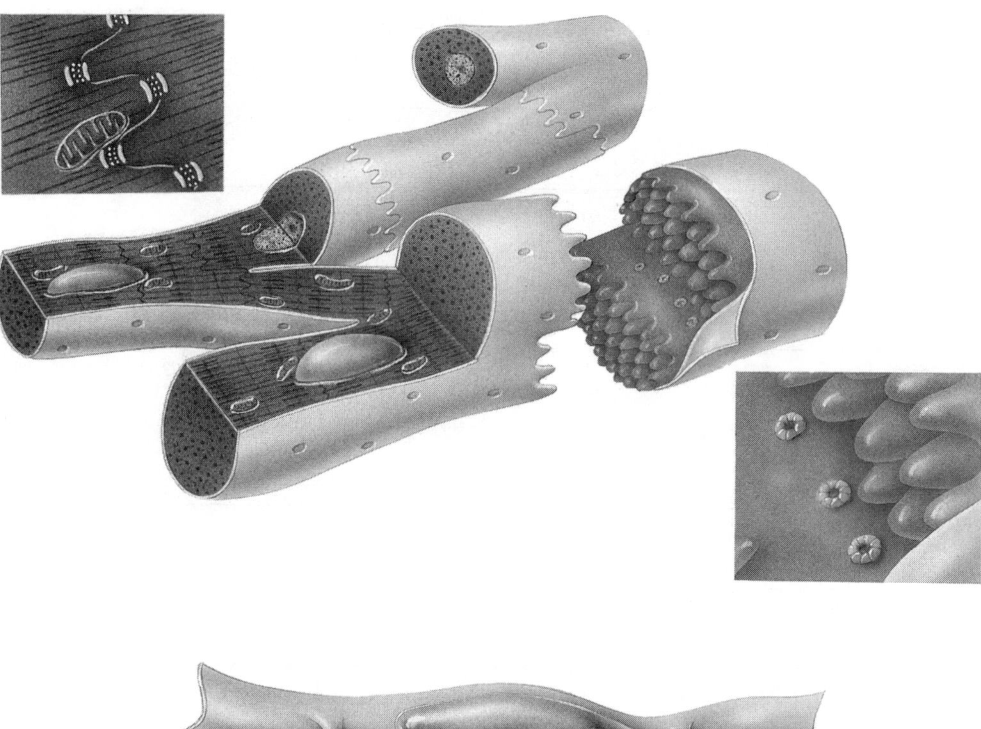

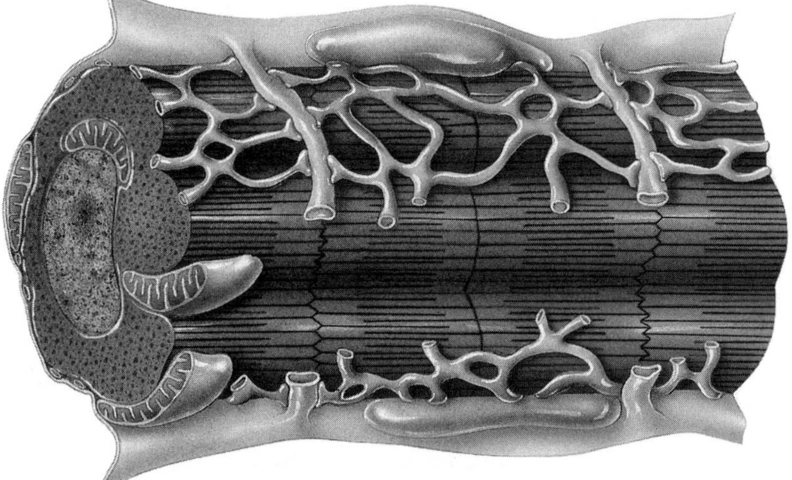

## NOTES

9

**Figure 9.9** Histology of smooth muscle tissue (page 245).

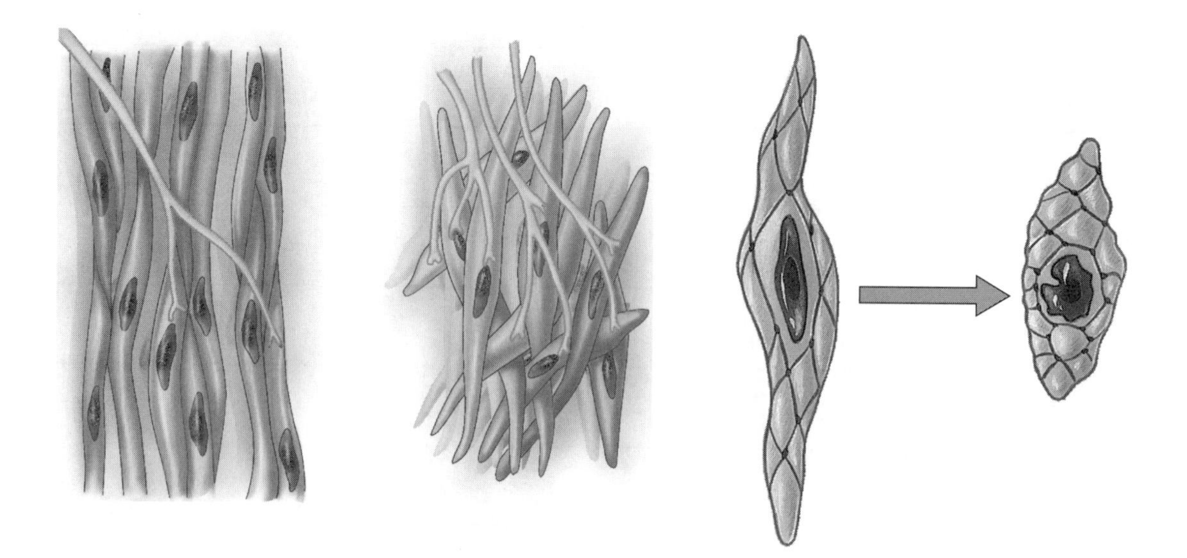

**Figure 9.10** Location and structure of somites (page 247).

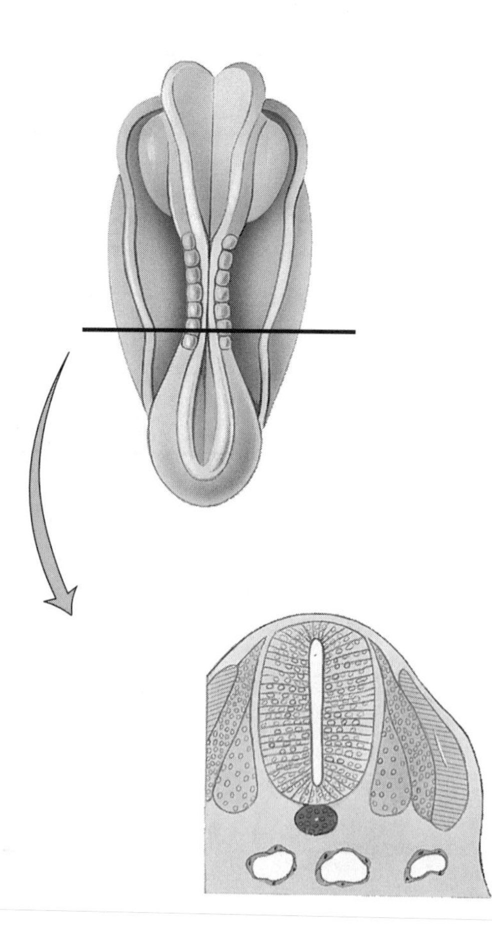

NOTES

9

**Figure 10.1** Relationship of skeletal muscles to bones (page 255).

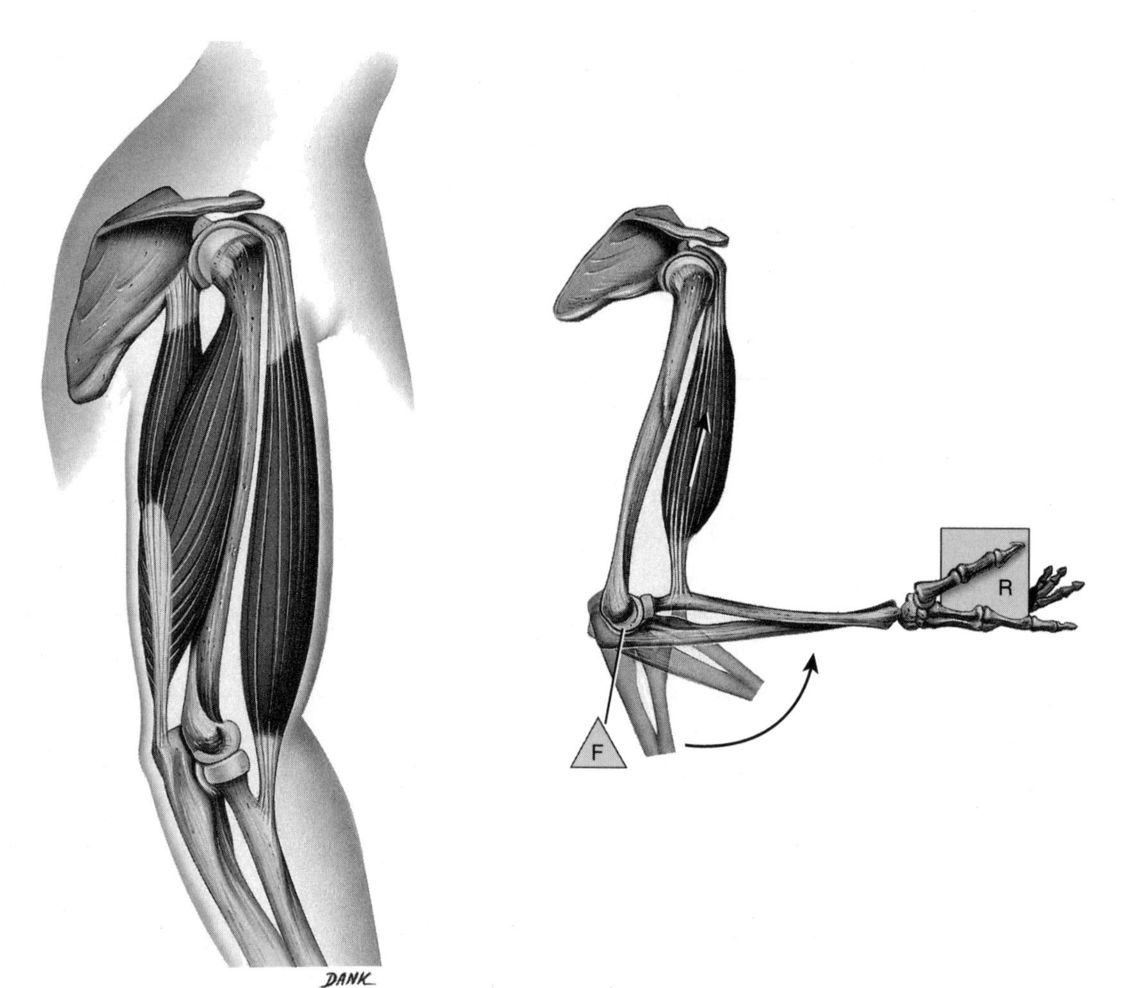

NOTES

10

**Figure 10.2** Types of levers (page 256).

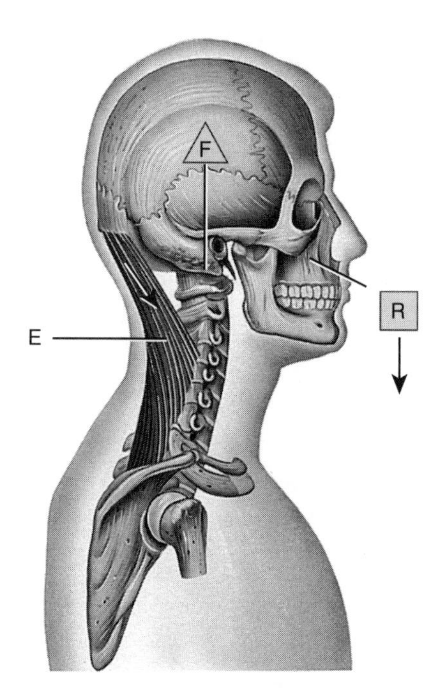


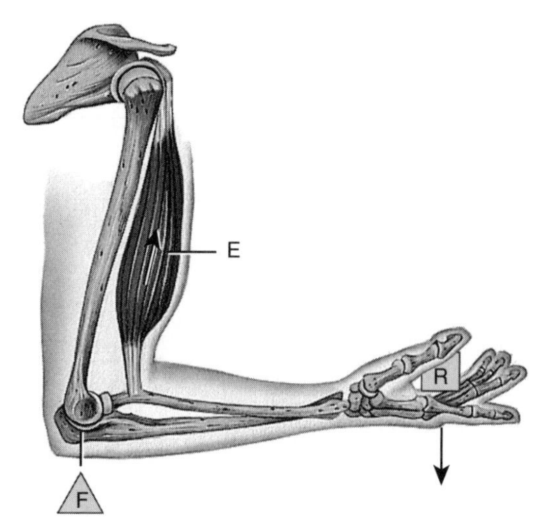


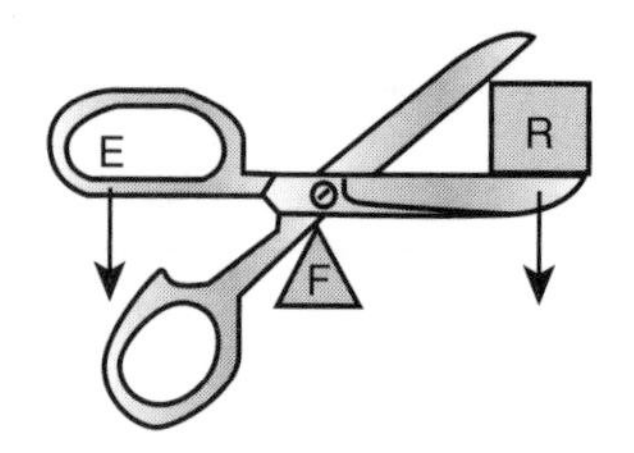


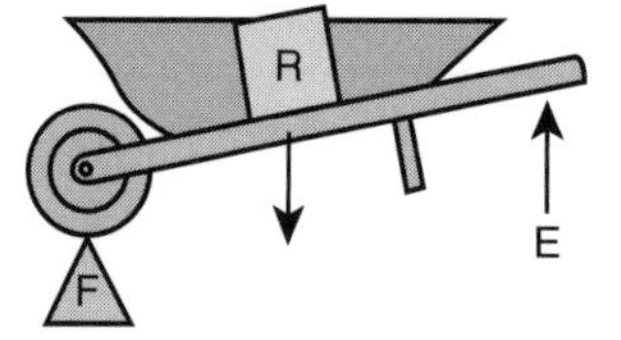


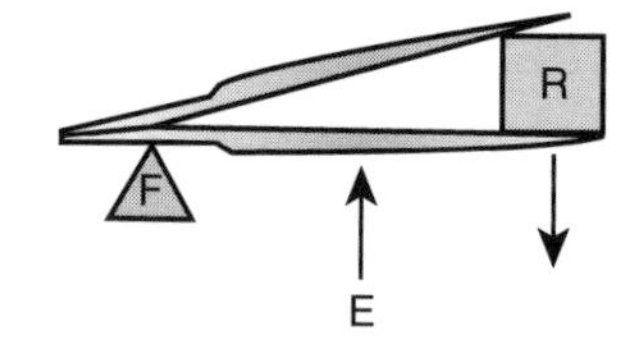

NOTES

10

**Figure 10.3a** Principal superficial skeletal muscles, anterior view (page 260).

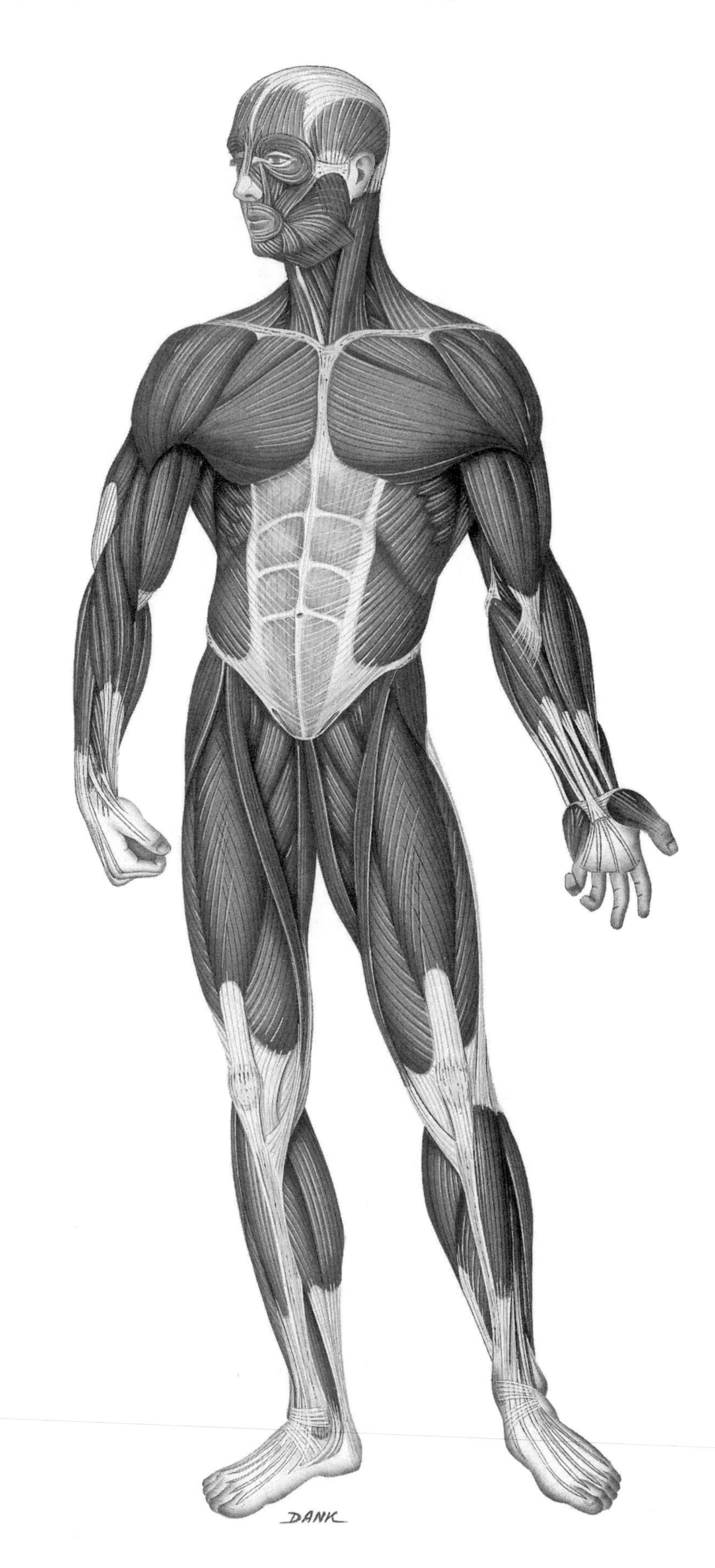

NOTES

10

**Figure 10.3b** Principal superficial skeletal muscles, posterior view (page 261).

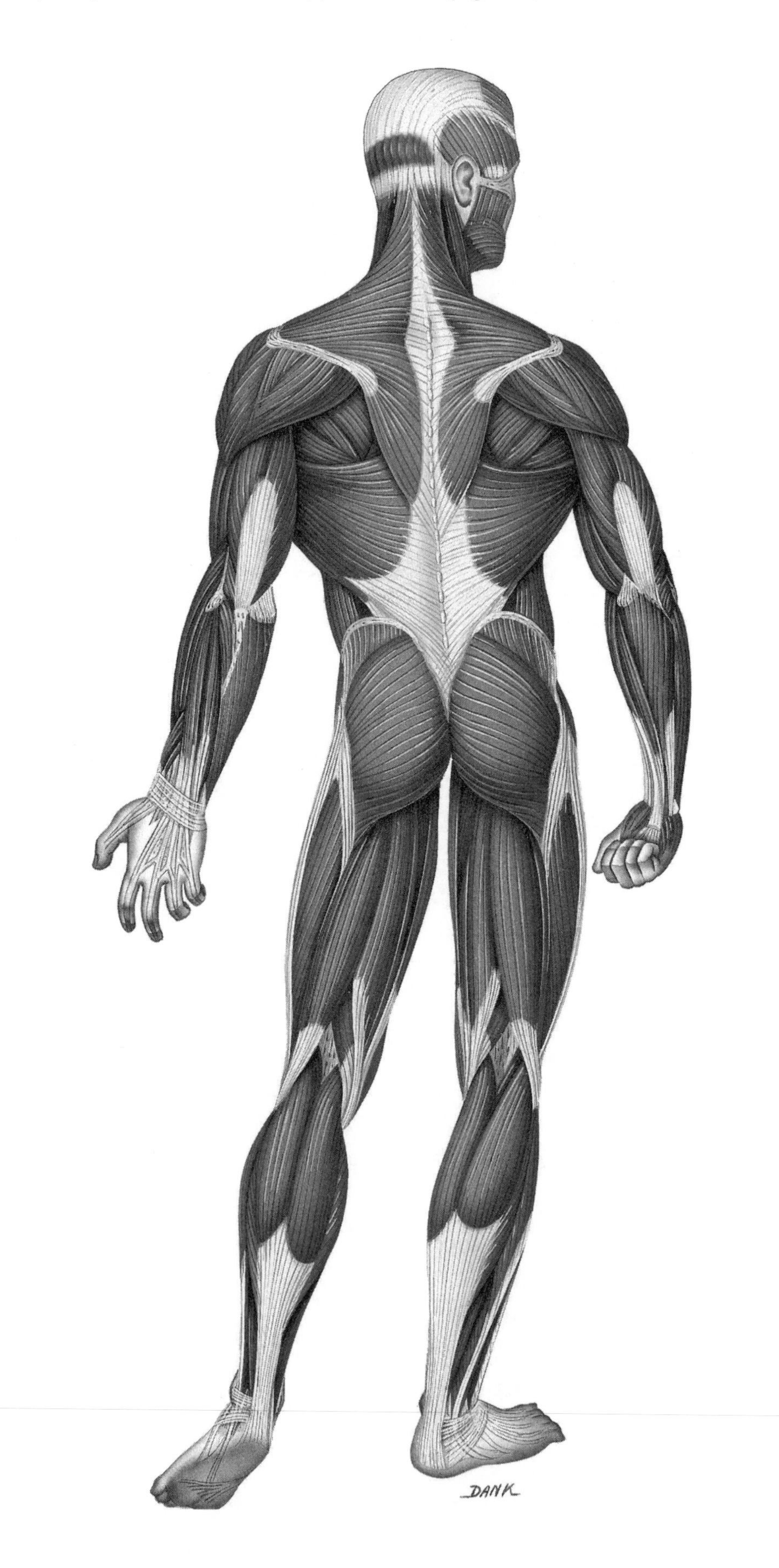

NOTES

10

**Figure 10.4a,b** Muscles of facial expression (page 265).

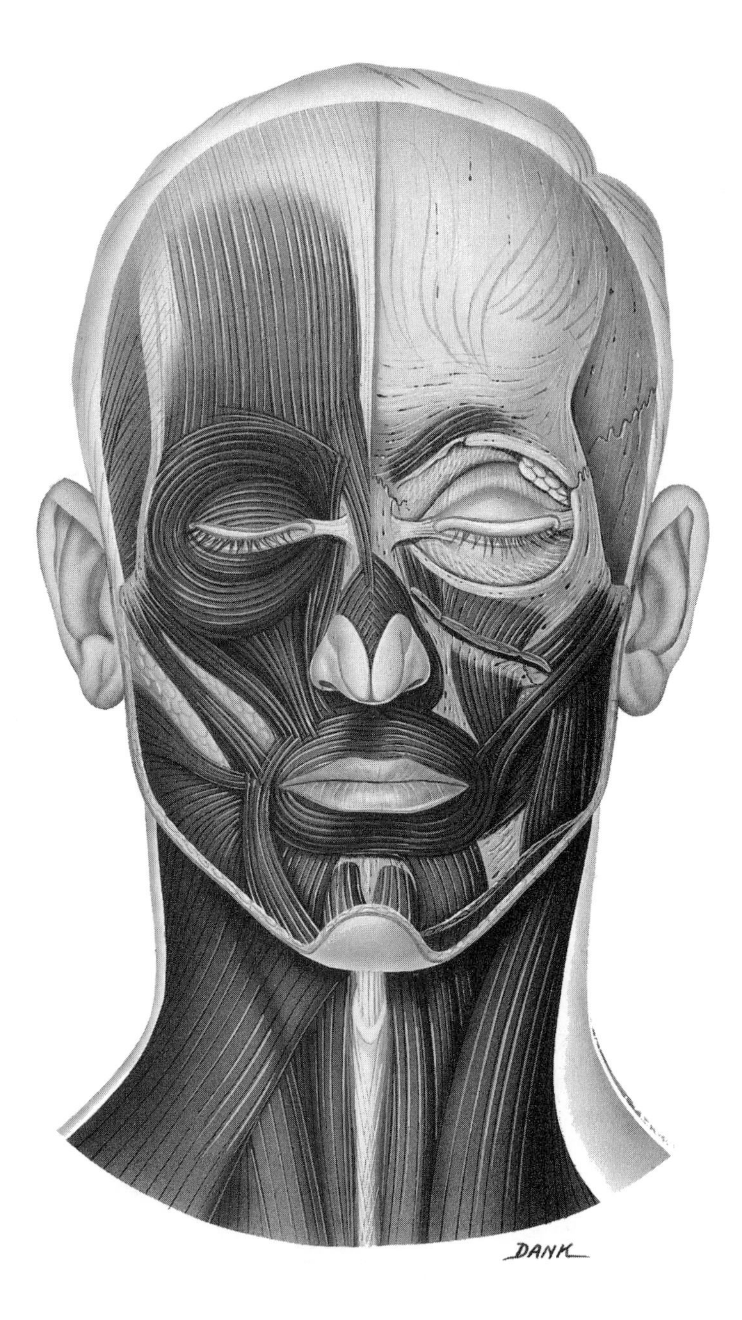

NOTES

10

**Figure 10.4c** Muscles of facial expression (page 266).

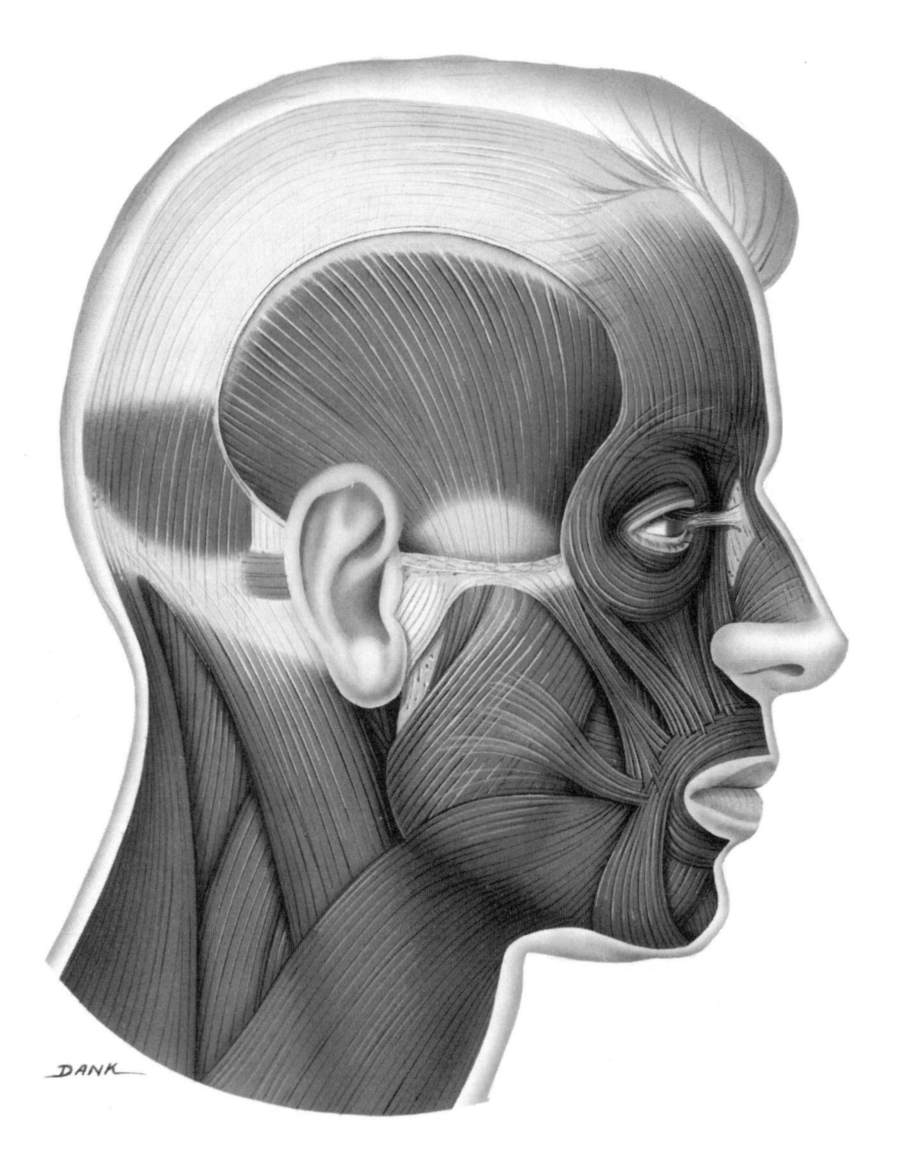

NOTES

10

**Figure 10.4d** Muscles of facial expression (page 267).

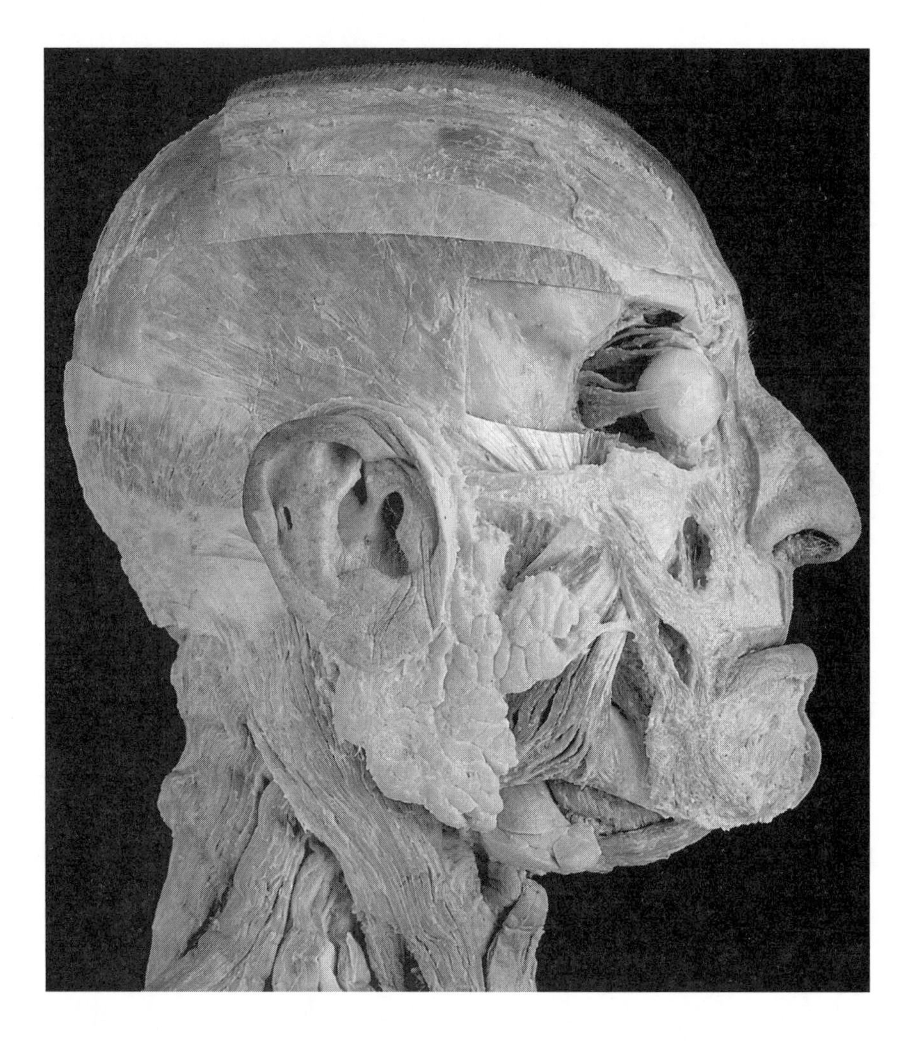

NOTES

10

**Figure 10.5a,b** Extrinsic muscles of the eyeball (page 269).

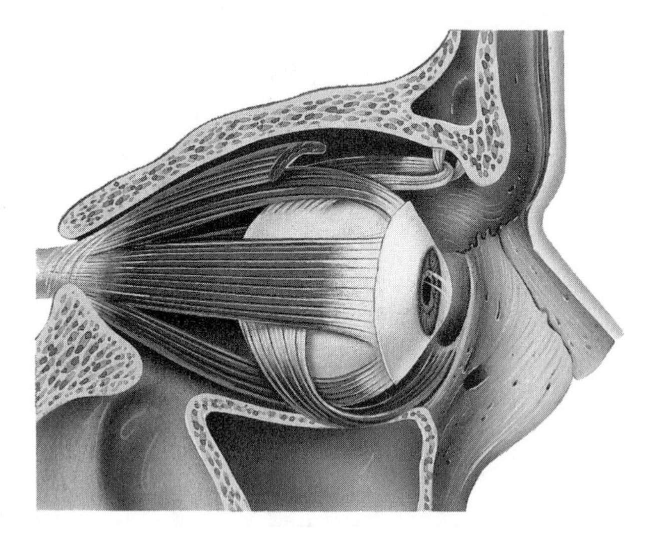

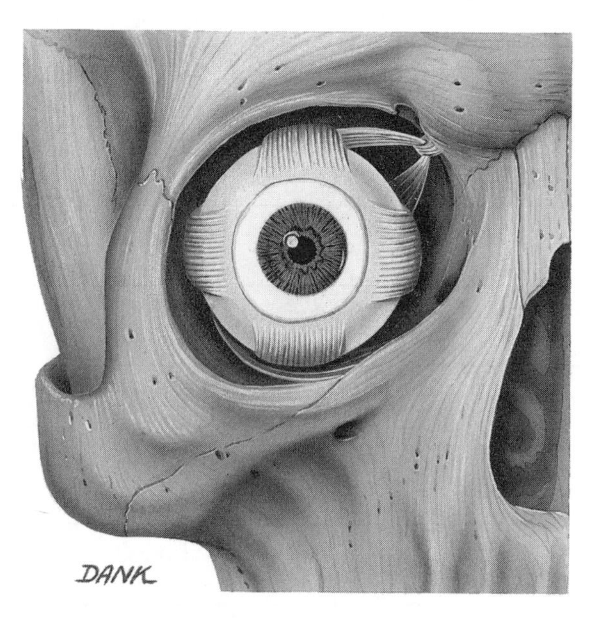


**Figure 10.5c** Extrinsic muscles of the eyeball (page 270).

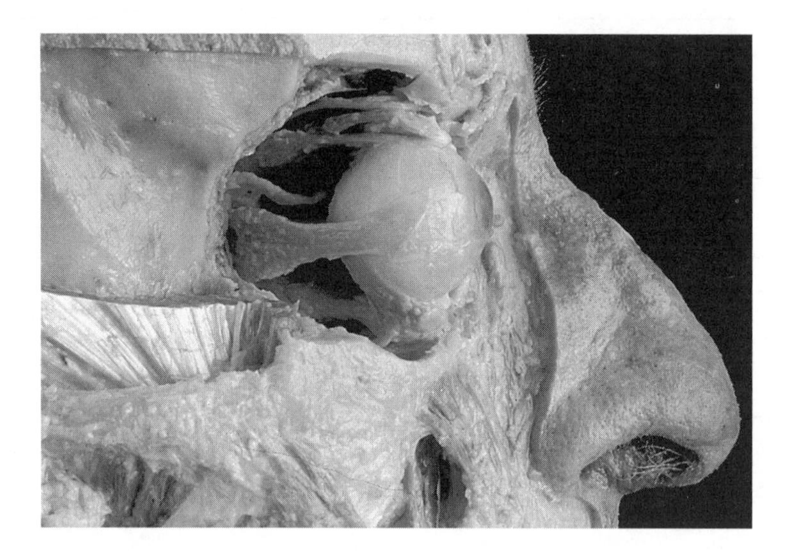

## NOTES

10

**Figure 10.6a** Muscles that move the mandible (lower jaw) (page 272).

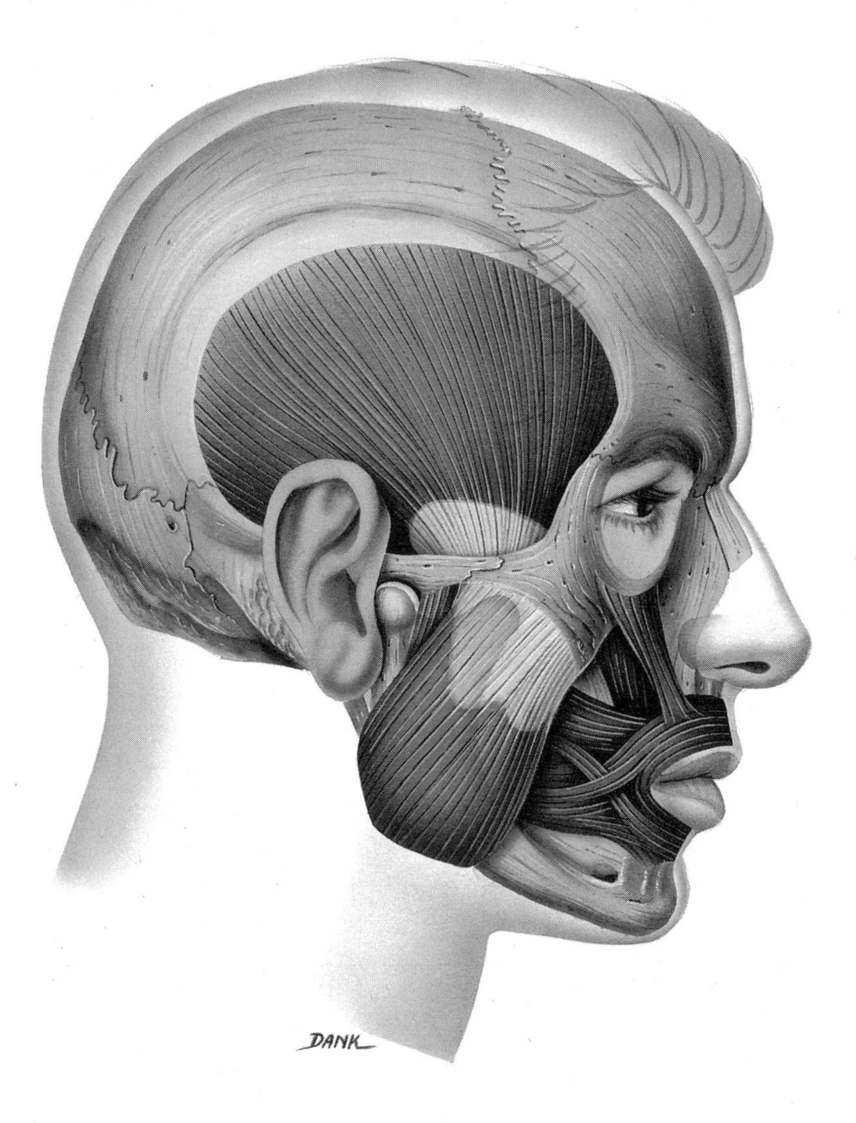

NOTES

10

**Figure 10.6b** Muscles that move the mandible (lower jaw) (page 273).

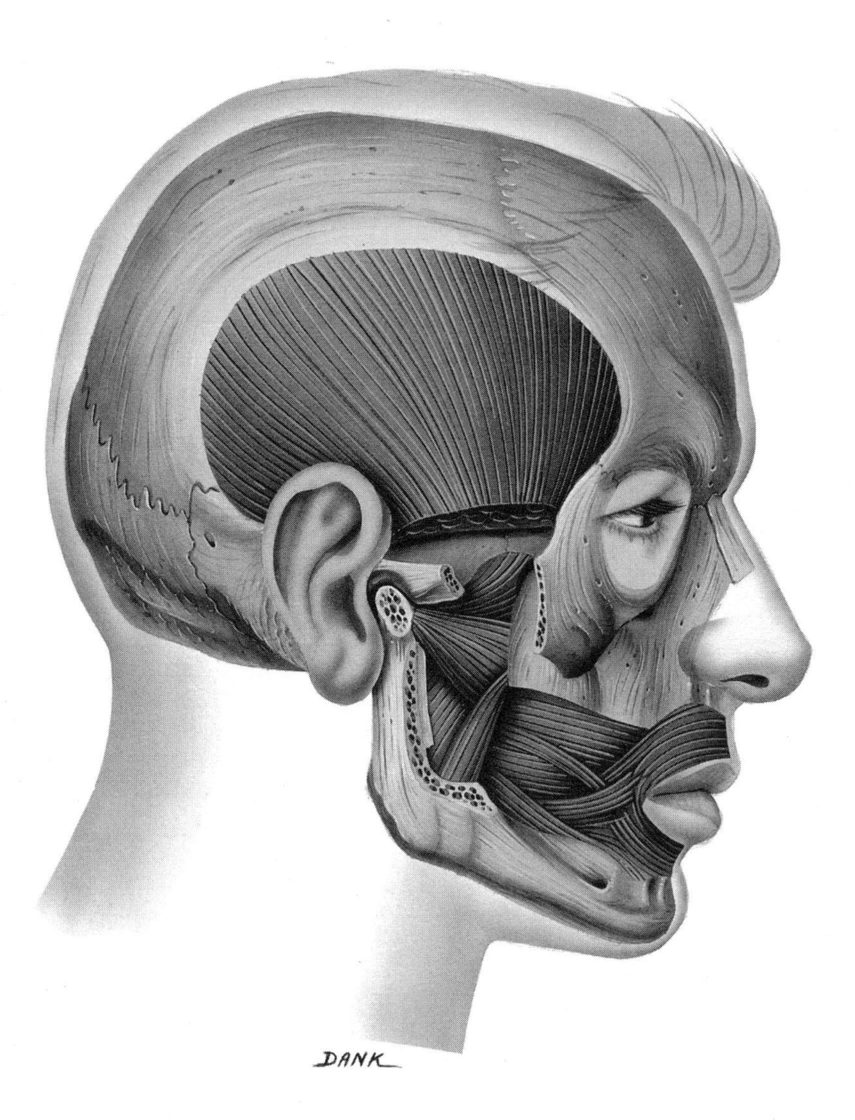

NOTES

10

**Figure 10.7** Muscles that move the tongue (page 275).

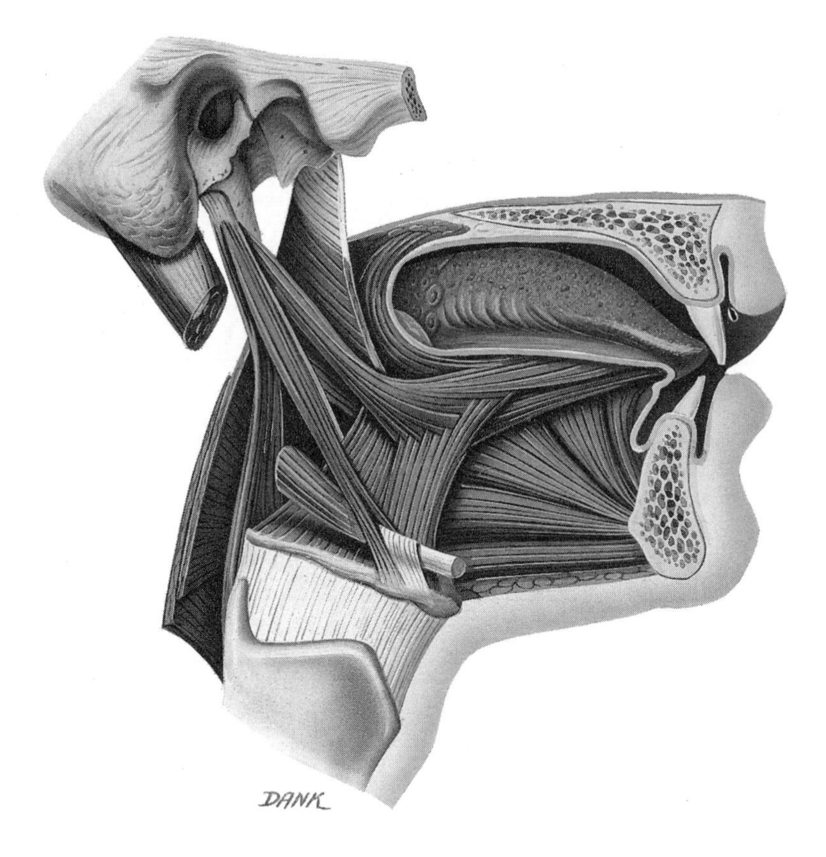


**Figure 10.8** Muscles of the floor of the oral cavity and front of the neck (page 277).

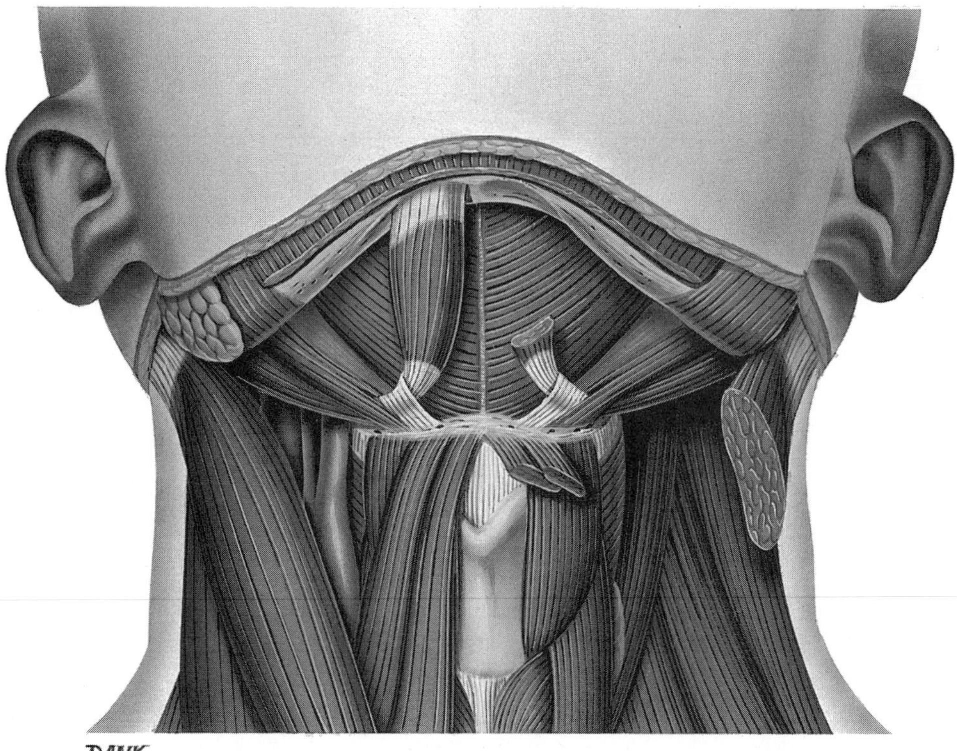

NOTES

10

**Figure 10.9** Triangles of the neck (page 279).

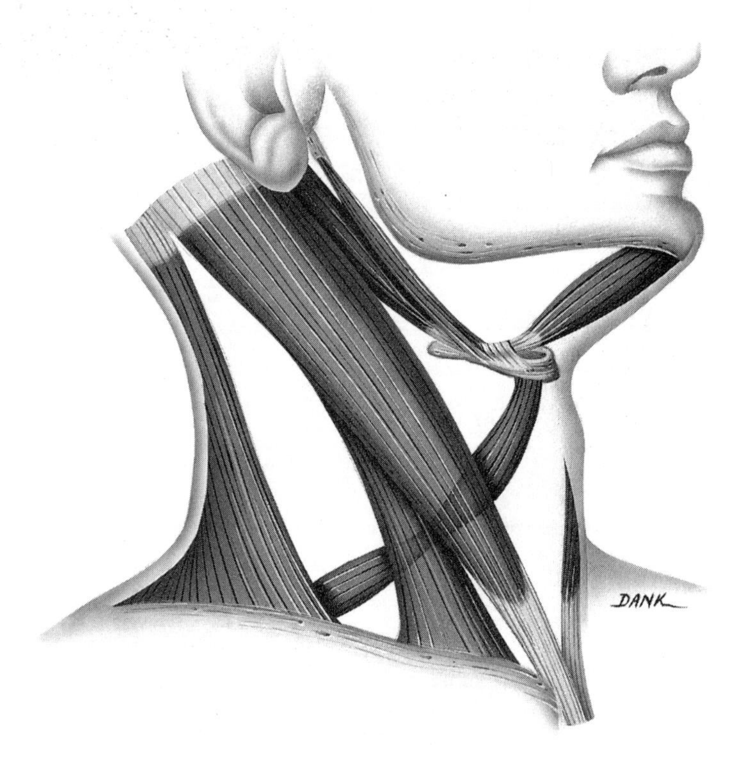

NOTES

10

**Figure 10.10** Muscles of the larynx (voice box) (page 281).

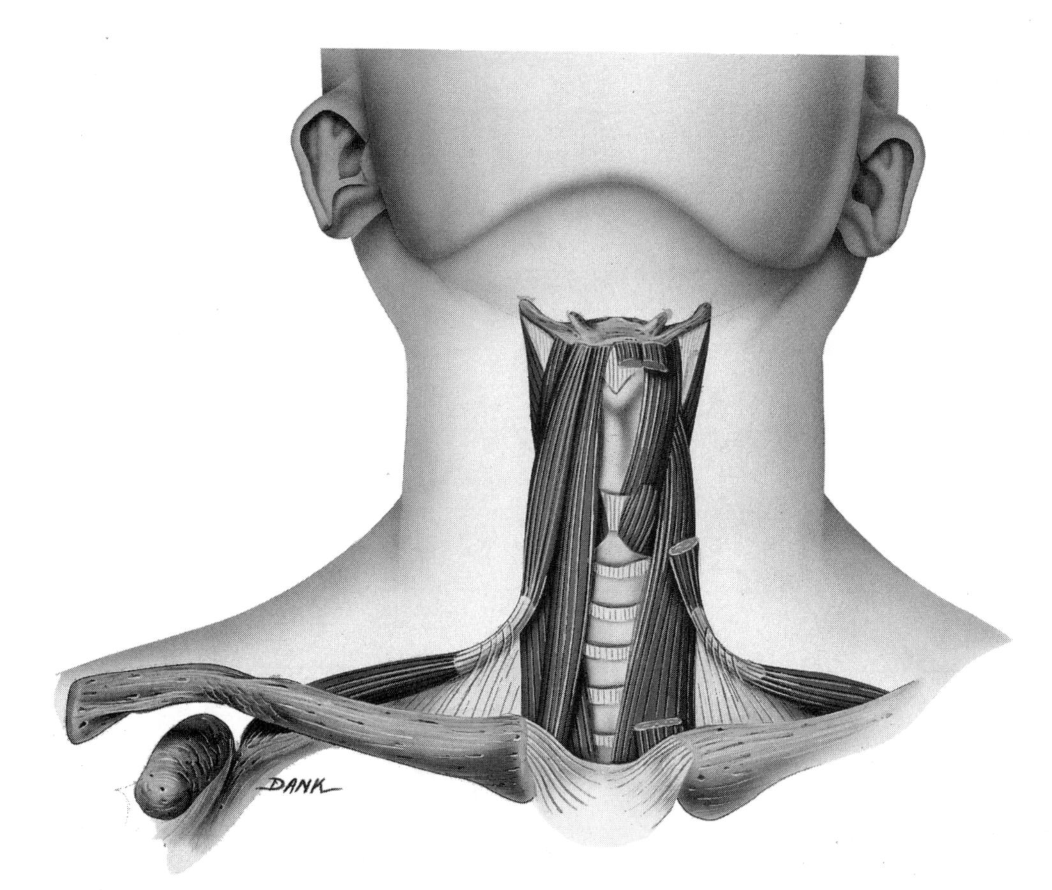


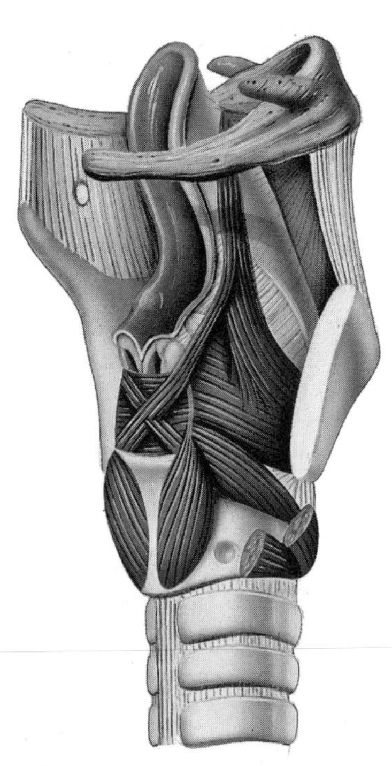

NOTES

10

**Figure 10.11** Muscles of the pharynx (page 283).

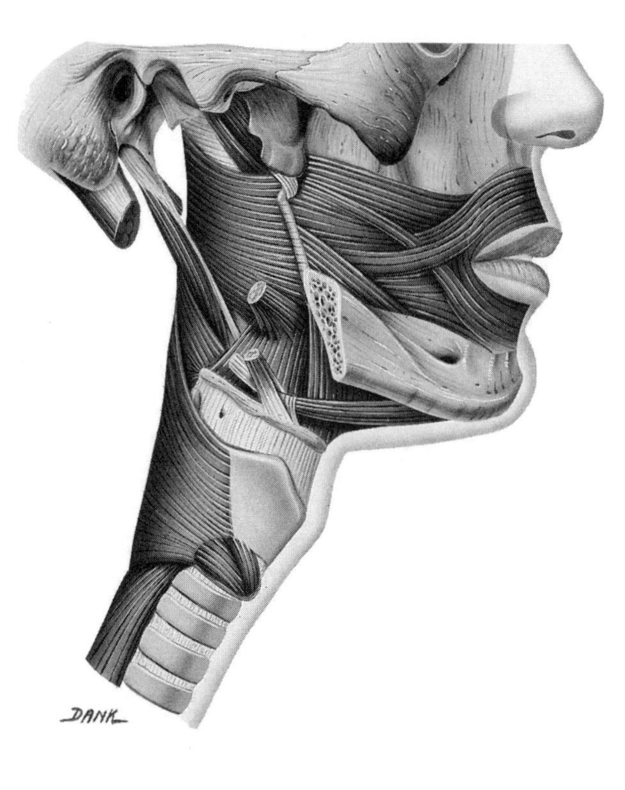


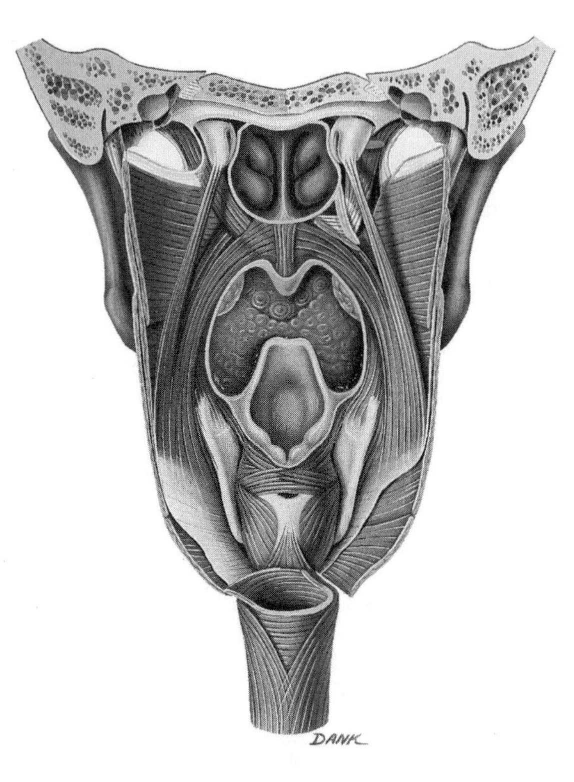

NOTES

10

**Figure 10.12a,b** Muscles of the male anterolateral abdominal wall (page 286).

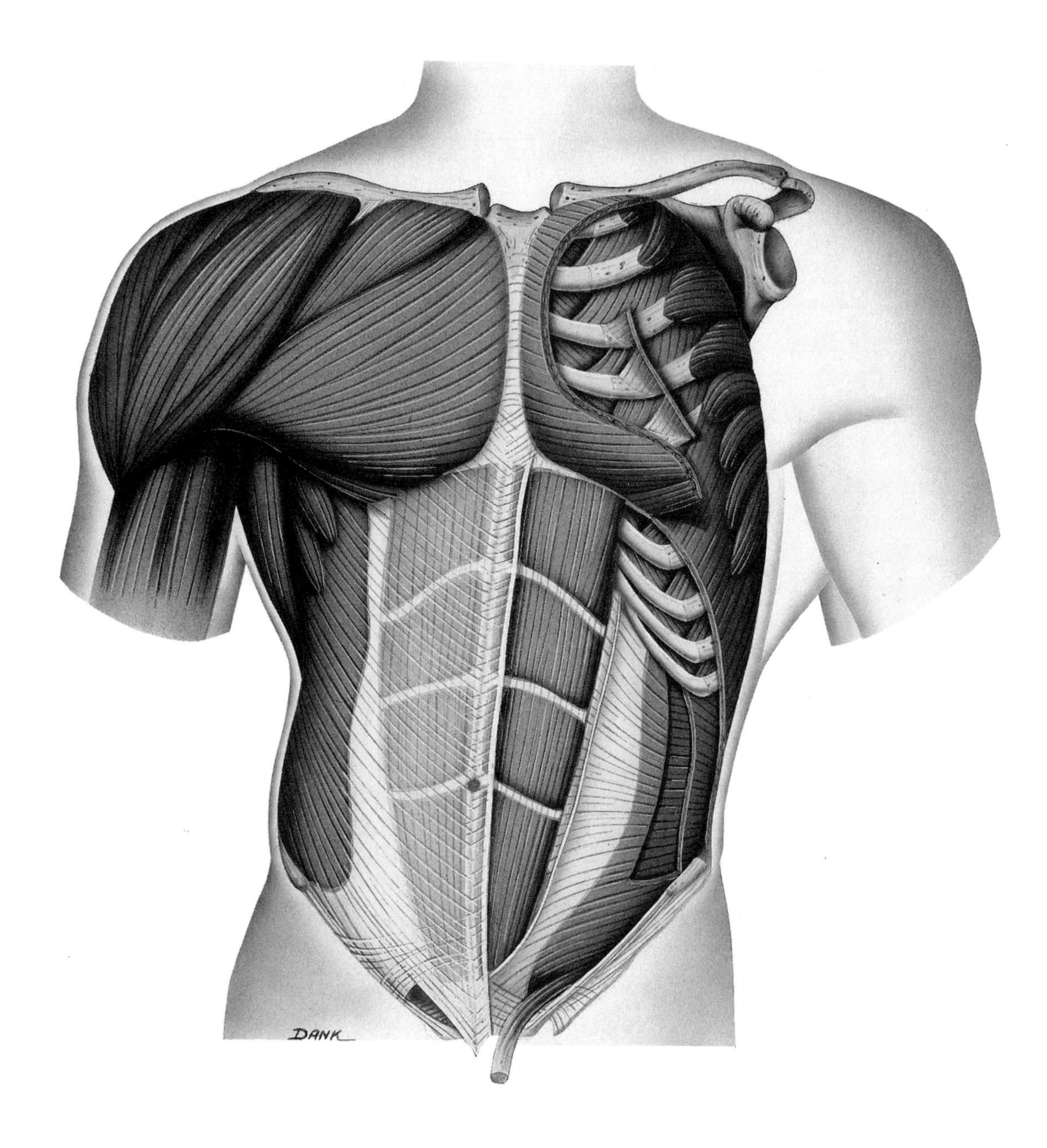

NOTES

10

**Figure 10.12c,d** Muscles of the male anterolateral abdominal wall (page 287).

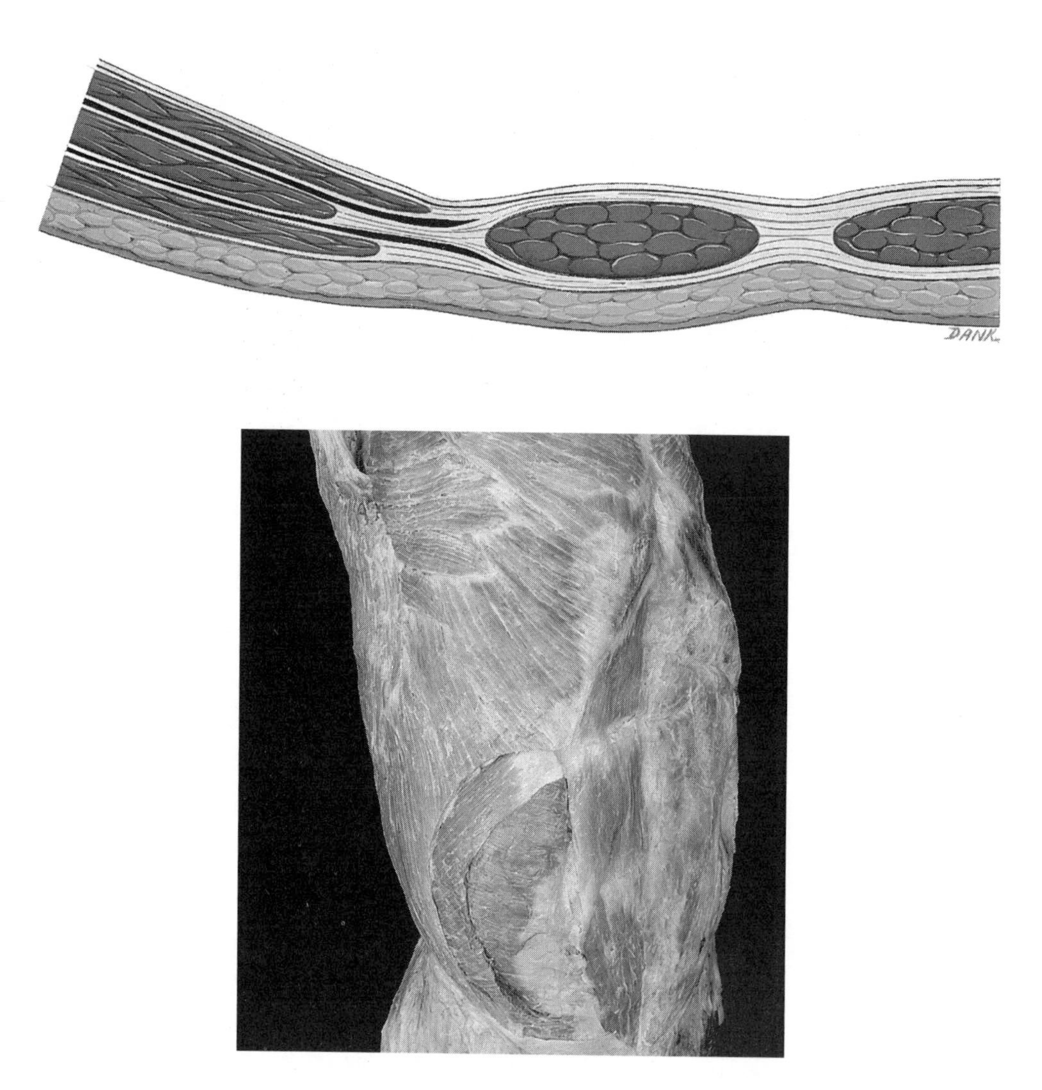

NOTES

10

**Figure 10.13** Muscles used in breathing, as seen in a male (page 289).

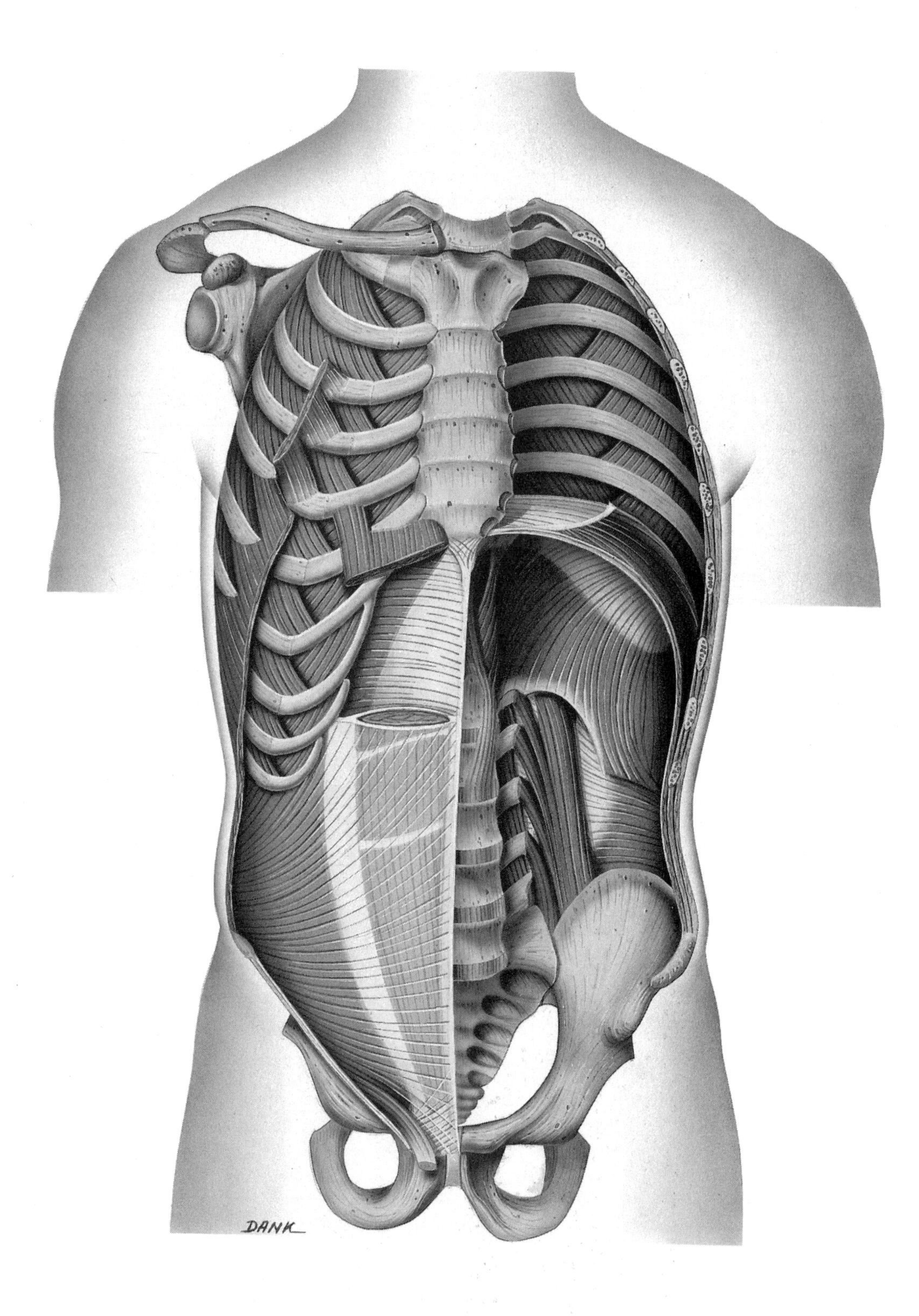

NOTES

10

**Figure 10.14** Muscles of the pelvic floor, as seen in the female perineum (page 291).

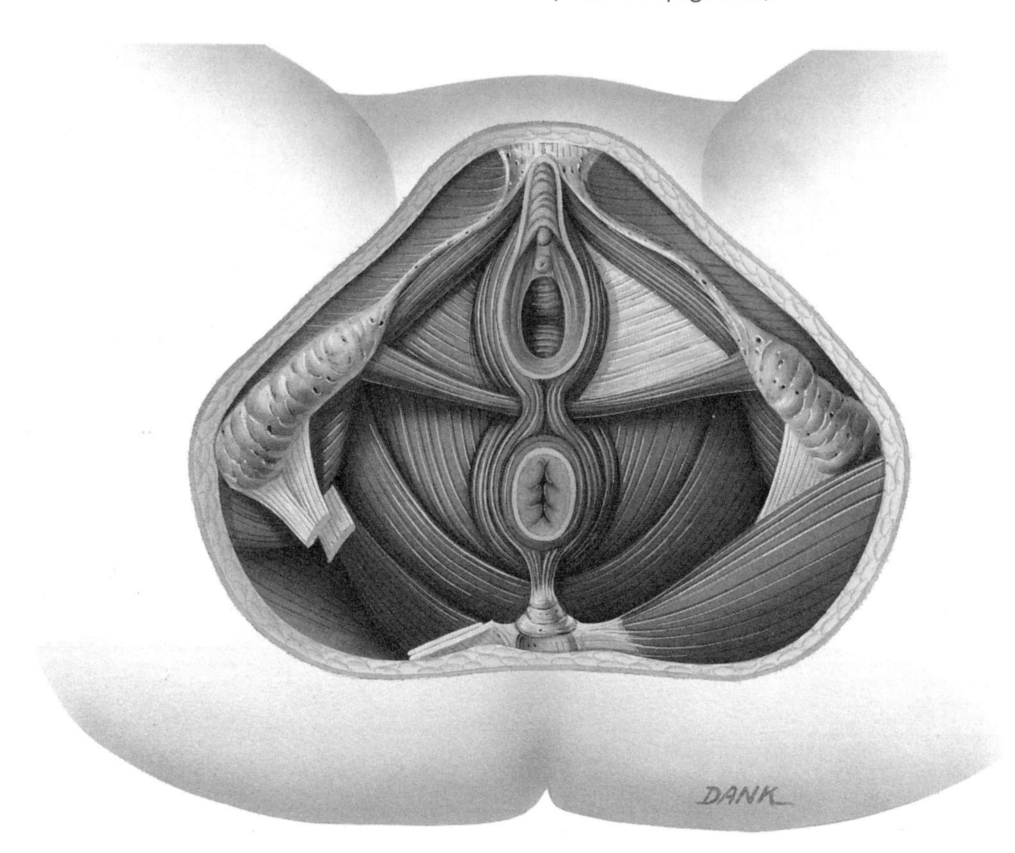

**Figure 10.15** Muscles of the male perineum (page 293).

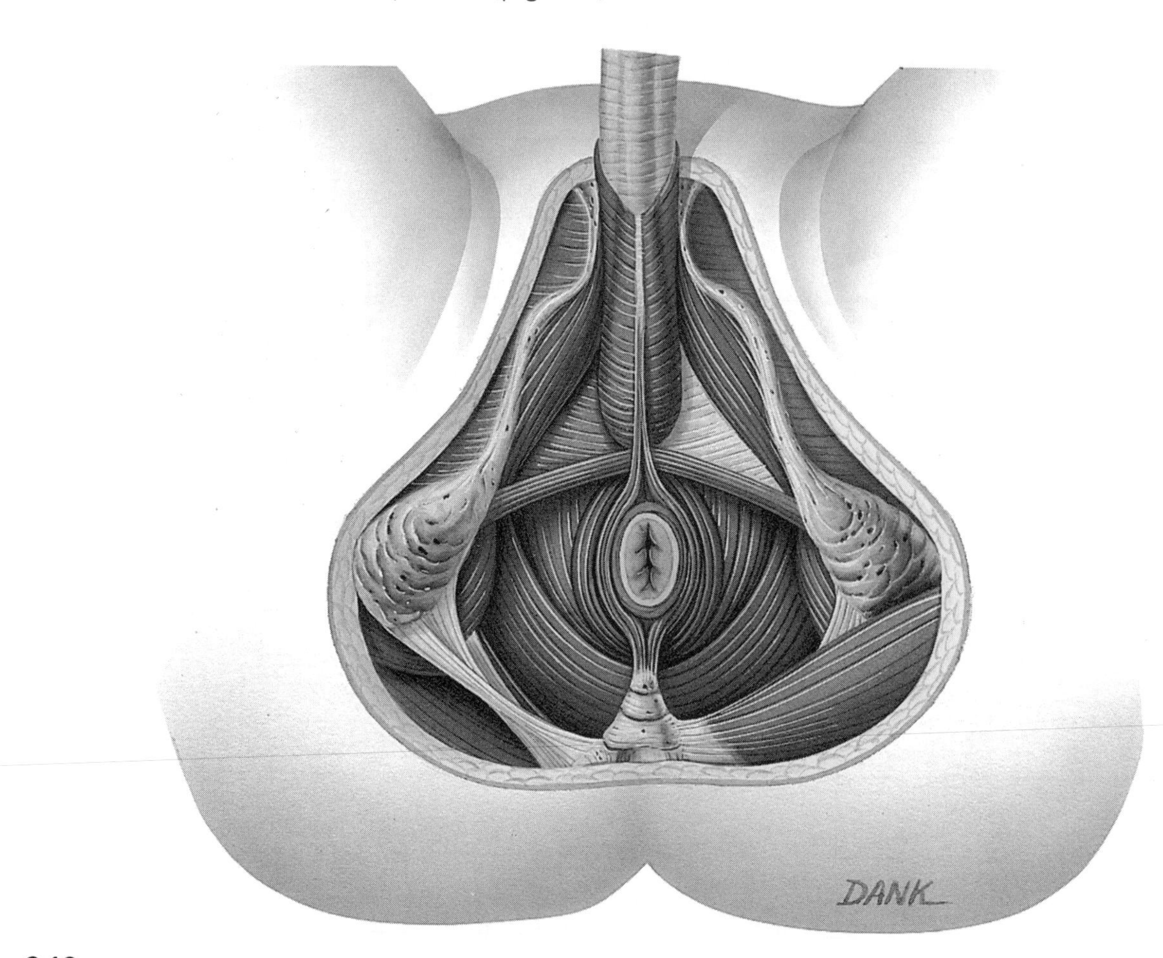

## NOTES

10

**Figure 10.16a,b** Muscles that move the pectoral (shoulder) girdle (page 296).

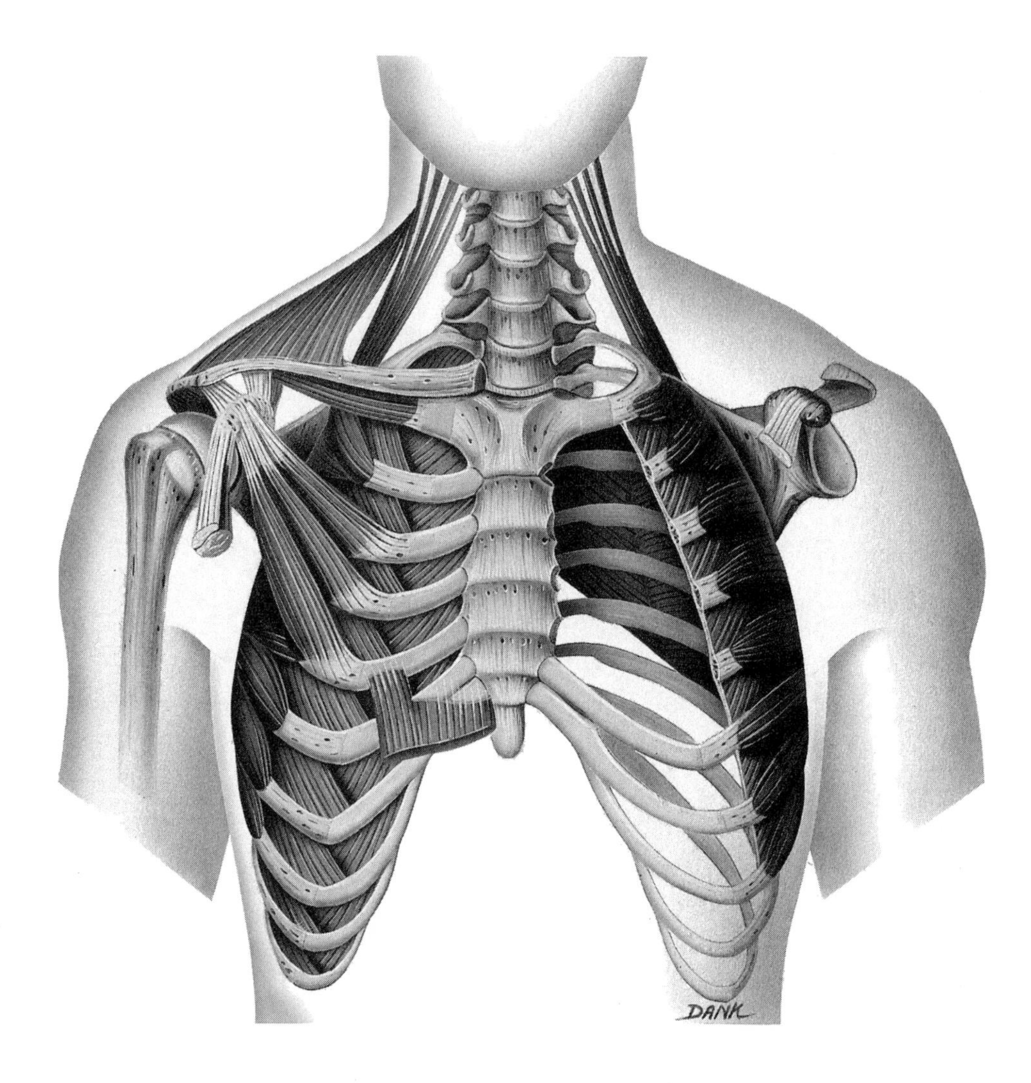

NOTES

10

**Figure 10.16c,d** Muscles that move the pectoral (shoulder) girdle: posterior view (page 297).

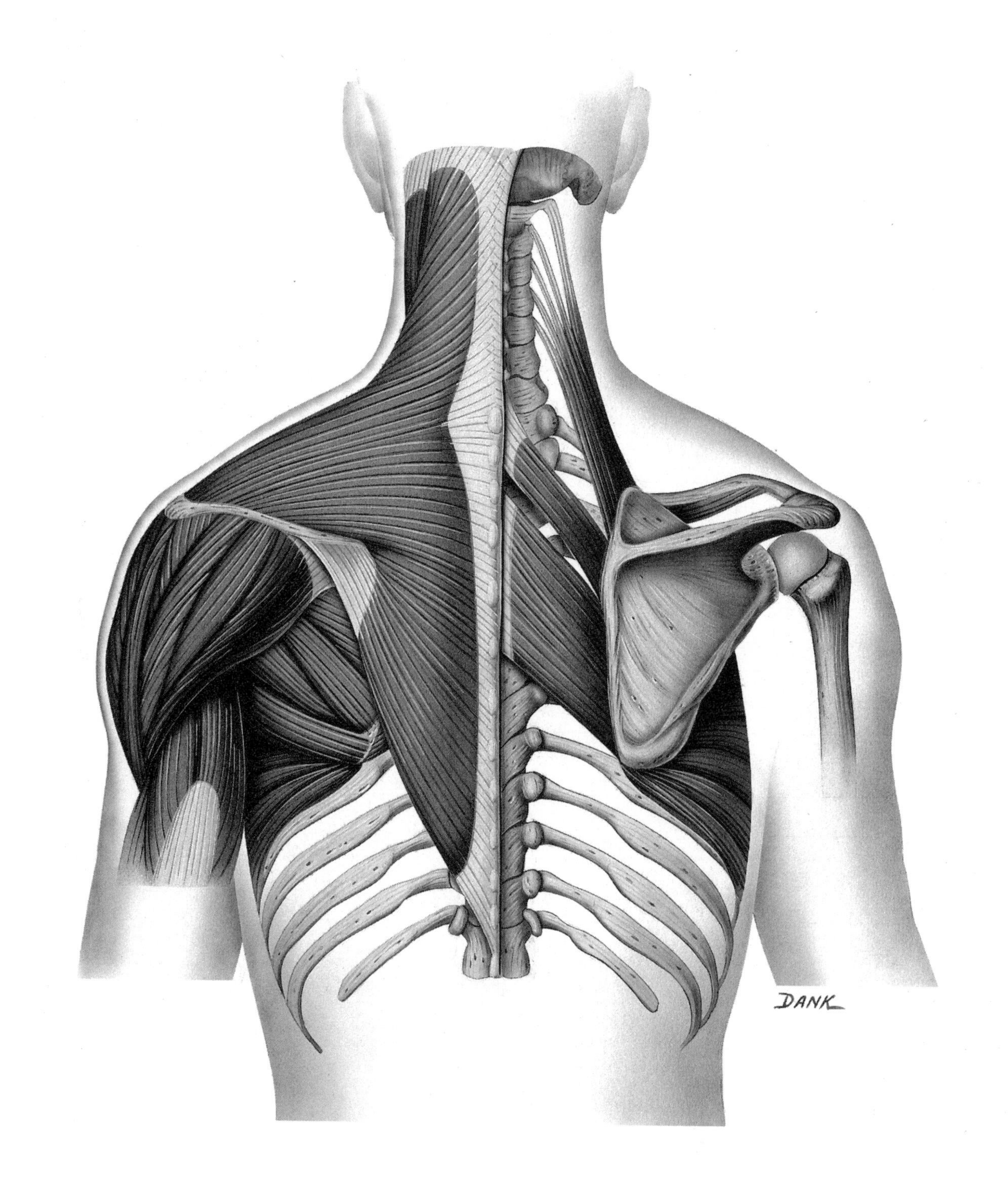

NOTES

10

**Figure 10.17a,b** Muscles that move the humerus (arm) (page 300).

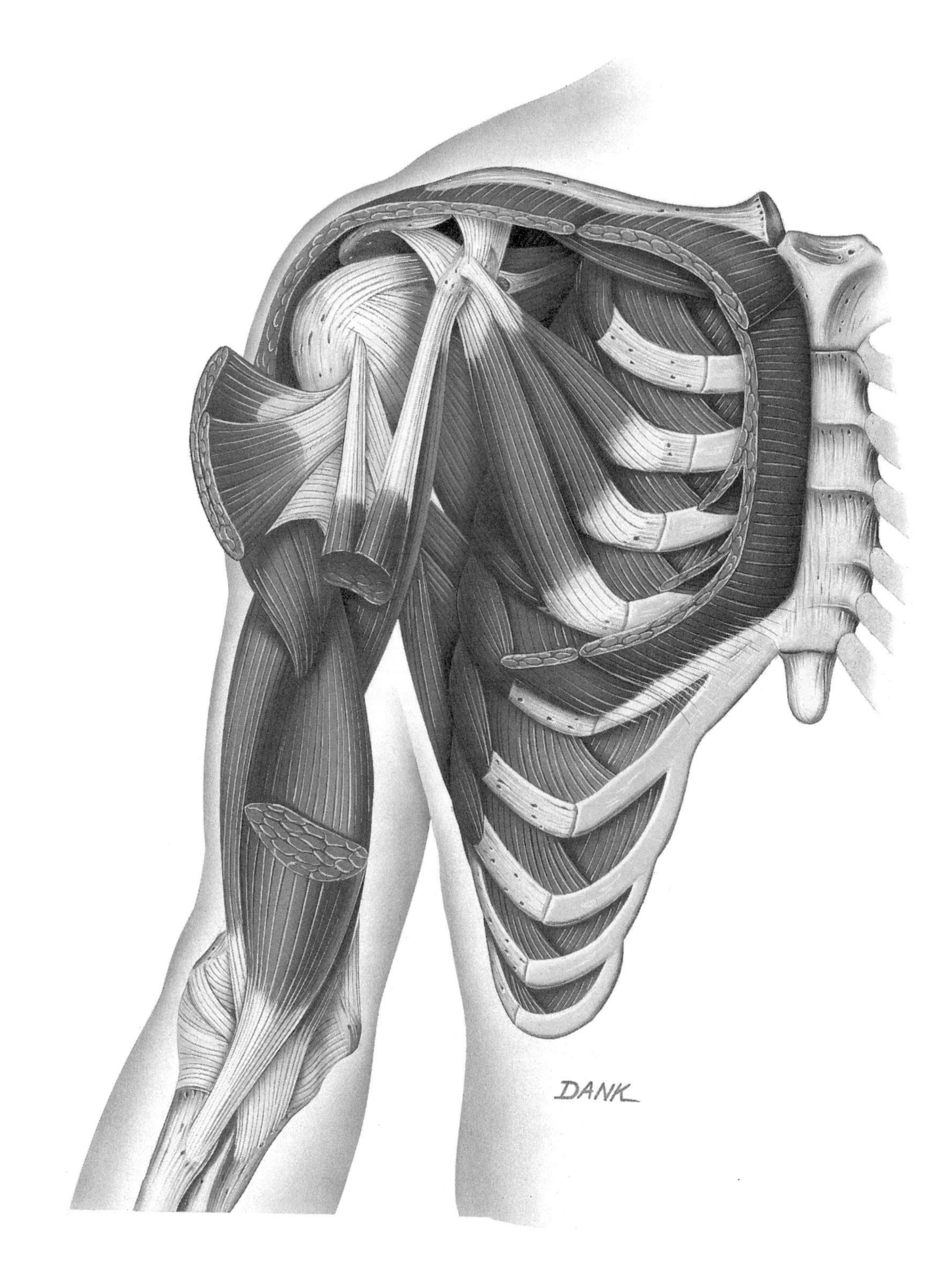

NOTES

10

**Figure 10.17c,d** Muscles that move the humerus (arm): posterior view (page 301).

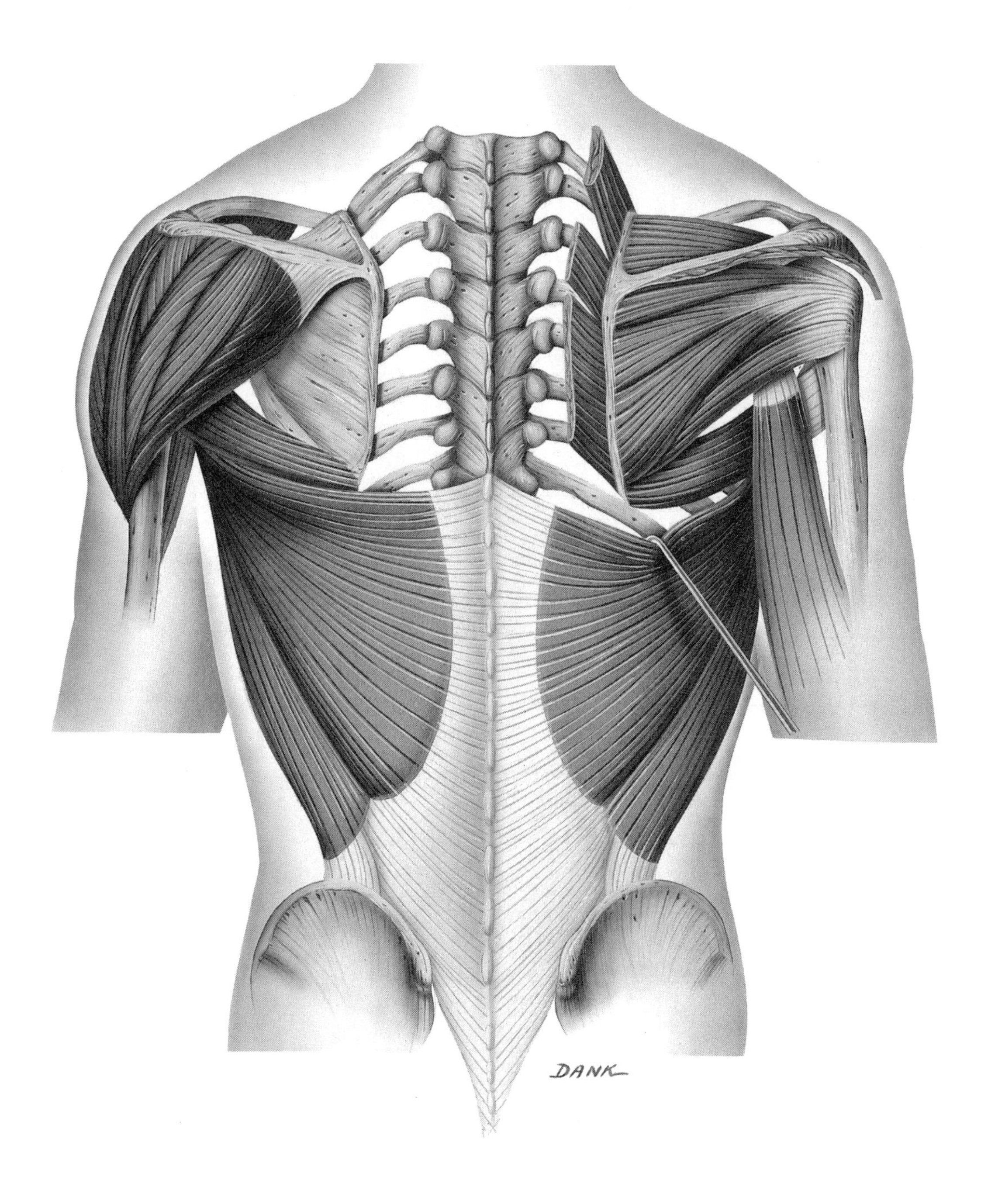

NOTES

10

**Figure 10.18a,b** Muscles that move the radius and ulna (forearm): anterior and posterior views (page 304).

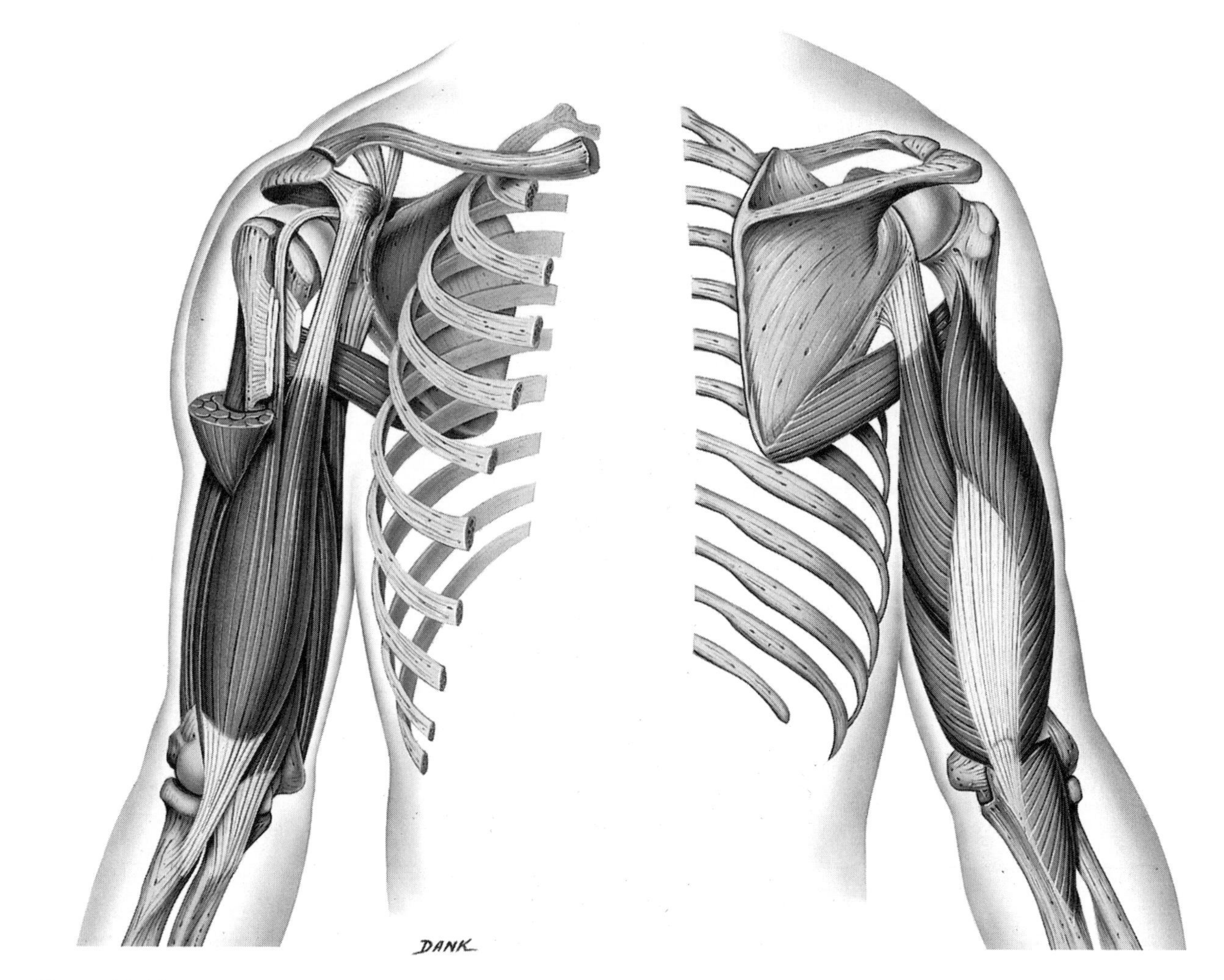


**Figure 10.18c** Muscles that move the radius and ulna (forearm): superior view (page 305).

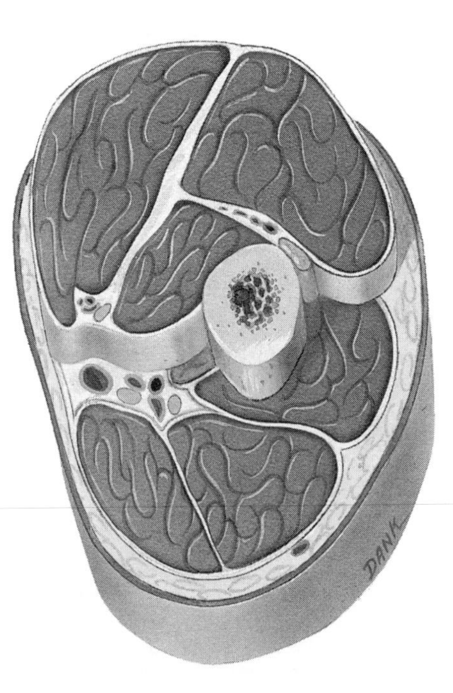

NOTES

10

**Figure 10.19a,b** Muscles that move the wrist, hand, and digits: anterior view (page 308).

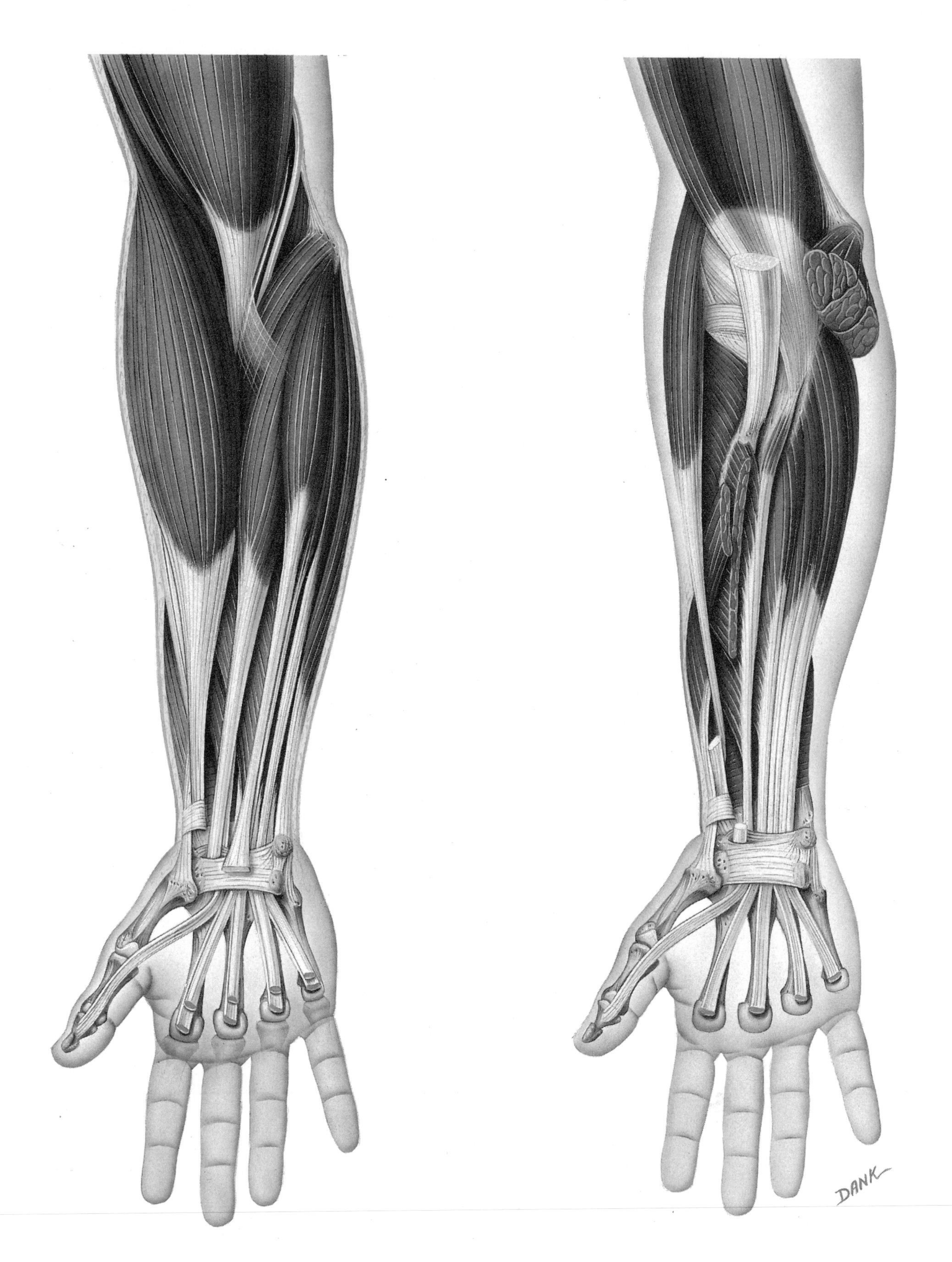

NOTES

**Figure 10.19c,d** Muscles that move the wrist, hand, and digits: posterior view (page 309).

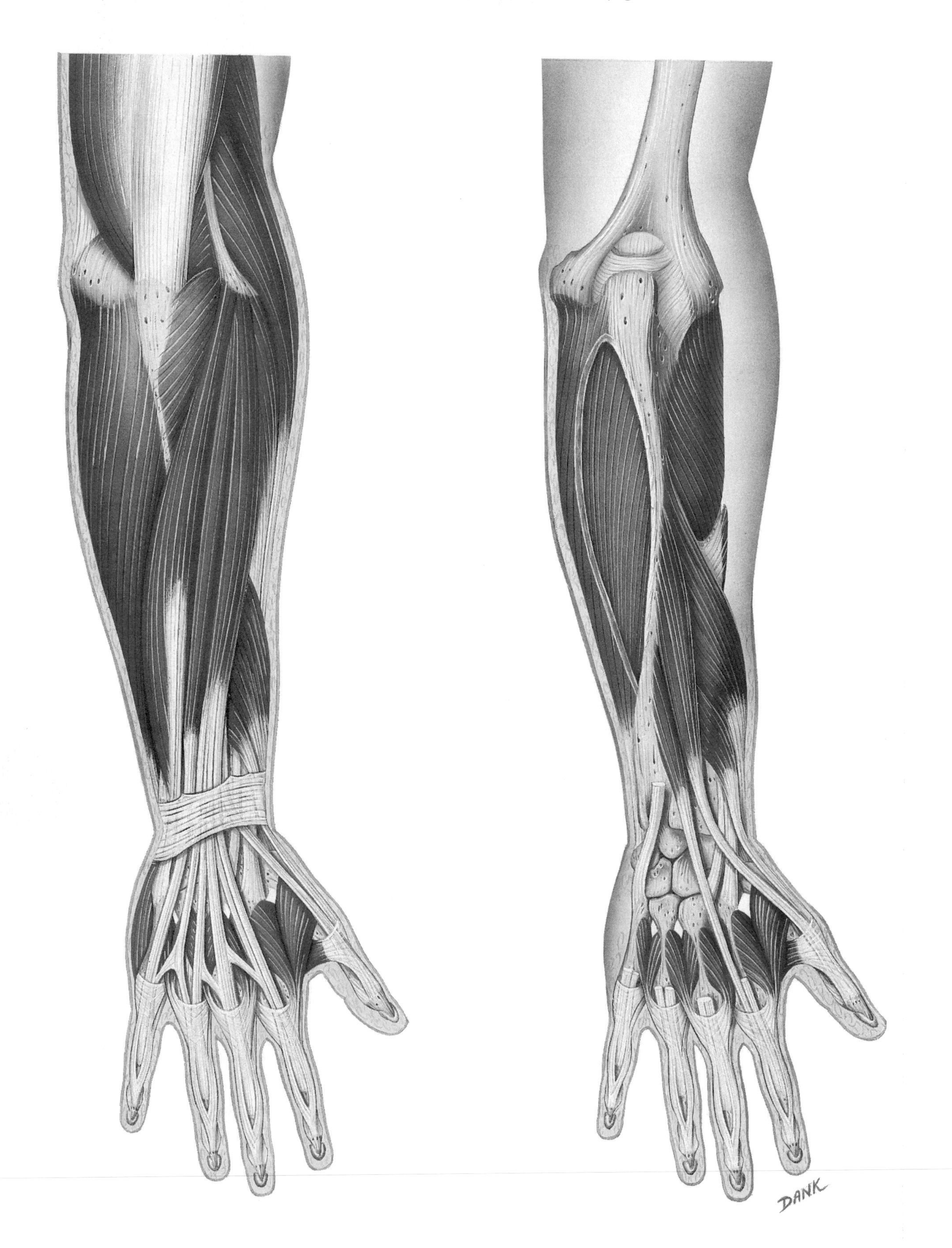

NOTES

10

**Figure 10.20a–c** Intrinsic muscles of the hand (page 314).

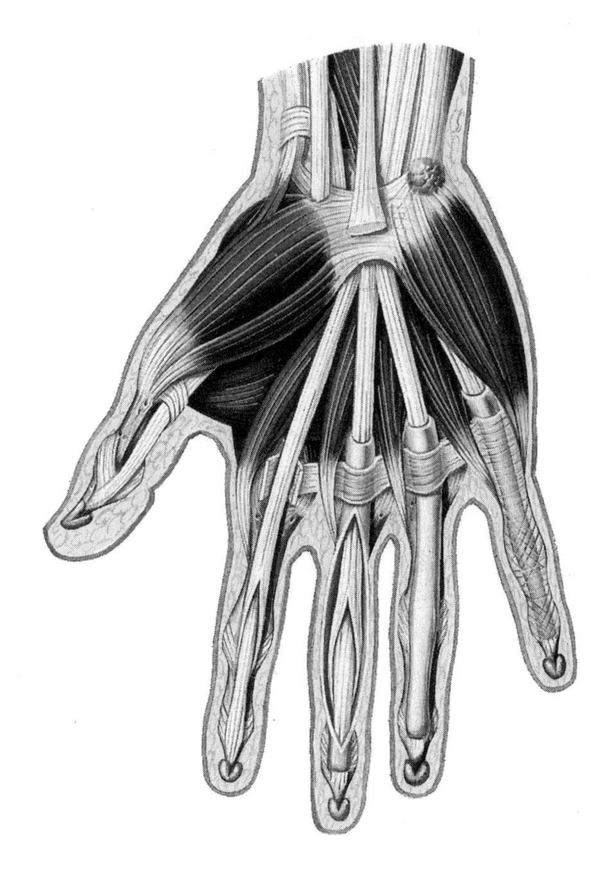

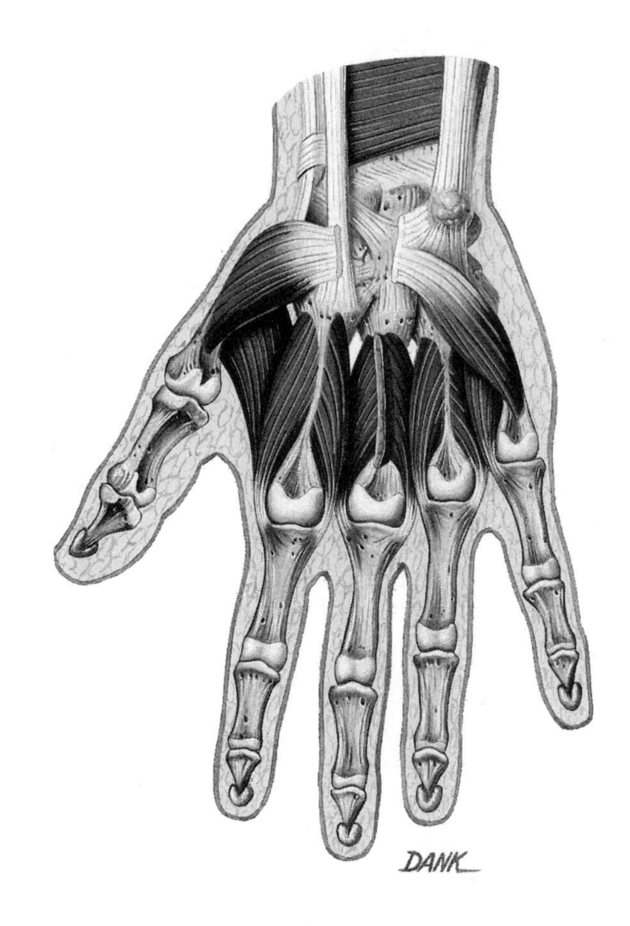


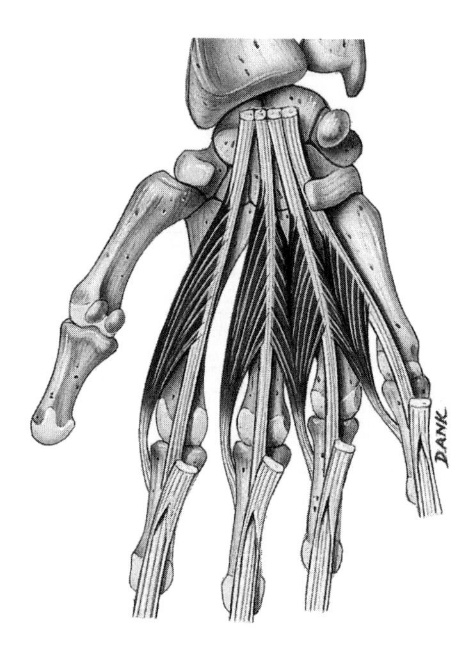

NOTES

10

**Figure 10.20d–f** Intrinsic muscles of the hand (page 315).

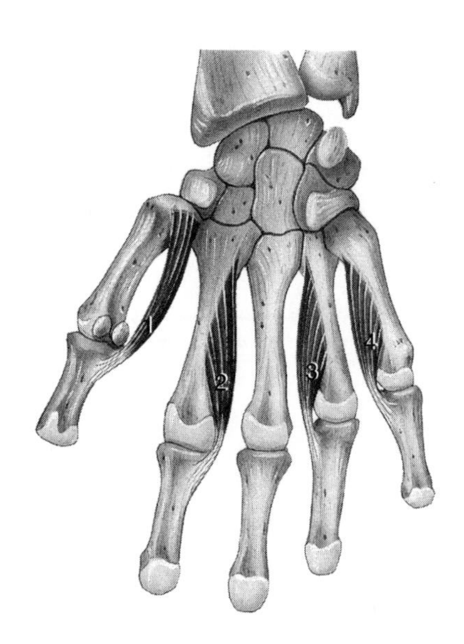


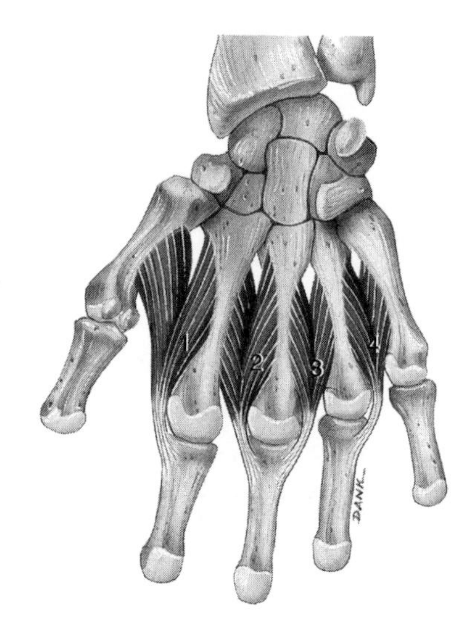


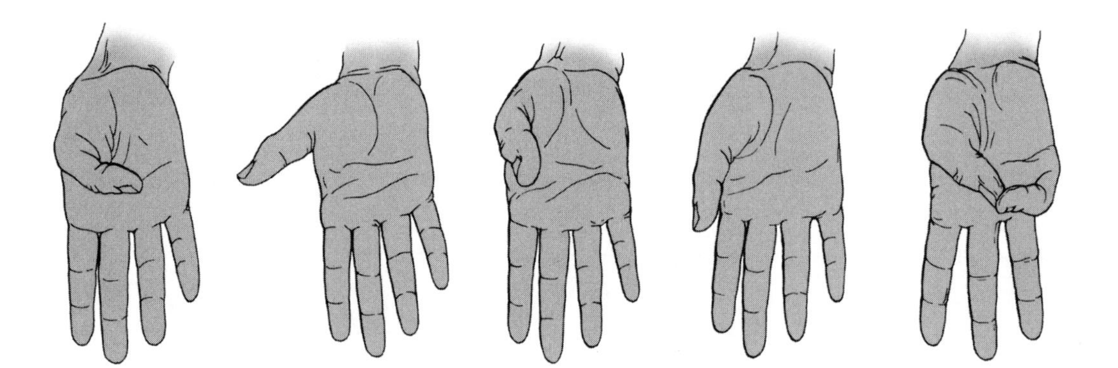

NOTES

10

**Figure 10.21a** Muscles that move the vertebral column (backbone) (page 319).

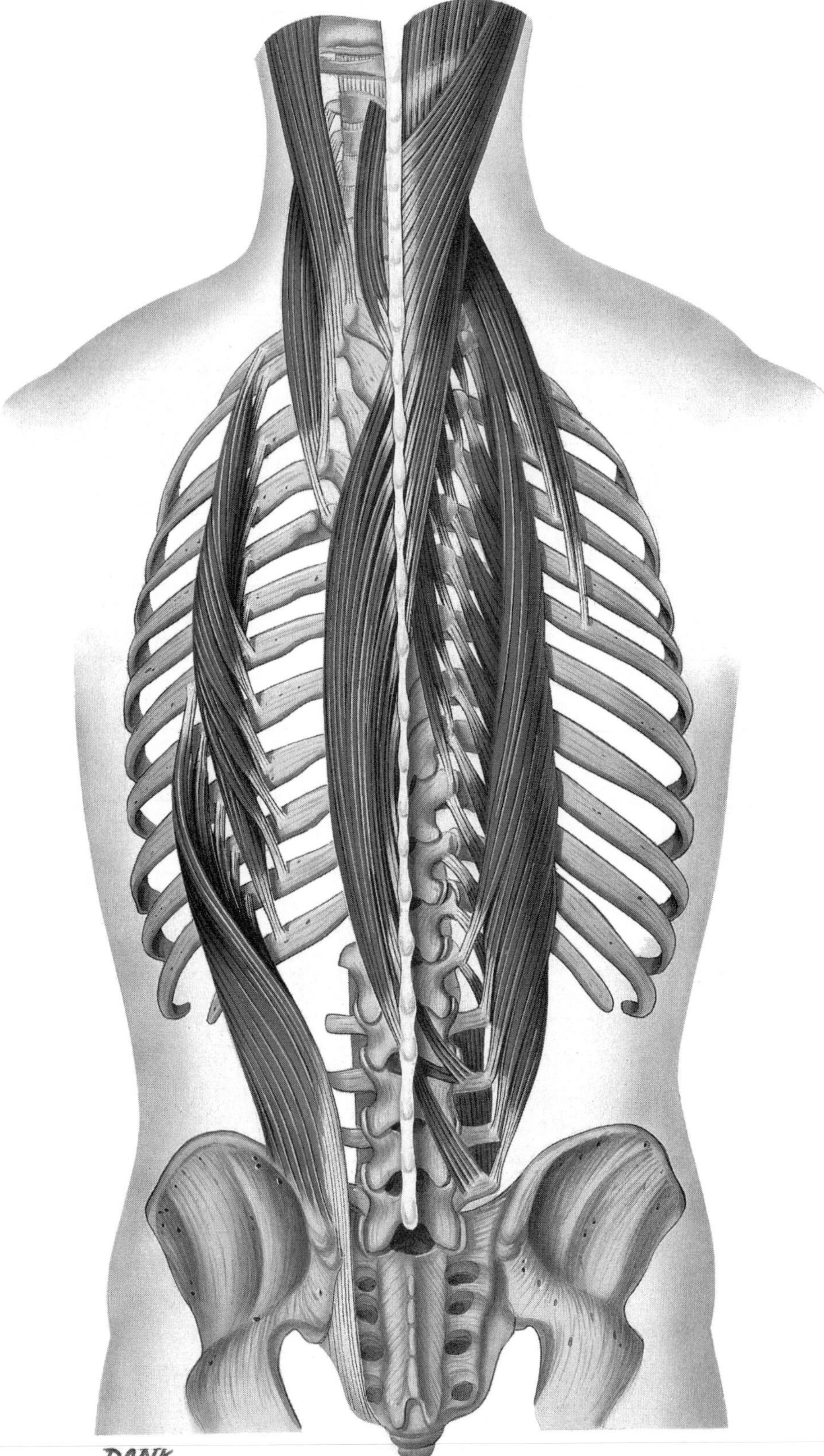

NOTES

10

**Figure 10.21b,c** Muscles that move the vertebral column (backbone) (page 320).

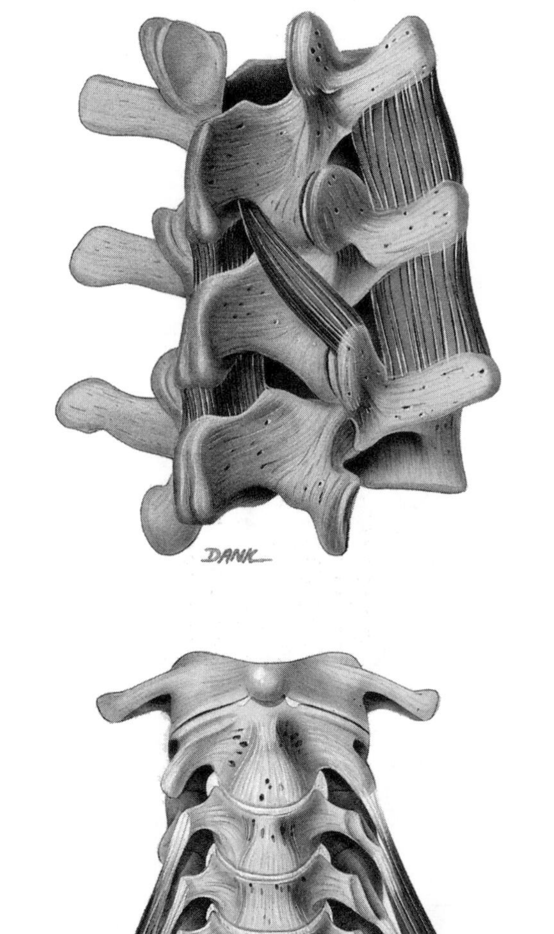


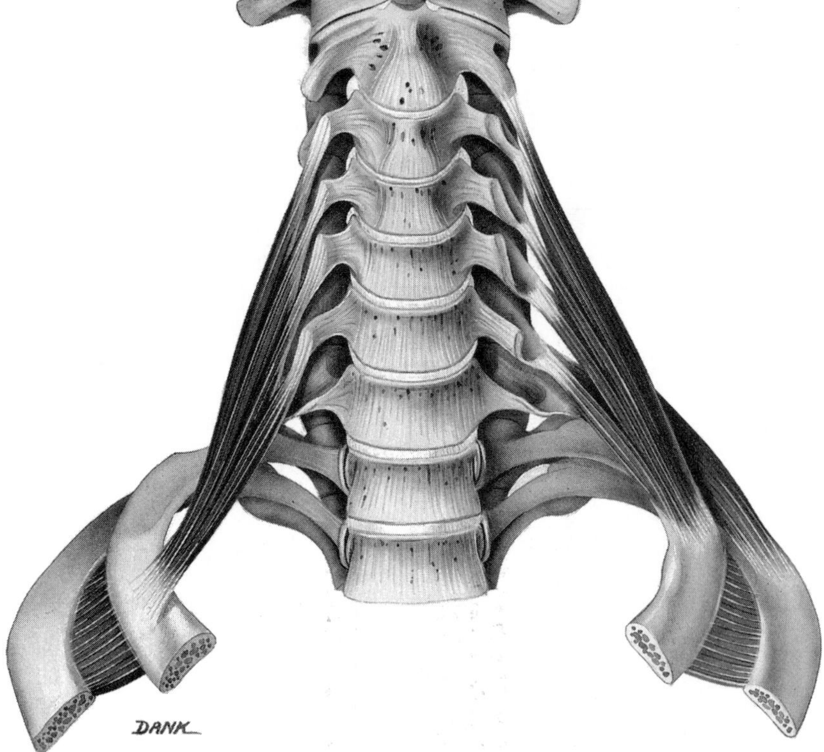

## NOTES

10

**Figure 10.22a** Muscles that move the femur (thigh): anterior view (page 323).

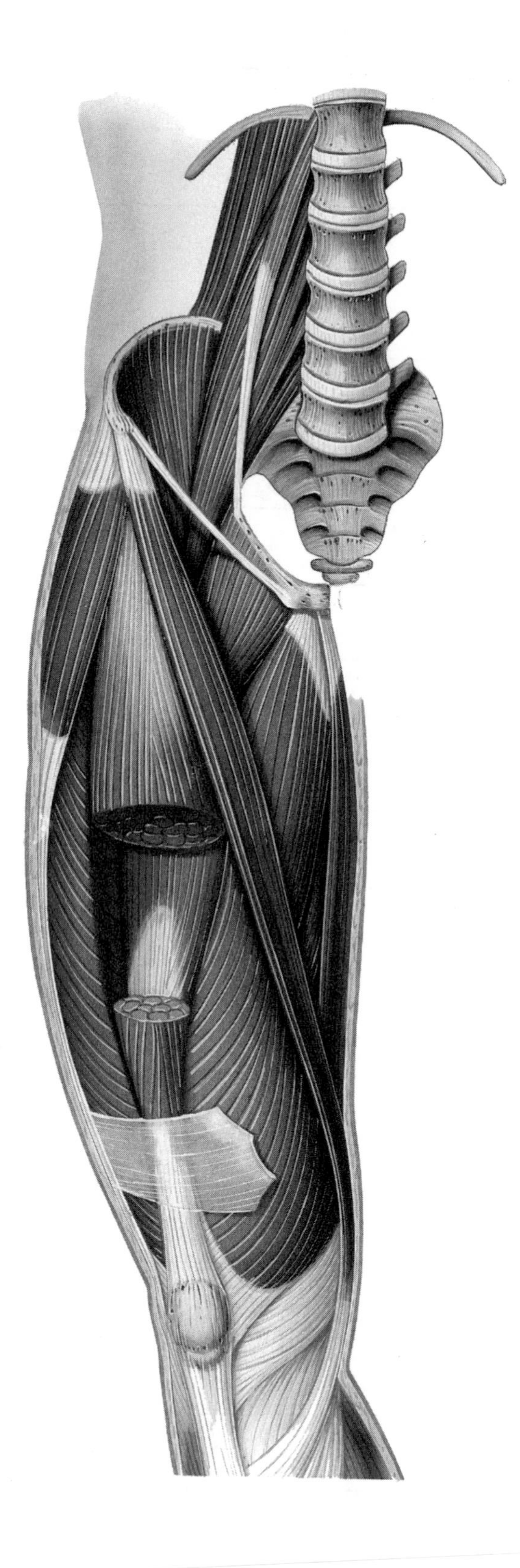

**Figure 10.22b** Muscles that move the femur (thigh): anterior view (page 324).

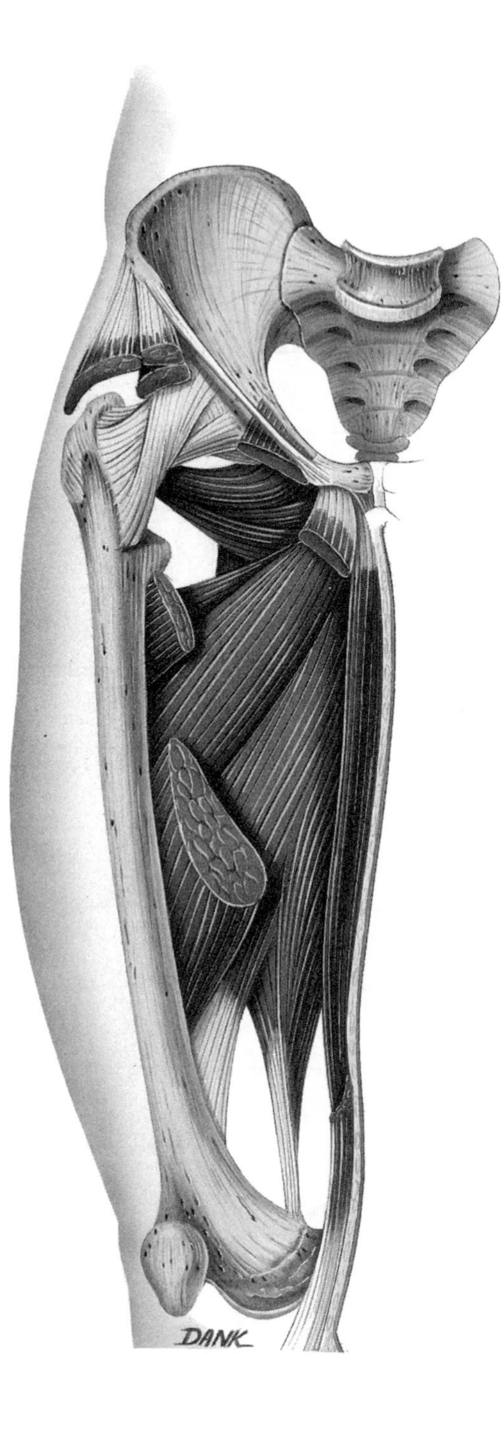

NOTES

10

**Figure 10.22c** Muscles that move the femur (thigh): posterior superficial view (page 325).

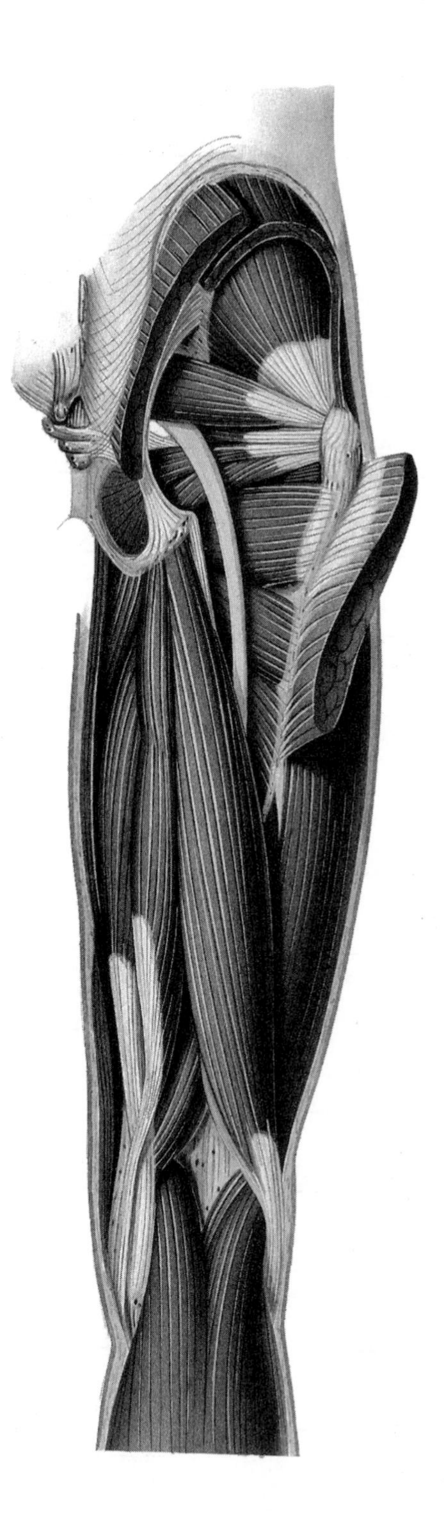

## NOTES

10

**Figure 10.23** Muscles that act on the femur (thigh) and tibia and fibula (leg) (page 329).

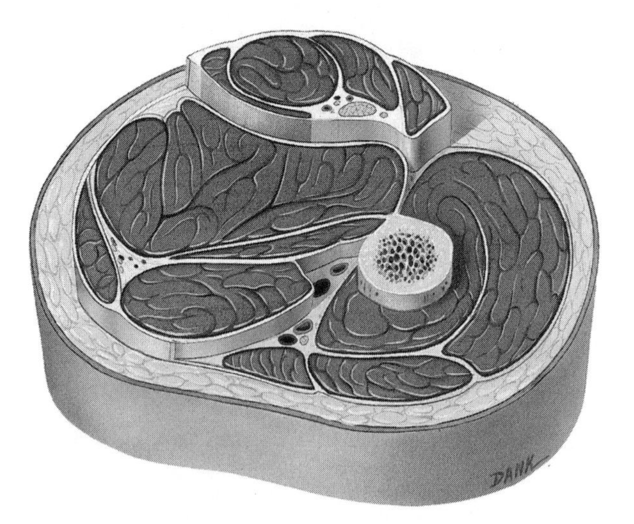

## NOTES

10

**Figure 10.24a,b** Muscles that move the foot and toes: anterior superficial view (page 332).

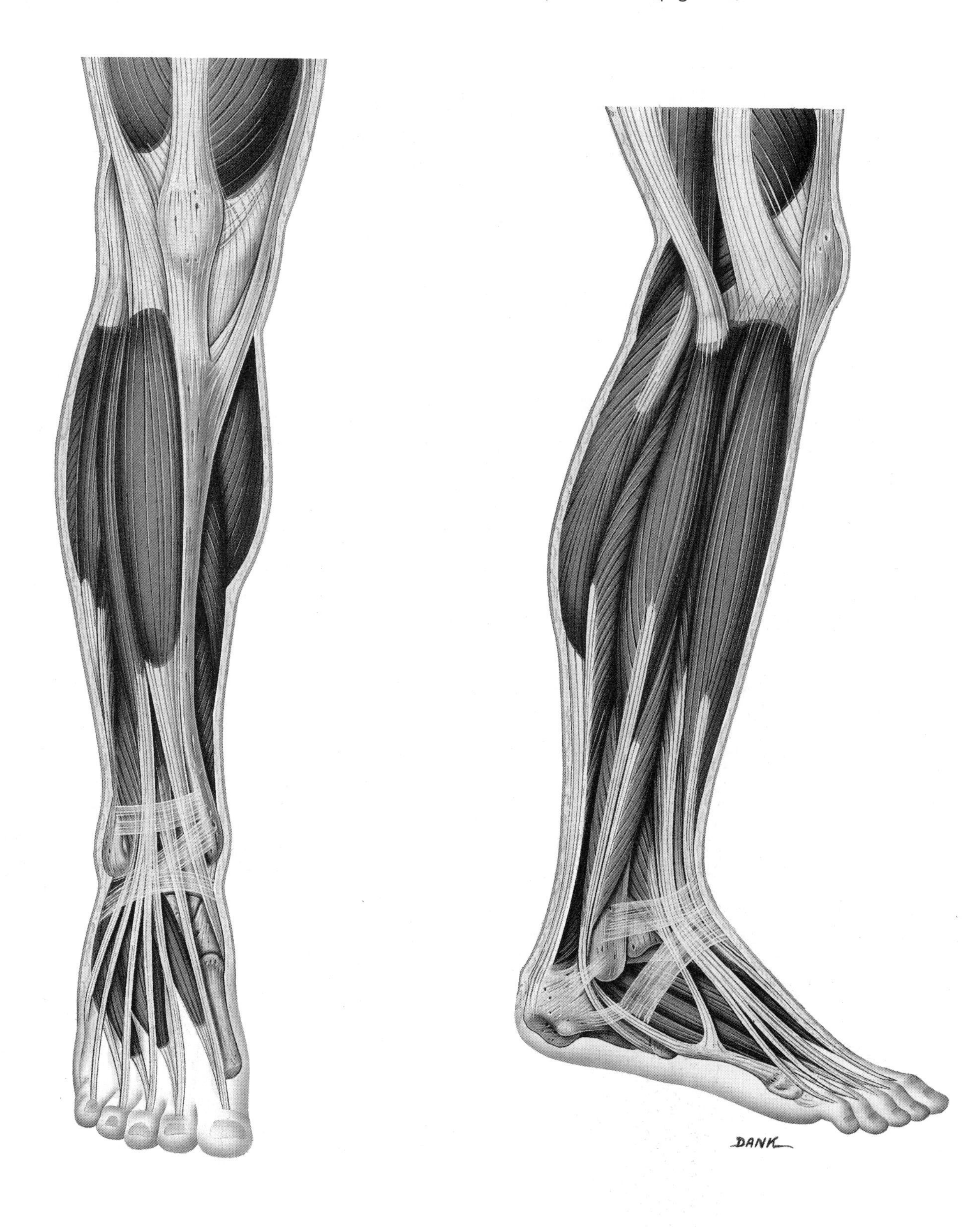

NOTES

10

**Figure 10.24c,d** Muscles that move the foot and toes: posterior view (page 333).

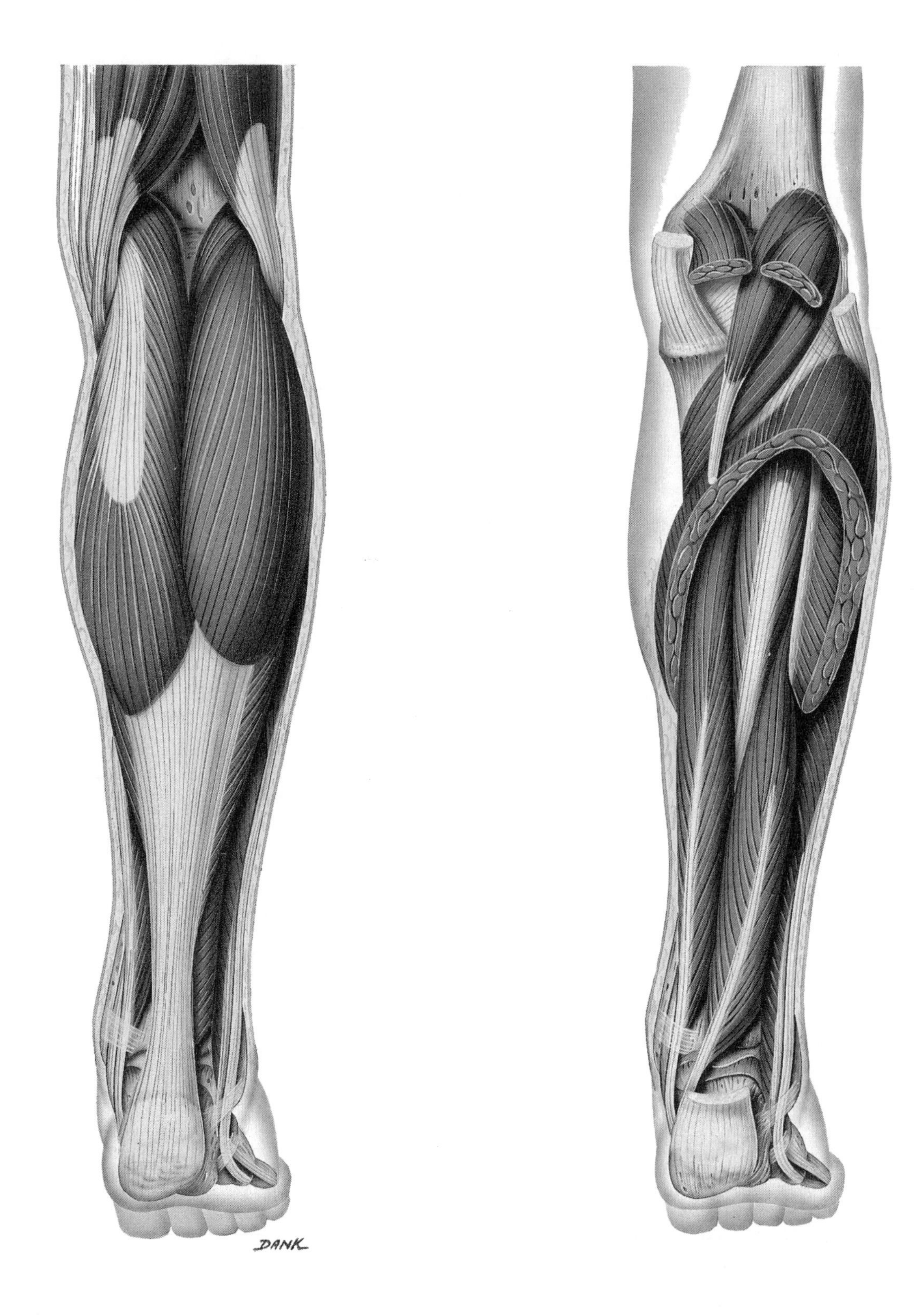

NOTES

10

**Figure 10.25a,b** Intrinsic muscles of the foot: Plantar superficial and deep view (page 338).

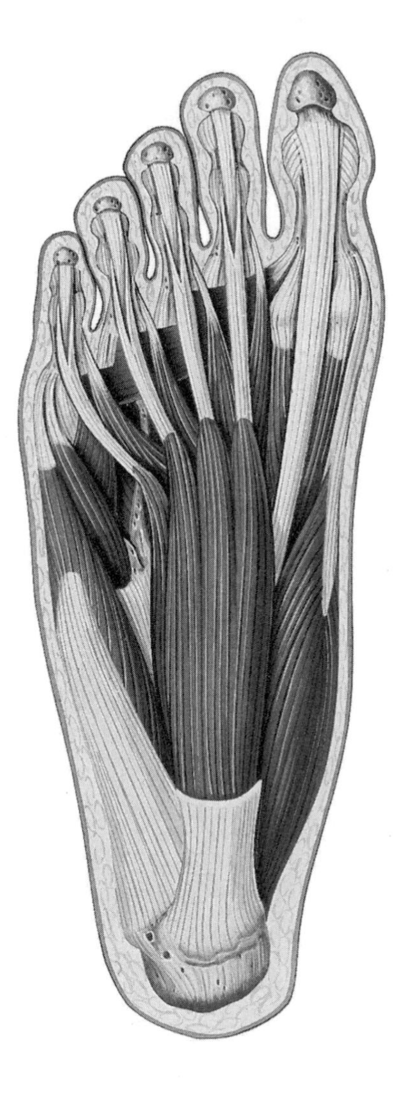

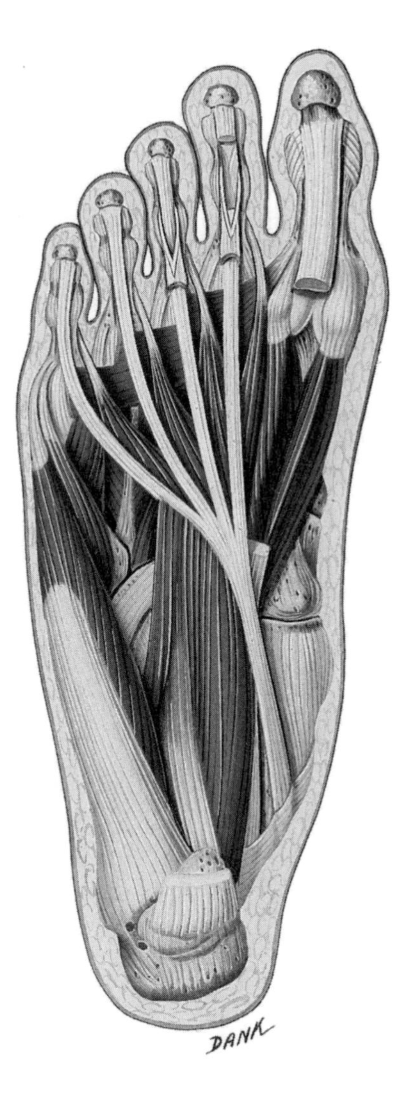


**Figure 10.25c,d** Intrinsic muscles of the foot: Plantar view (page 339).

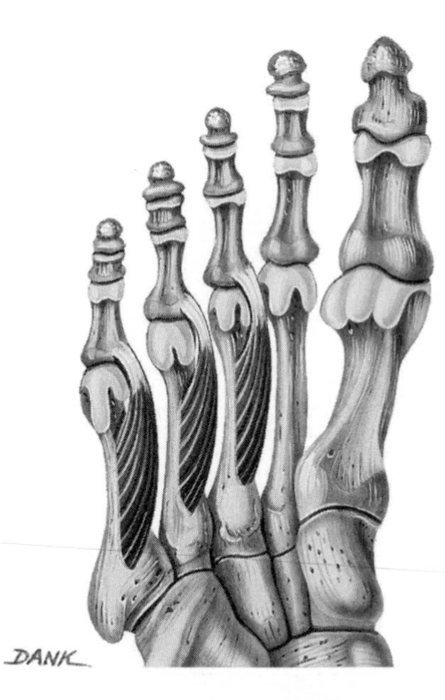


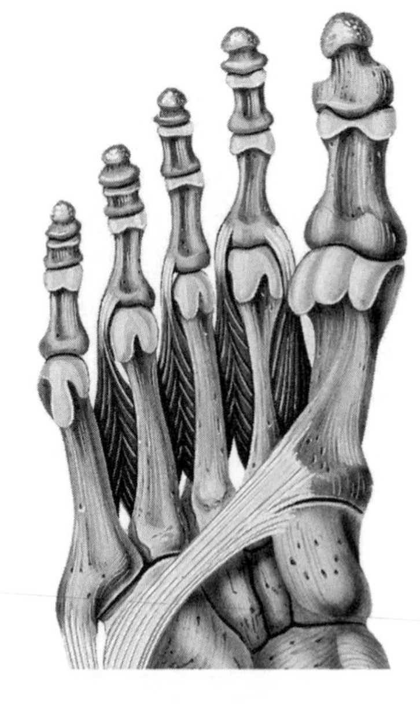

NOTES

10

**Figure 11.1** Principal regions of the cranium and face (page 346).

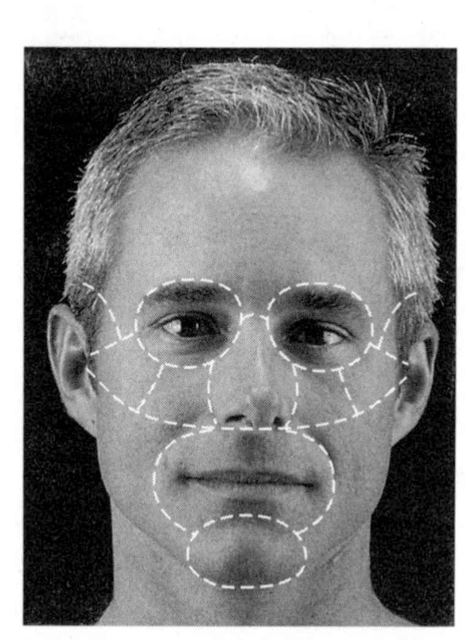

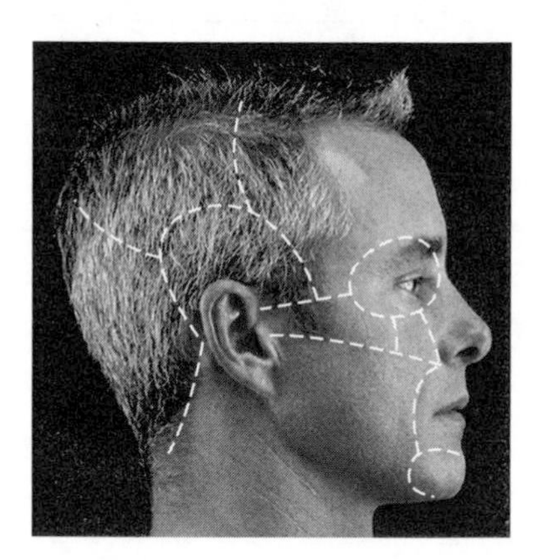

**Figure 11.2** Surface anatomy of the head (page 347).

NOTES

11

**Figure 11.3** Surface anatomy of the right eye (page 348).

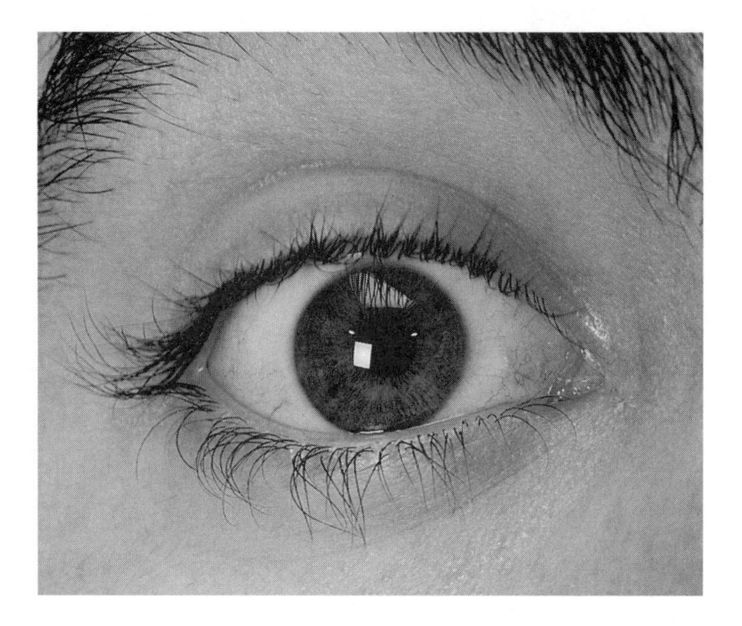

**Figure 11.4** Surface anatomy of the right ear (page 349).

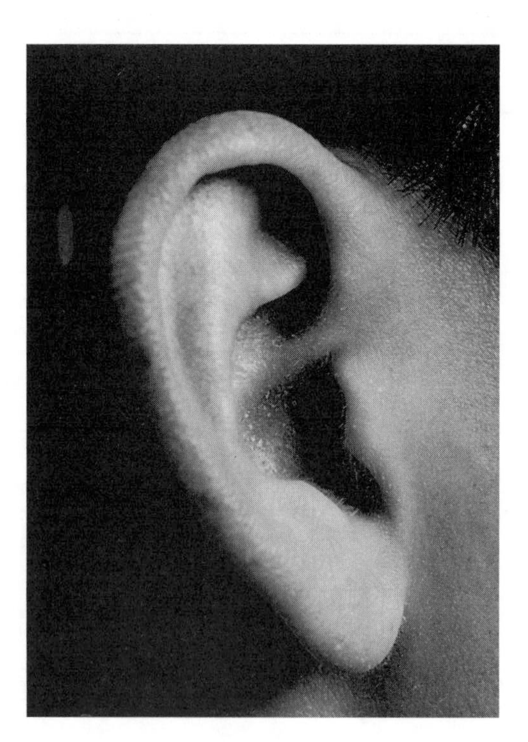

## NOTES

11

**Figure 11.5** Surface anatomy of the nose and lips (page 350).

**Figure 11.6** Surface anatomy of the neck (page 351).

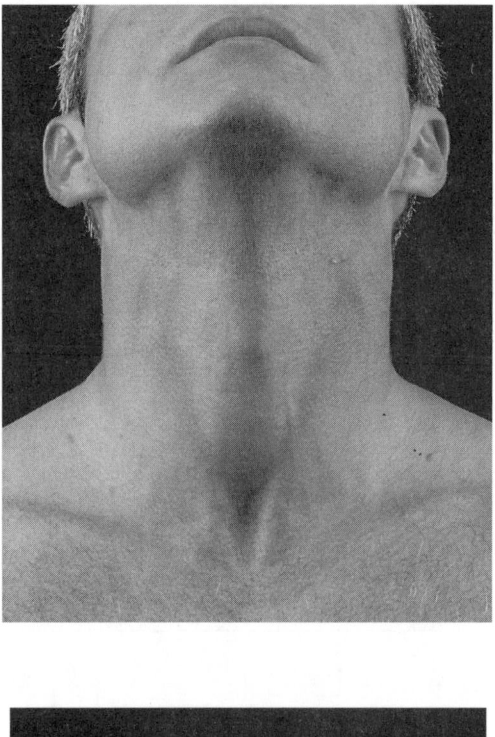

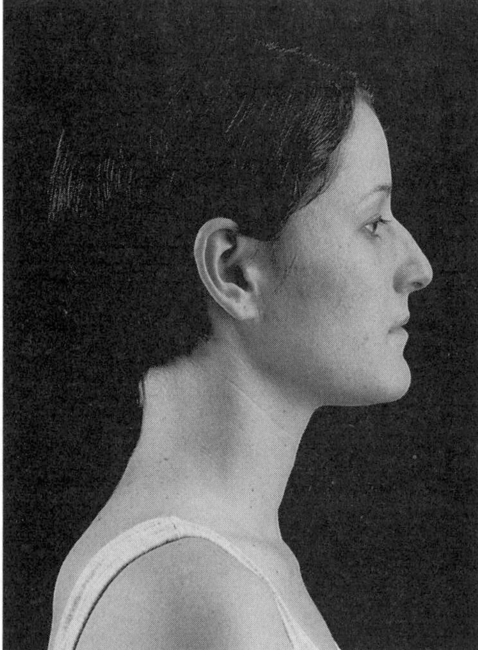

## NOTES

11

**Figure 11.7** Surface anatomy of the back (page 352).

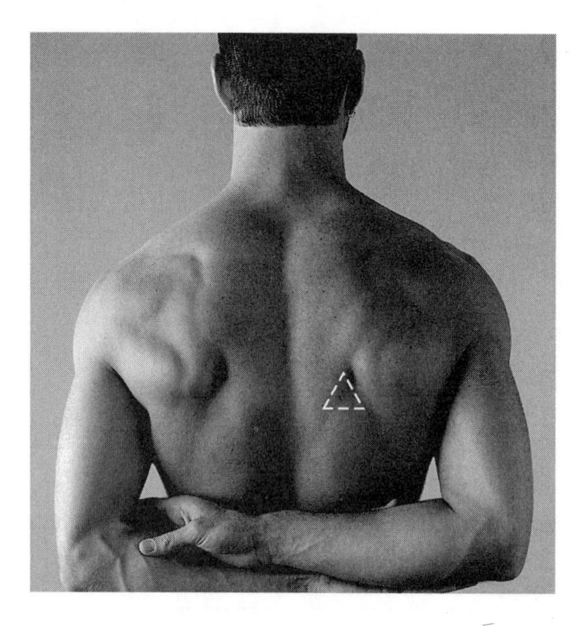

**Figure 11.8** Surface anatomy of the chest (page 353).

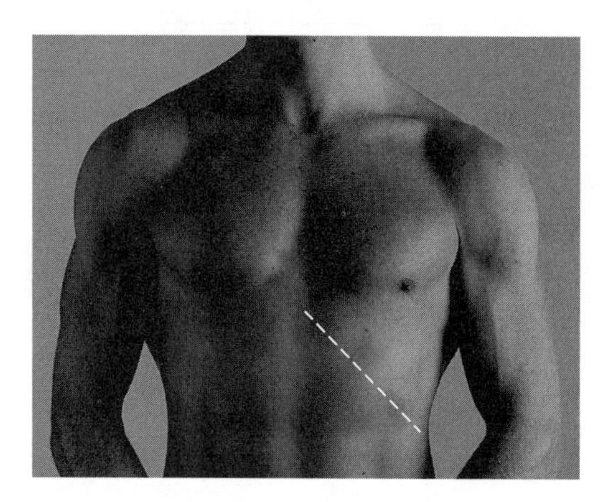

## NOTES

11

**Figure 11.9a** Surface anatomy of the abdomen and pelvis (page 356).

**Figure 11.9b,c** Surface anatomy of the abdomen and pelvis (page 357).

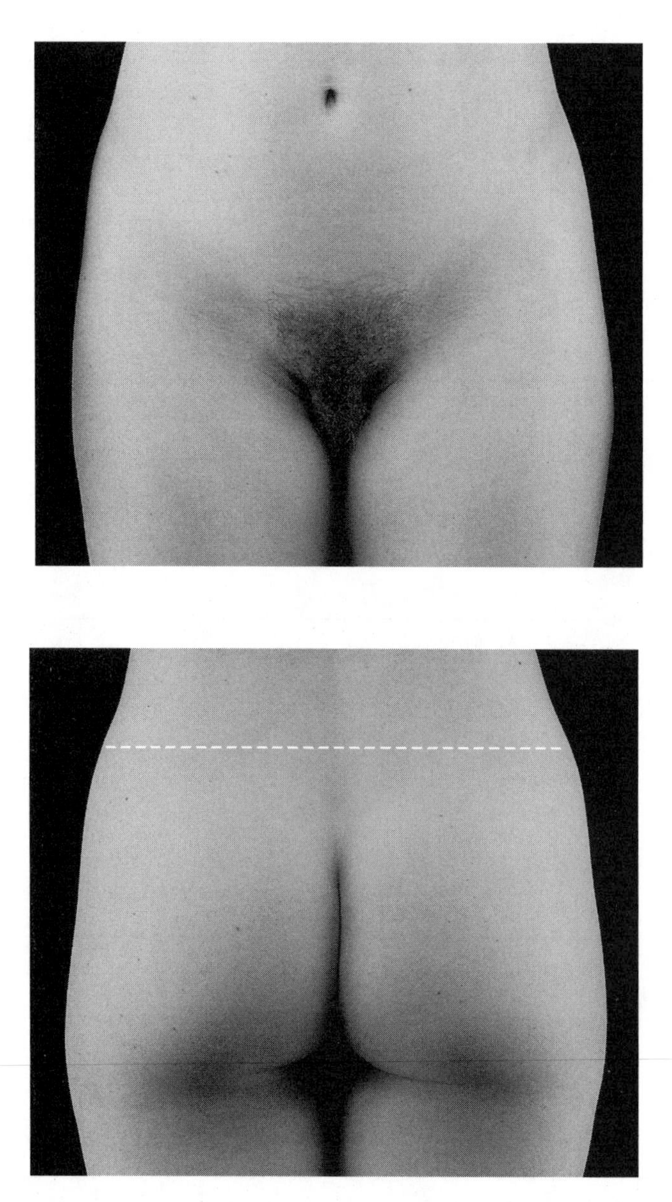

## NOTES

11

**Figure 11.10** Surface anatomy of the shoulder (page 358).

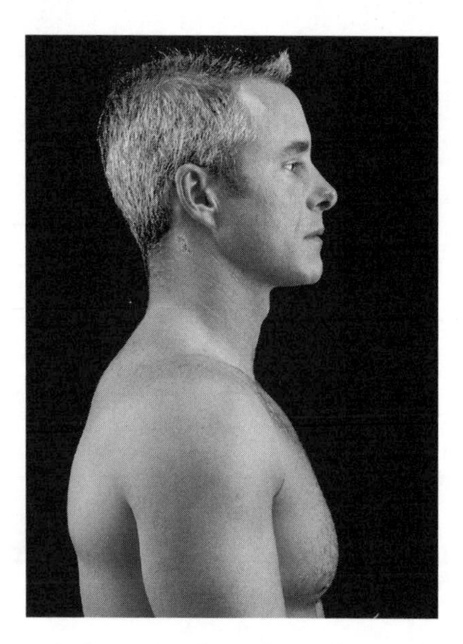

## NOTES

11

**Figure 11.11** Surface anatomy of the arm and elbow (page 359).

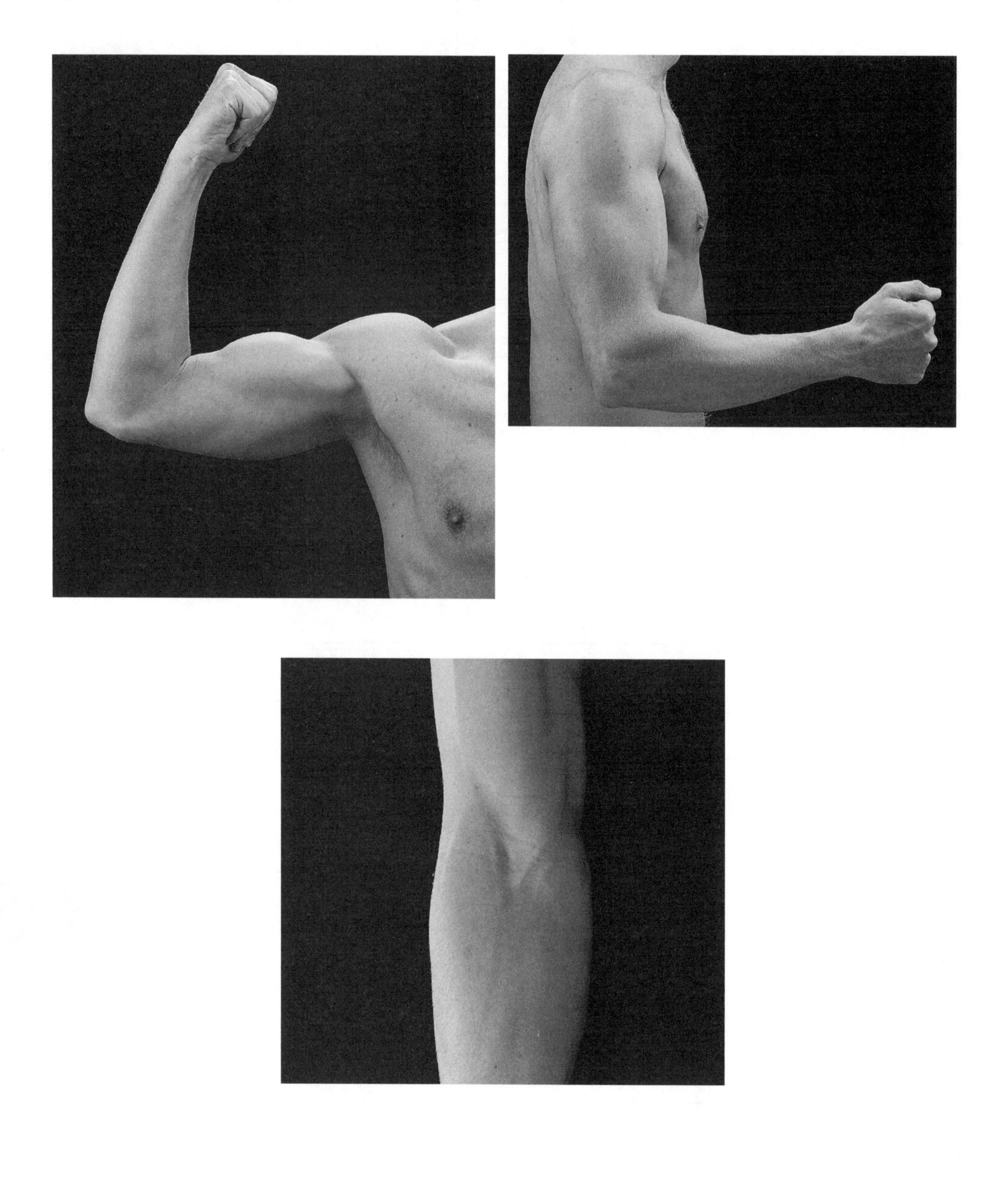

NOTES

11

**Figure 11.12** Surface anatomy of the forearm and wrist (page 361).

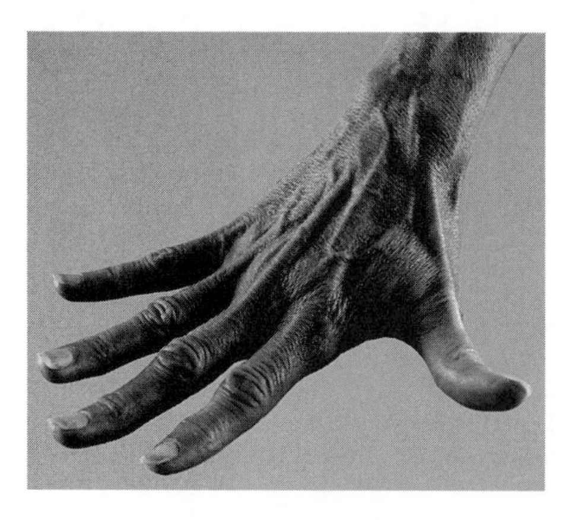

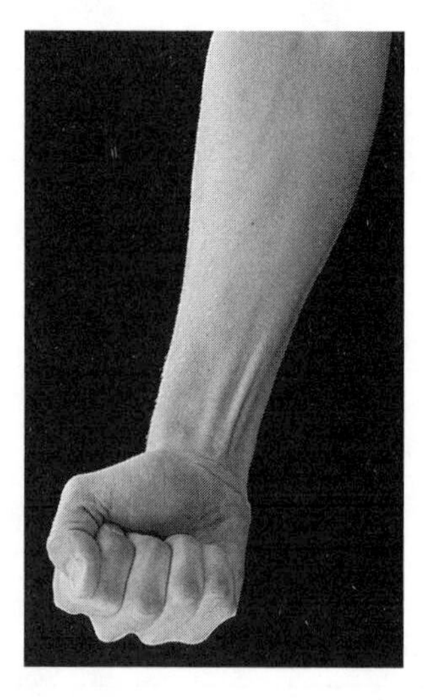

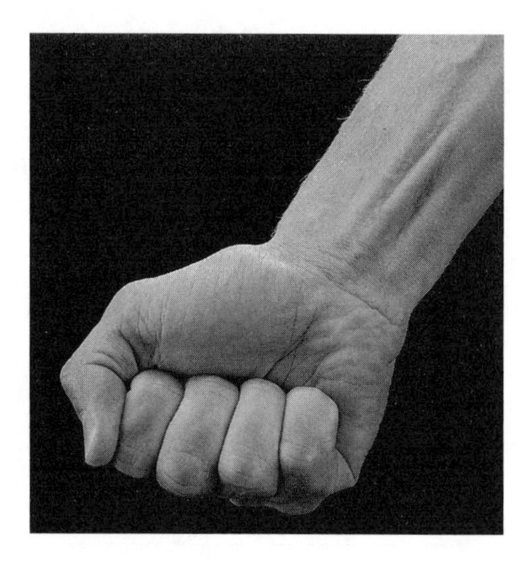

NOTES

11

**Figure 11.13** Surface anatomy of the hand (page 362).

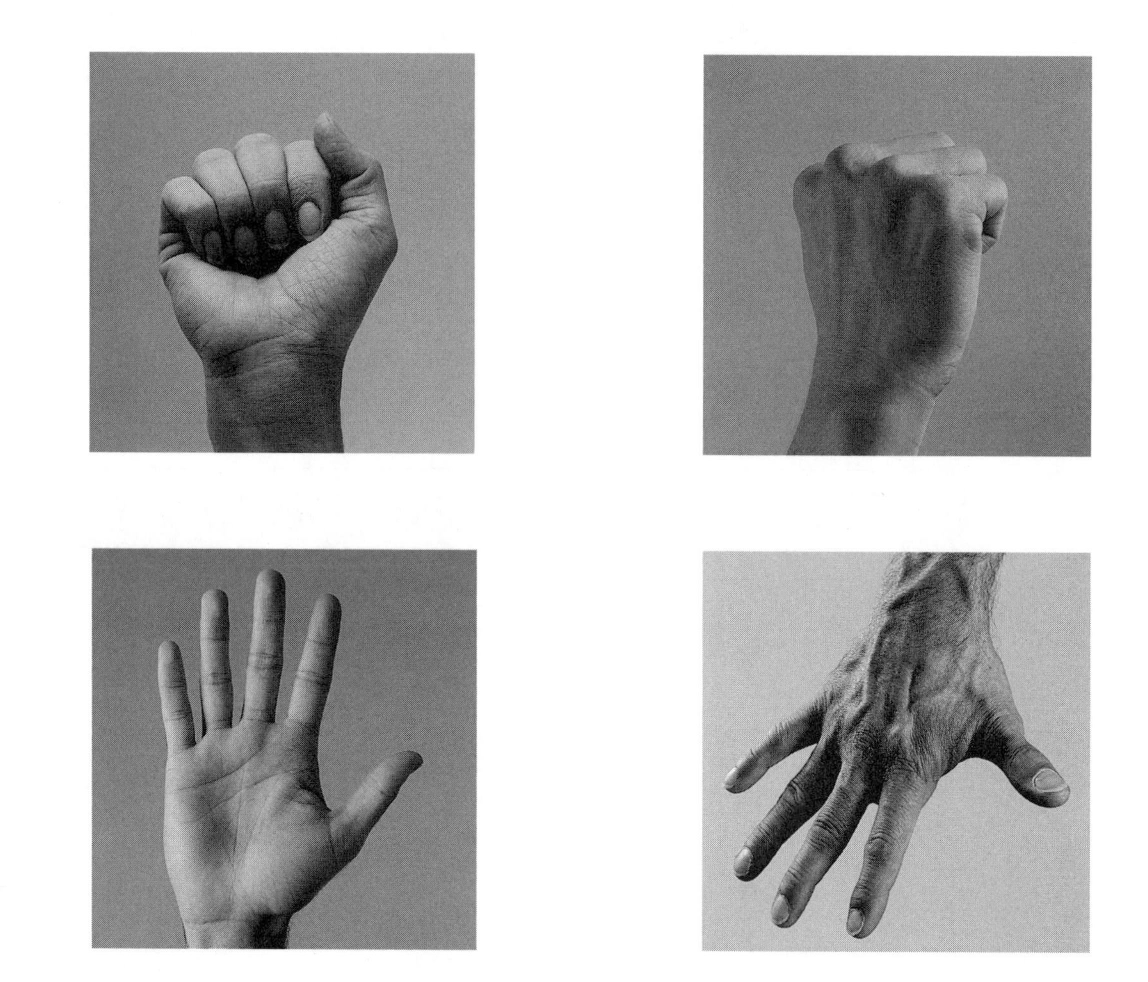

**Figure 11.14** Surface anatomy of the buttocks (page 363).

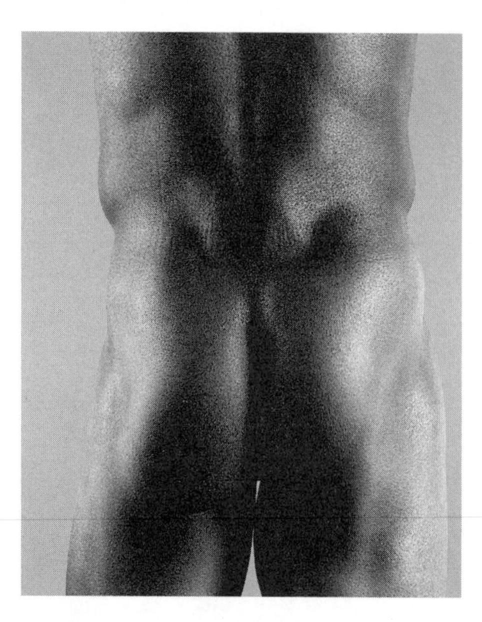

NOTES

11

**Figure 11.15** Surface anatomy of the thigh and knee (page 364).

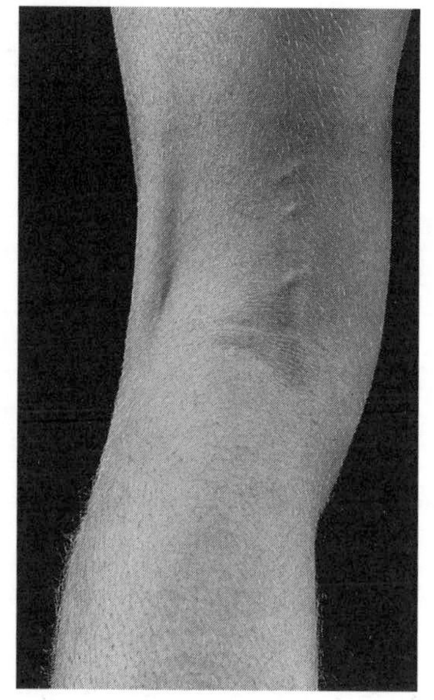

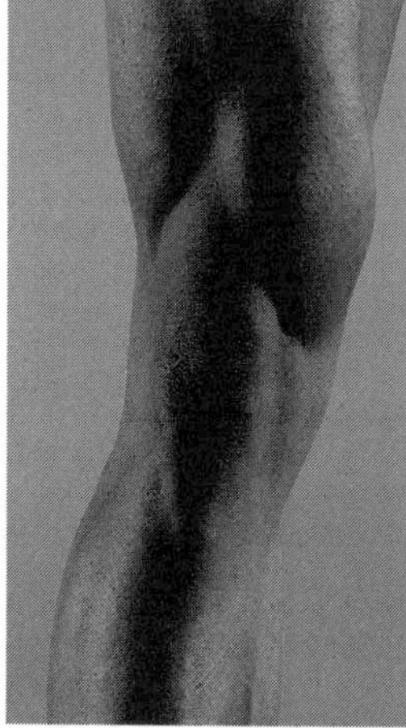

NOTES

11

**Figure 11.16a,b** Surface anatomy of the leg, ankle, and foot (page 365).

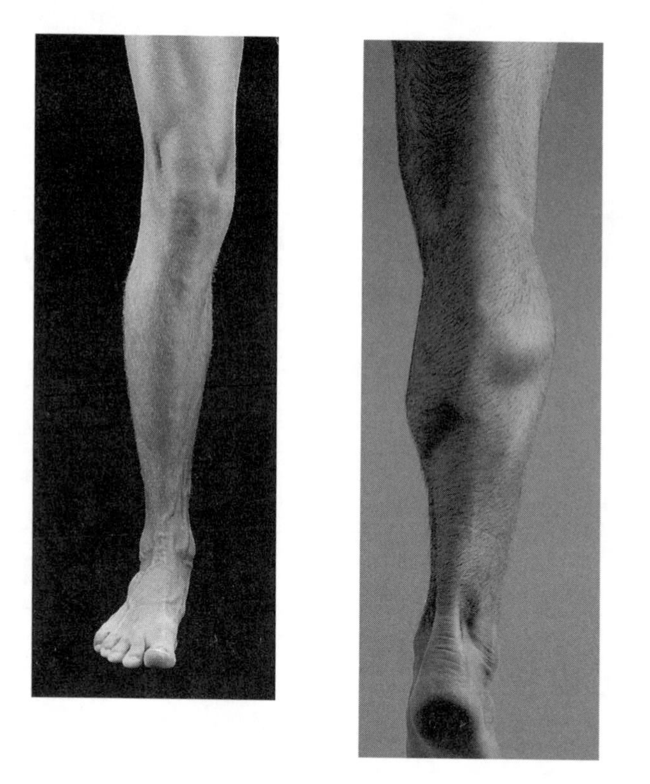

**Figure 11.16c,d** Surface anatomy of the leg, ankle, and foot (page 366).

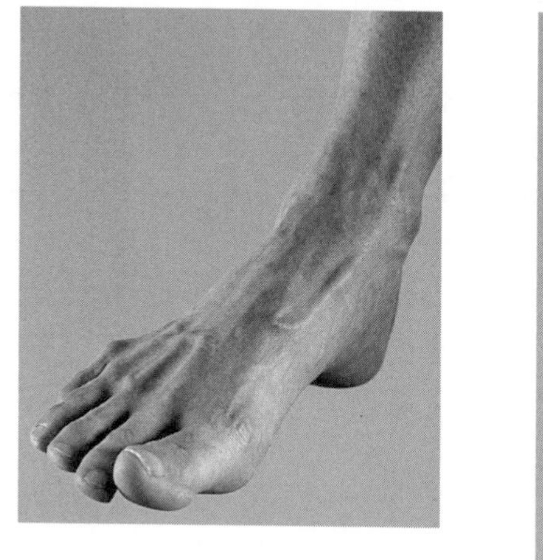

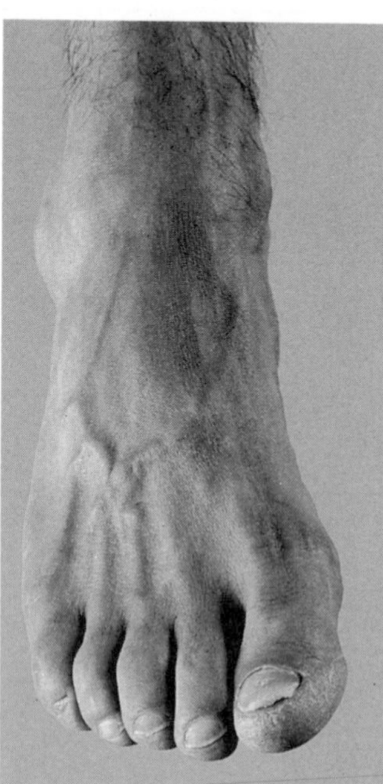

## NOTES

11

**Figure 12.1** Components of blood in a normal adult (page 372).

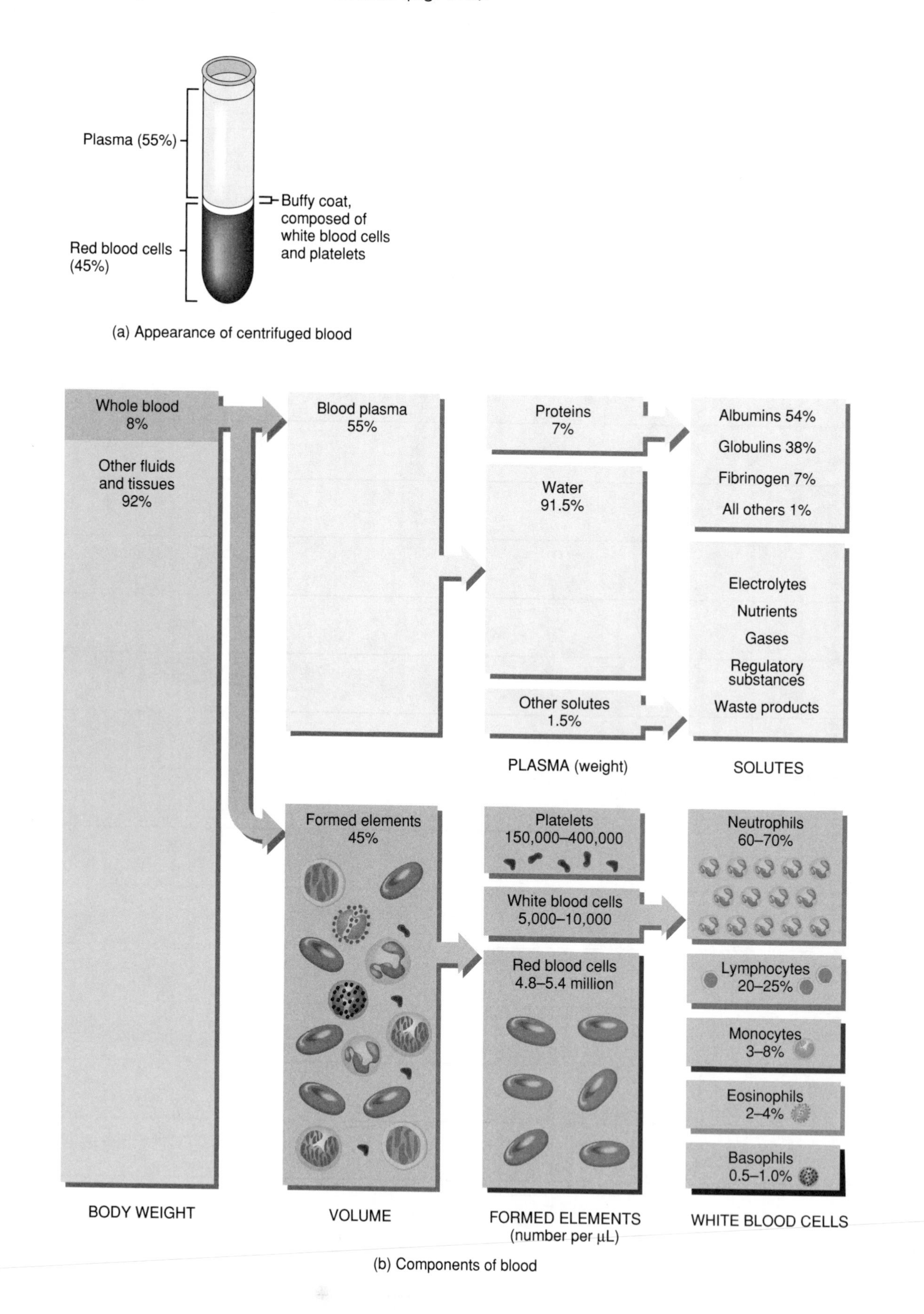


(b) Components of blood

NOTES

12

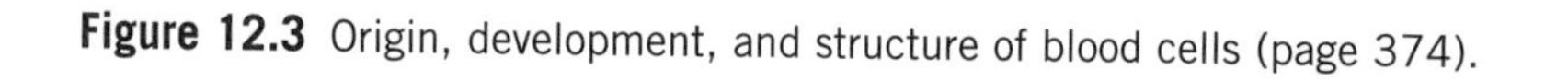

**Figure 12.3** Origin, development, and structure of blood cells (page 374).

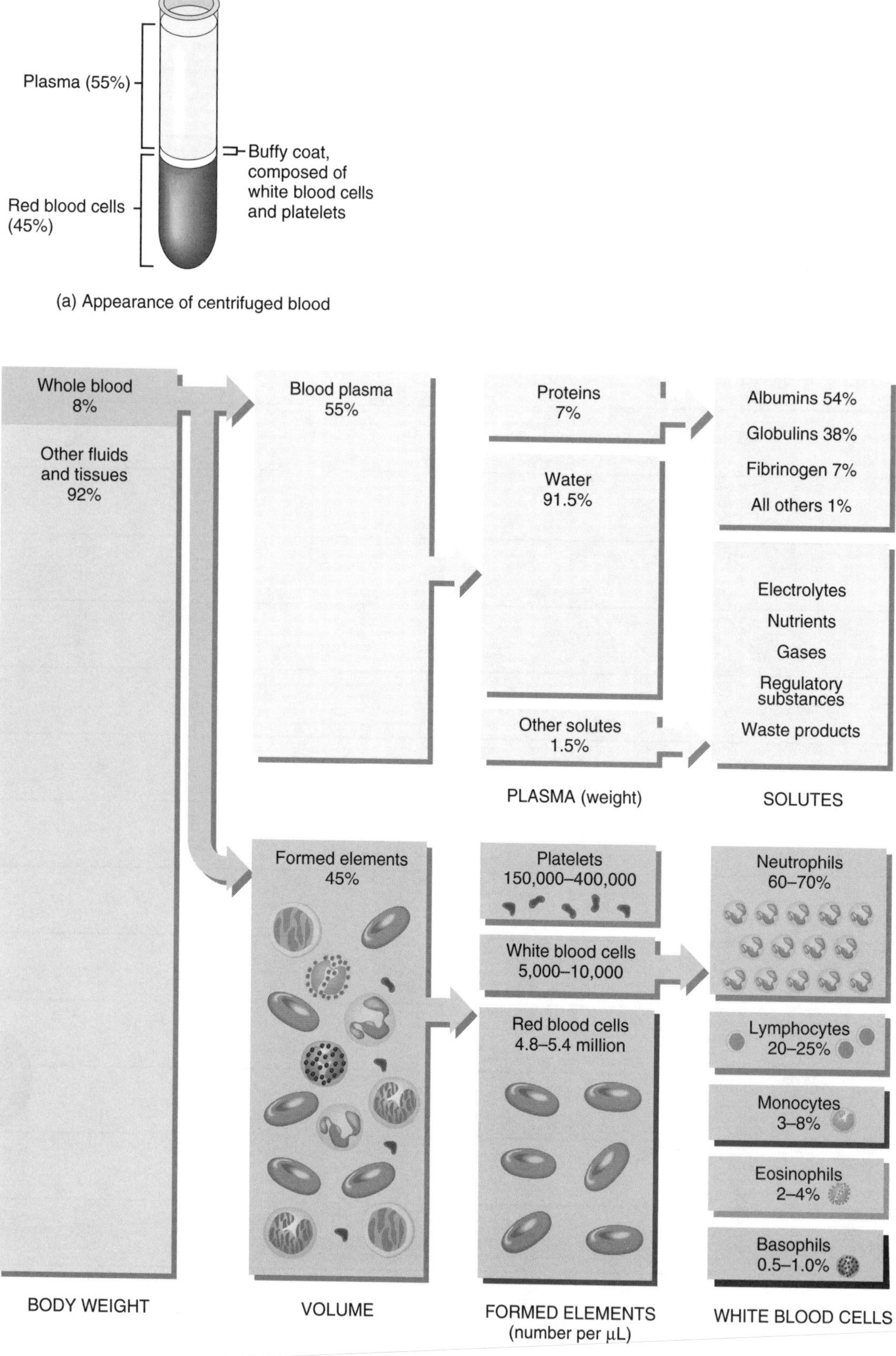

NOTES

12

**Figure 12.4** The shapes of a red blood cell (RBC) and a hemoglobin molecule (page 375).

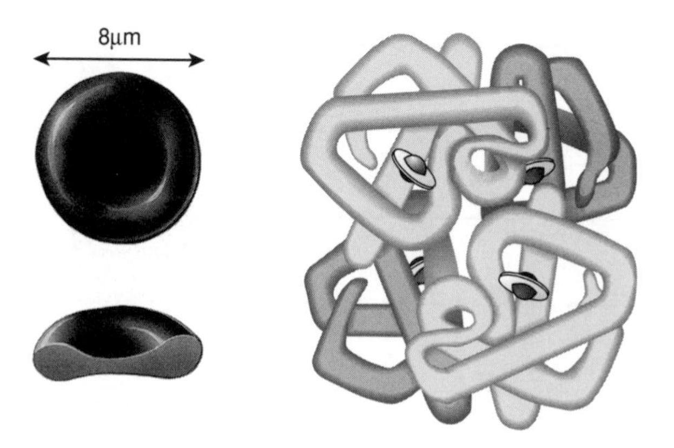


**Figure 12.5** Structure of white blood cells (page 378).

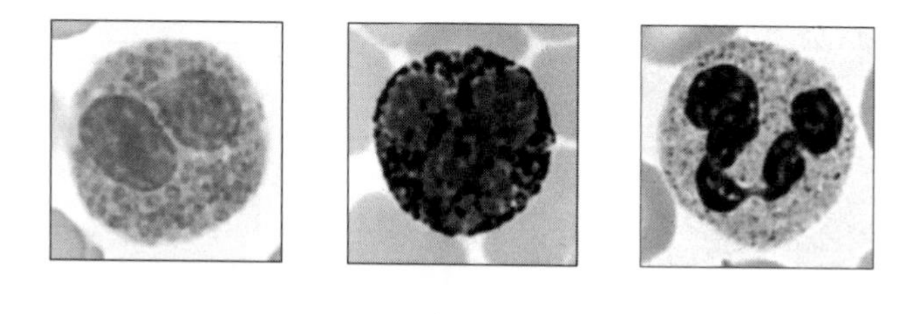

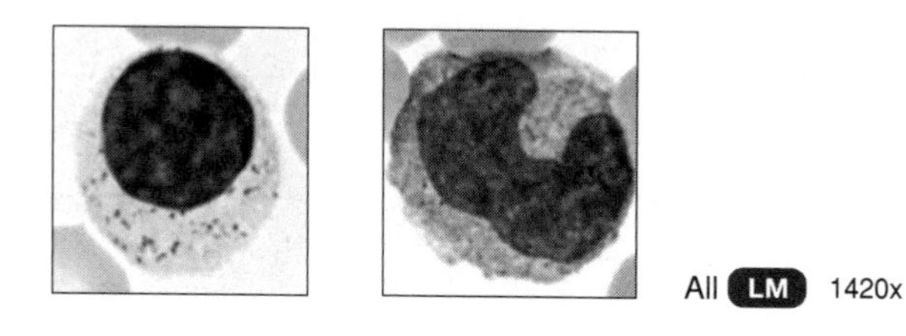

## NOTES

**Figure 13.1** Position of the heart and associated structures in the mediastinum (page 389).

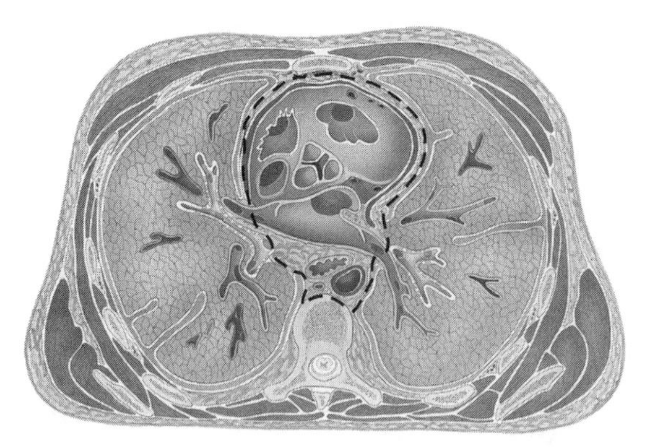

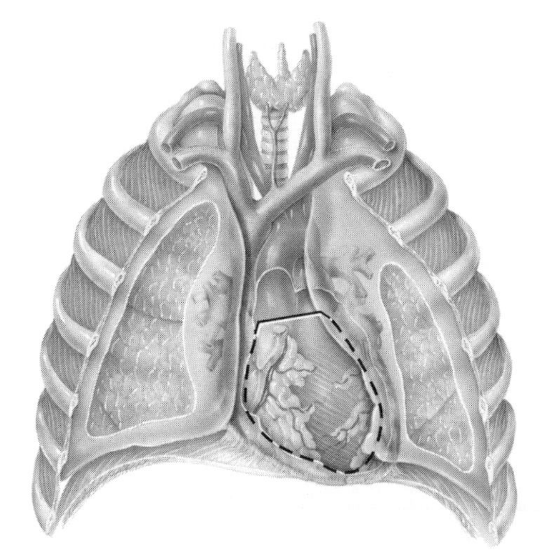

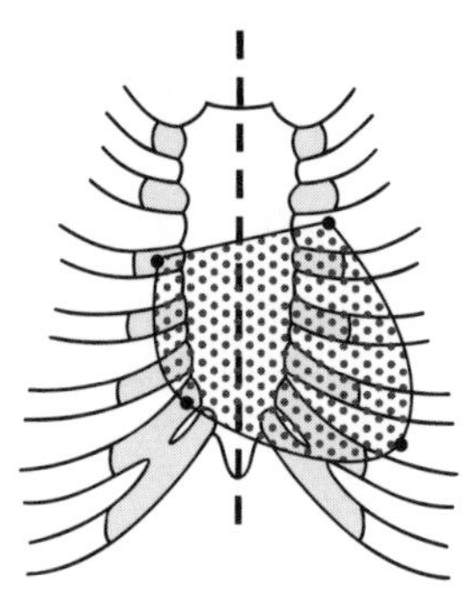

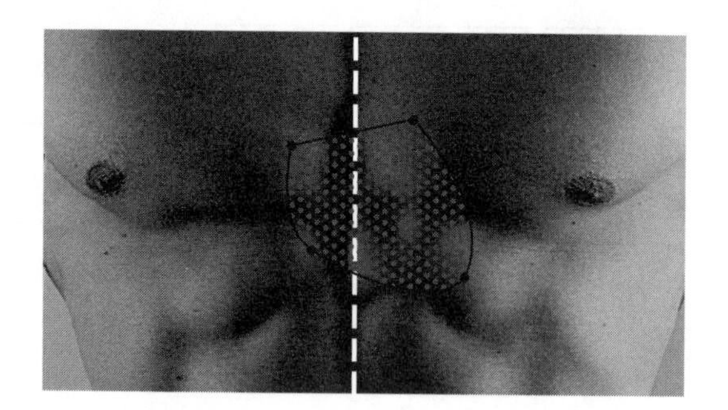

NOTES

13

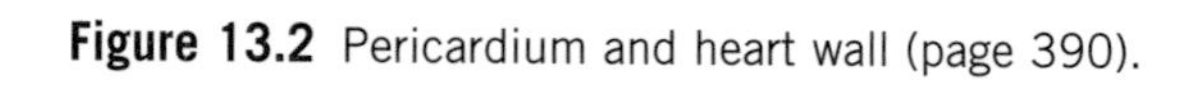

**Figure 13.2** Pericardium and heart wall (page 390).

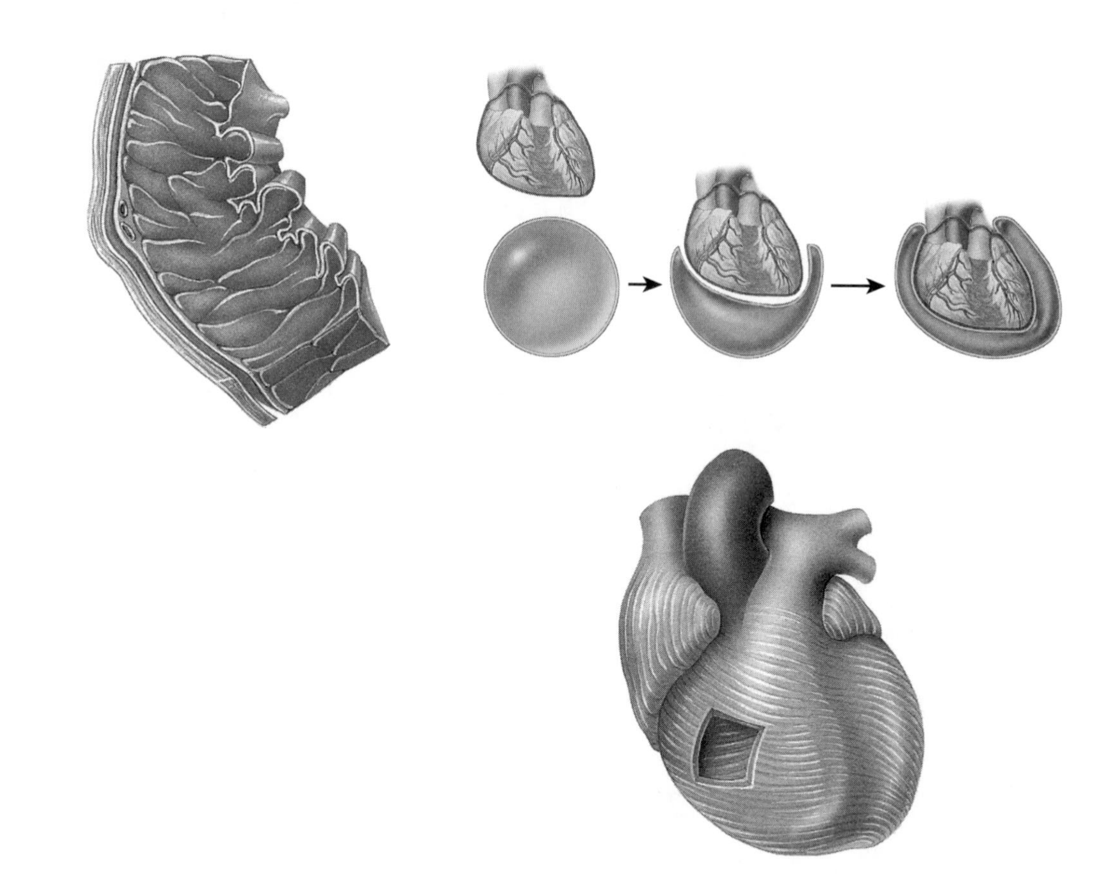

**Figure 13.3a** Structure of the heart: surface features (page 392).

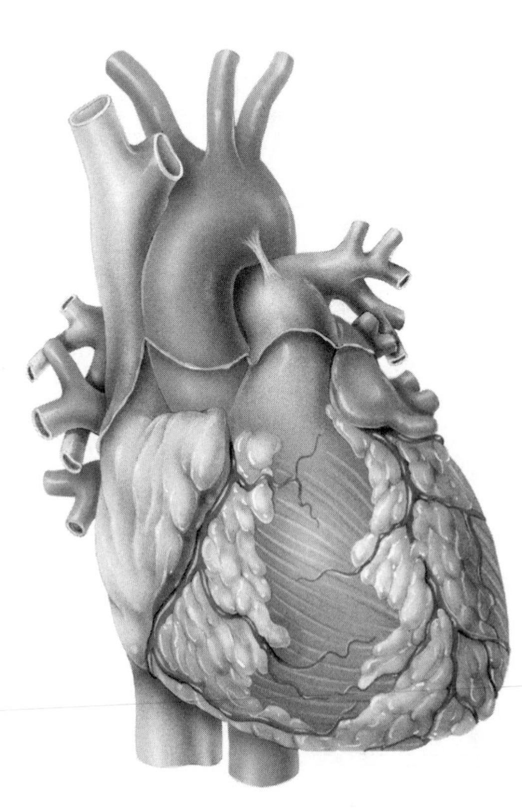

NOTES

13

**Figure 13.3b,c** Structure of the heart: surface features (page 393).

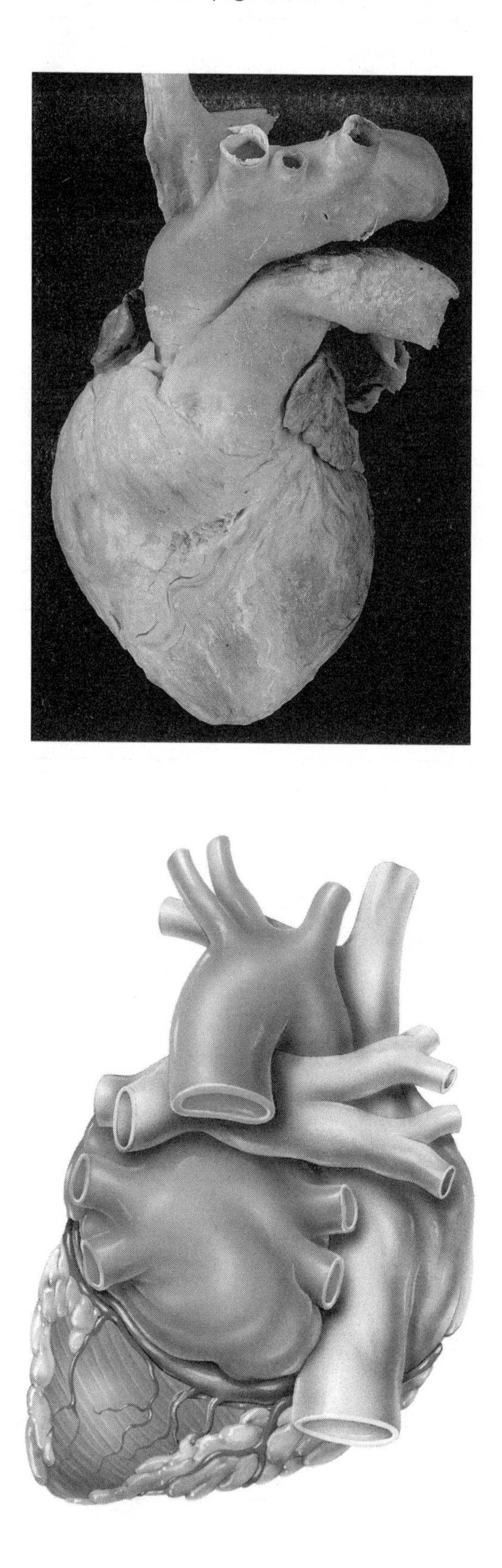

NOTES

13

**Figure 13.4a** Structure of the heart: internal anatomy (page 394).

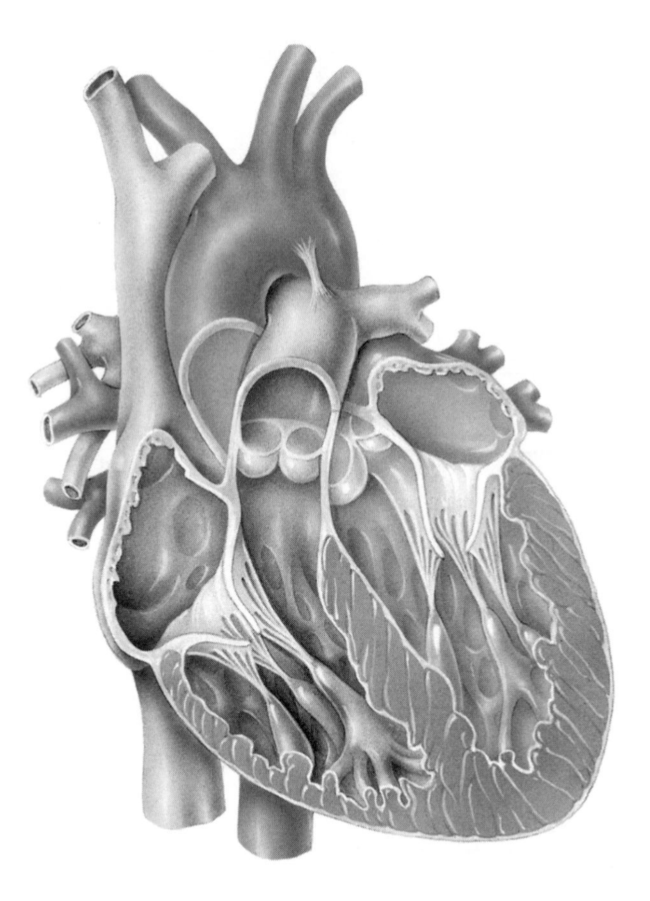

NOTES

13

**Figure 13.4b,c** Structure of the heart: internal anatomy (page 395).

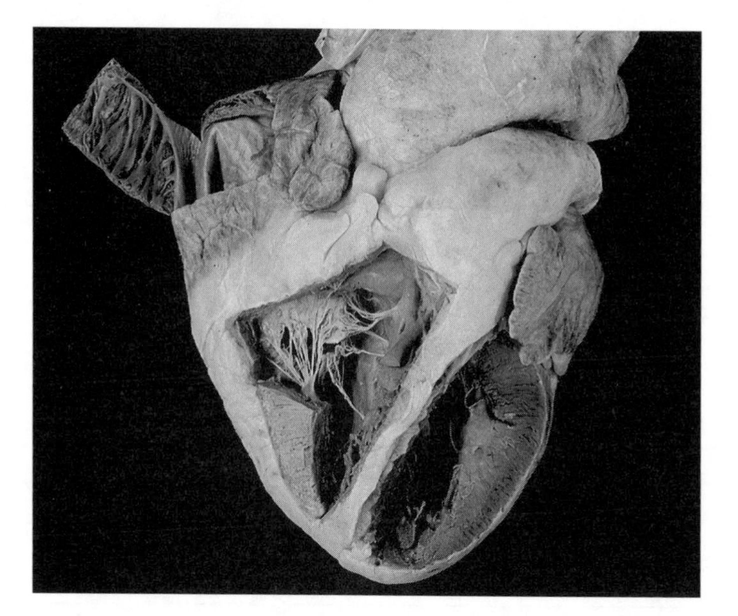

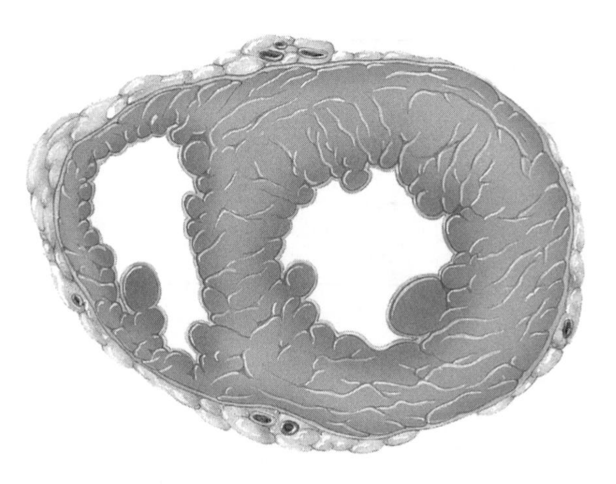

**Figure 13.5** Fibrous skeleton of the heart (page 396).

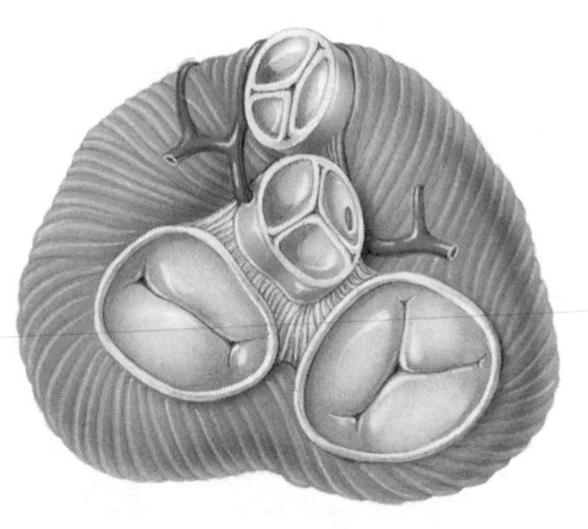

NOTES

13

**Figure 13.6** Valves of the heart (page 397).

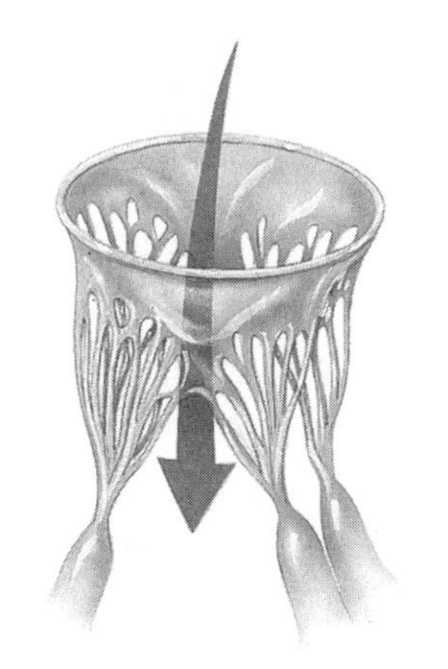

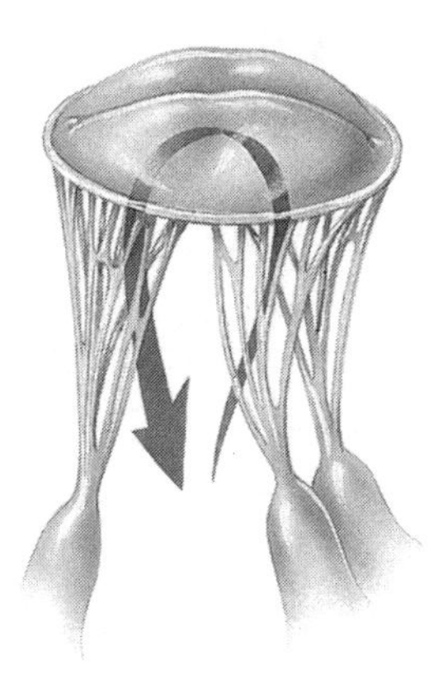

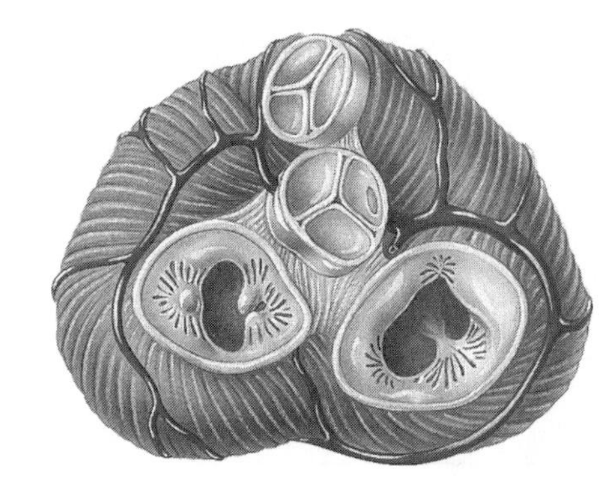

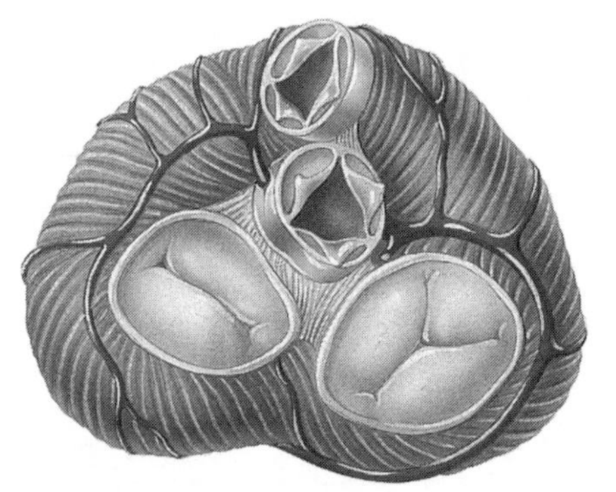

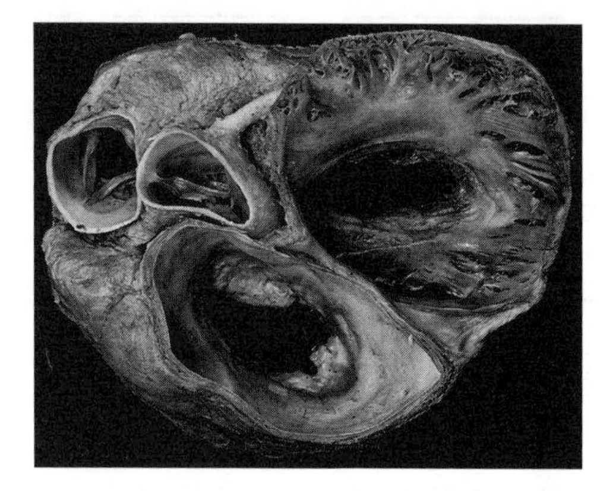

NOTES

13

**Figure 13.7a** Systemic and pulmonary circulations (page 398).

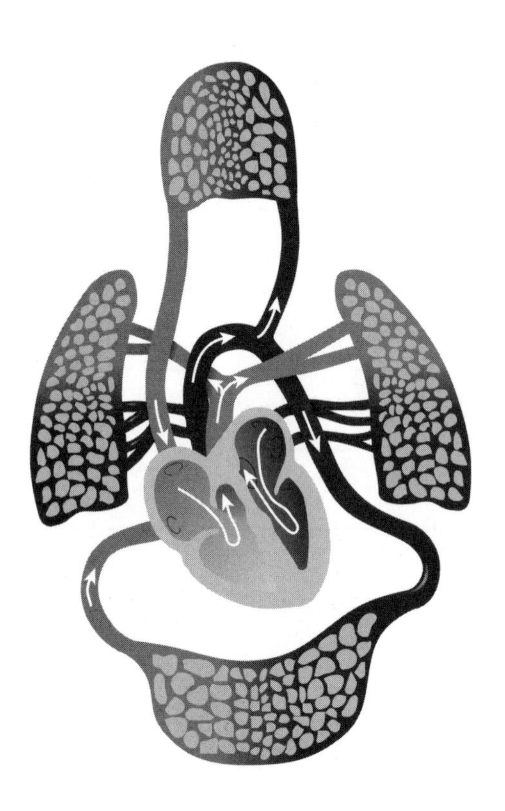

**Figure 13.7b** Systemic and pulmonary circulations (page 339).

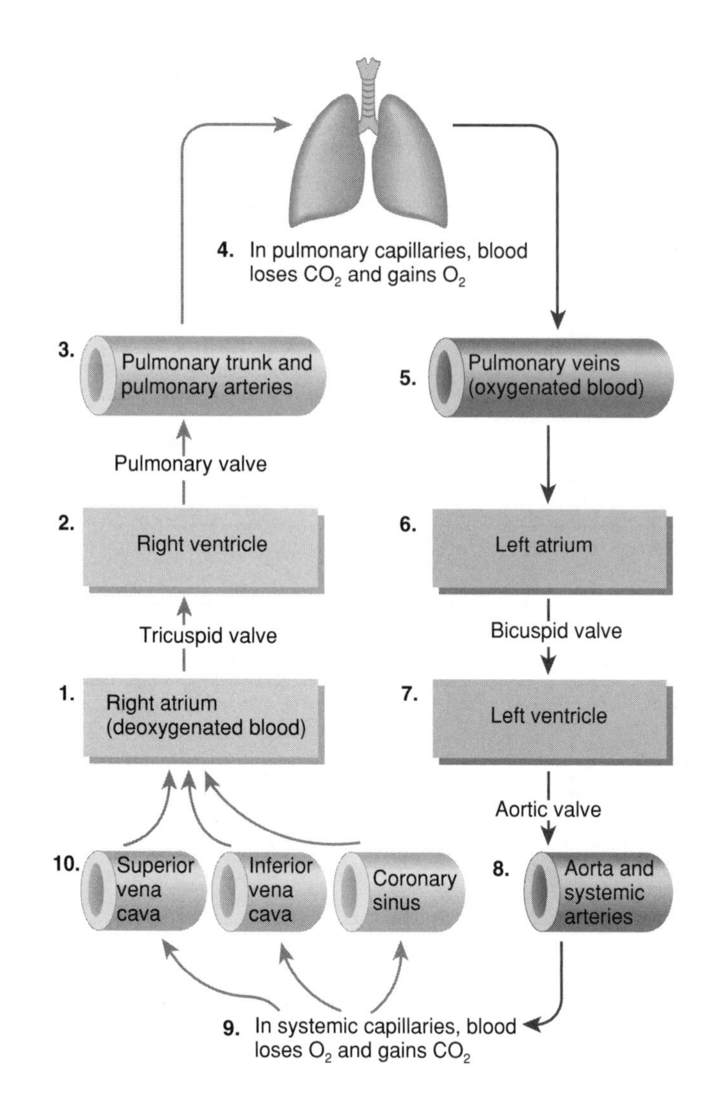

NOTES

13

**Figure 13.8** Coronary (cardiac) circulation (page 400).

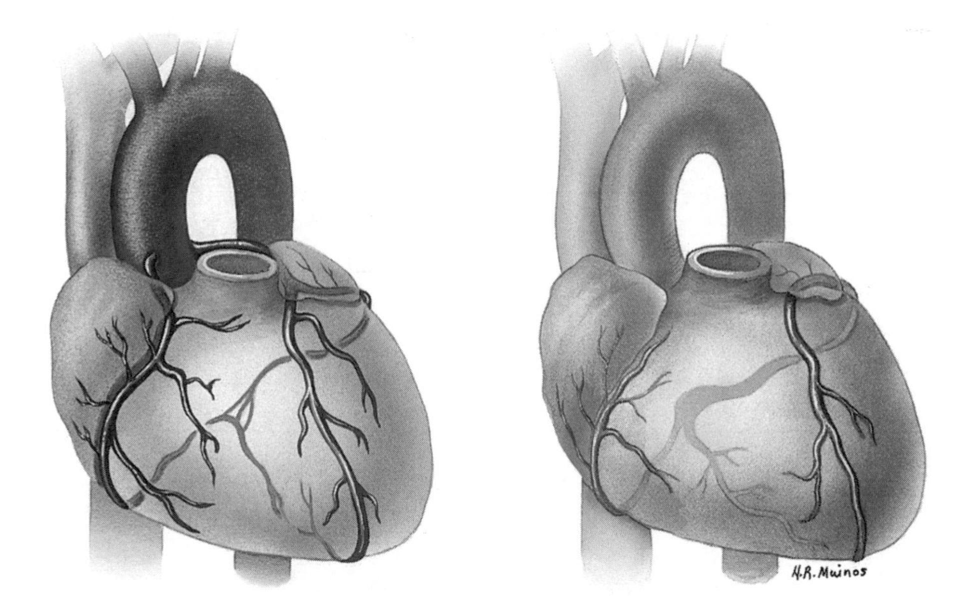

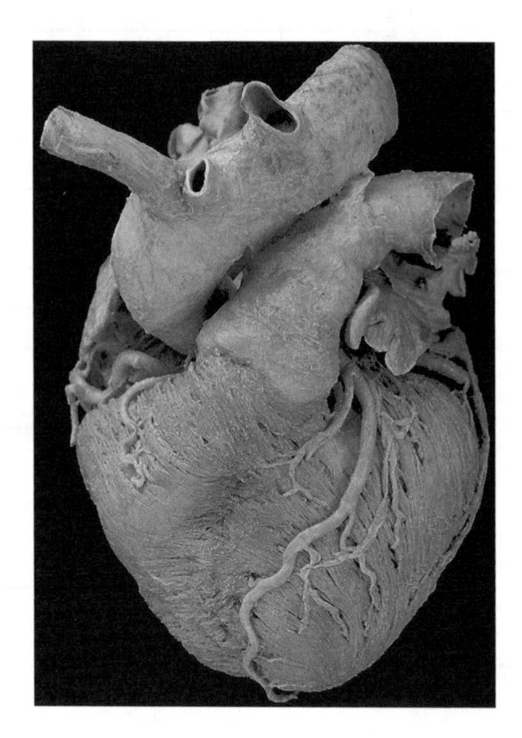

NOTES

13

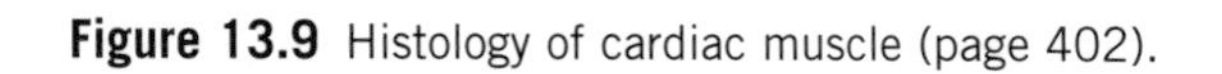

**Figure 13.9** Histology of cardiac muscle (page 402).

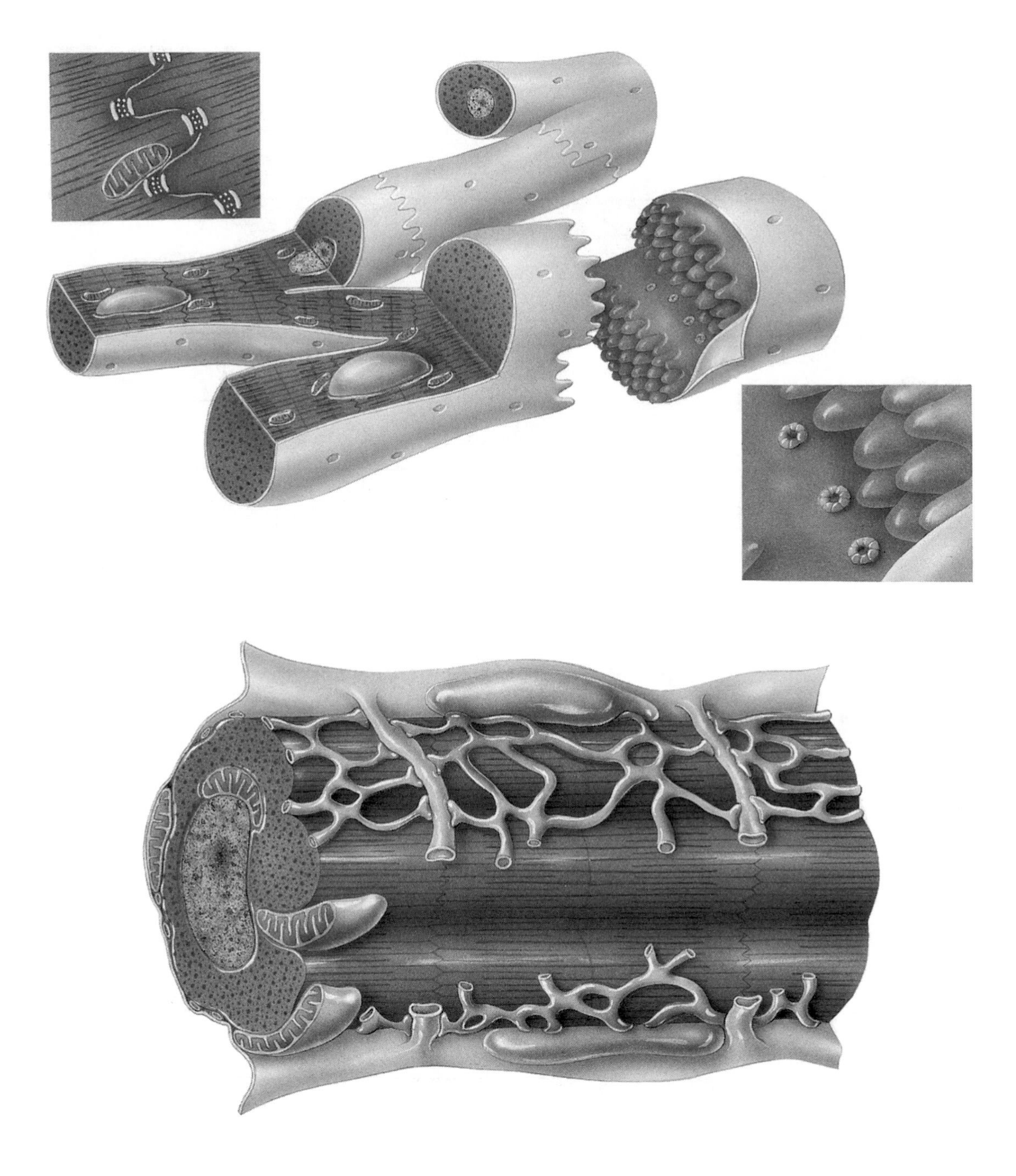

NOTES

13

**Figure 13.10** Conduction system of the heart and a normal electrocardiogram (page 403).

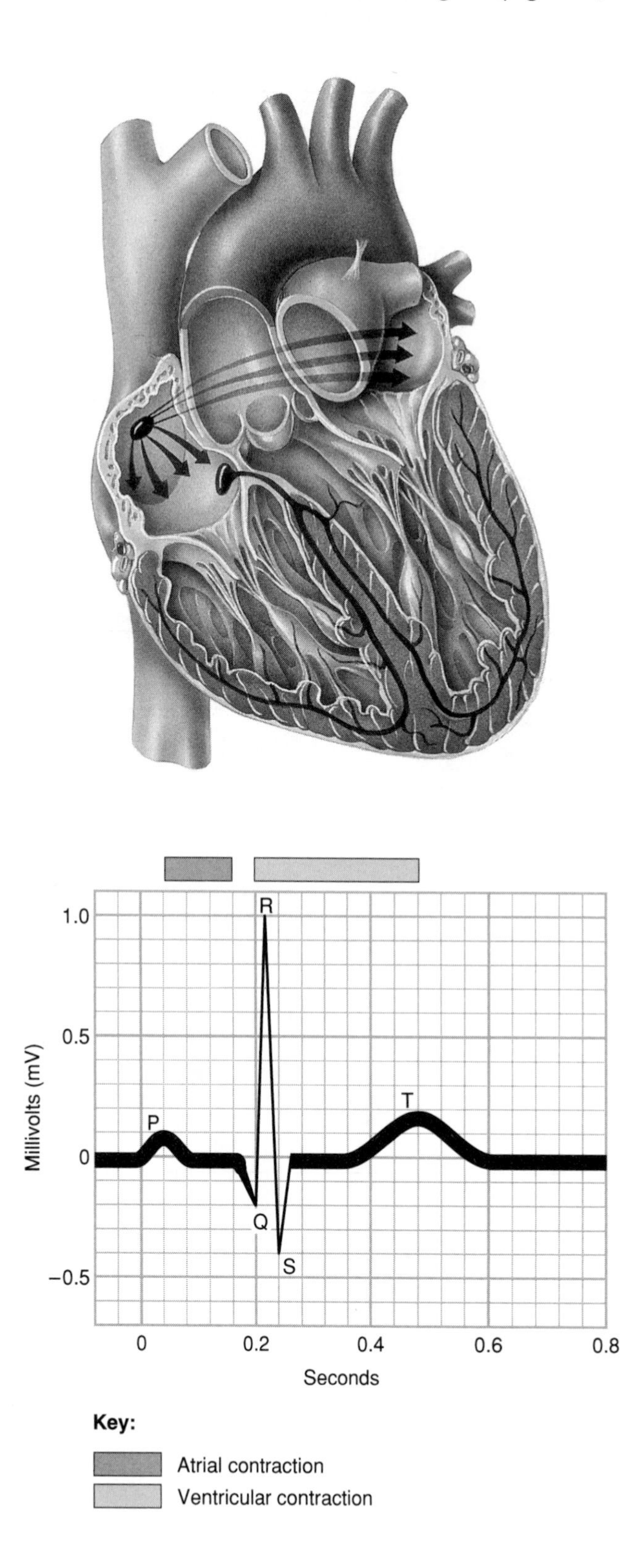

NOTES

13

**Figure 13.11** The cardiac cycle (heartbeat) (page 405).

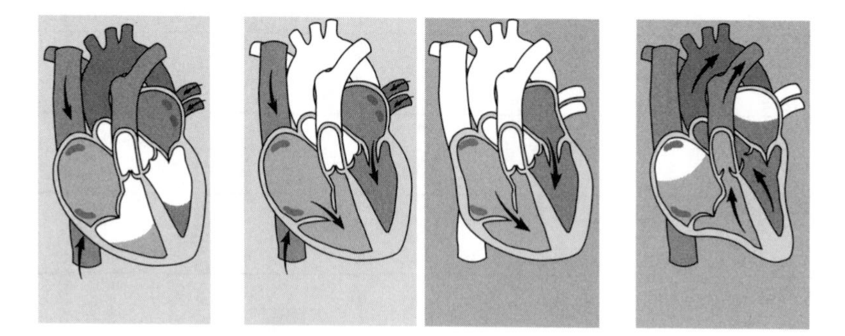

**Figure 13.12** Location of valves and auscultation sites for heart sounds (page 406).

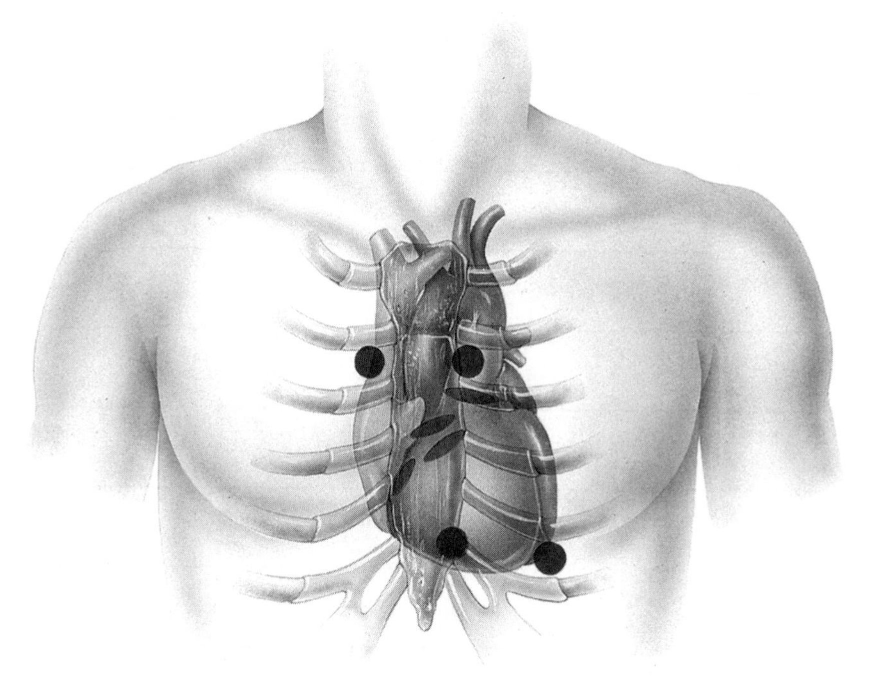

NOTES

13

**Figure 13.13** Development of the heart (page 407).

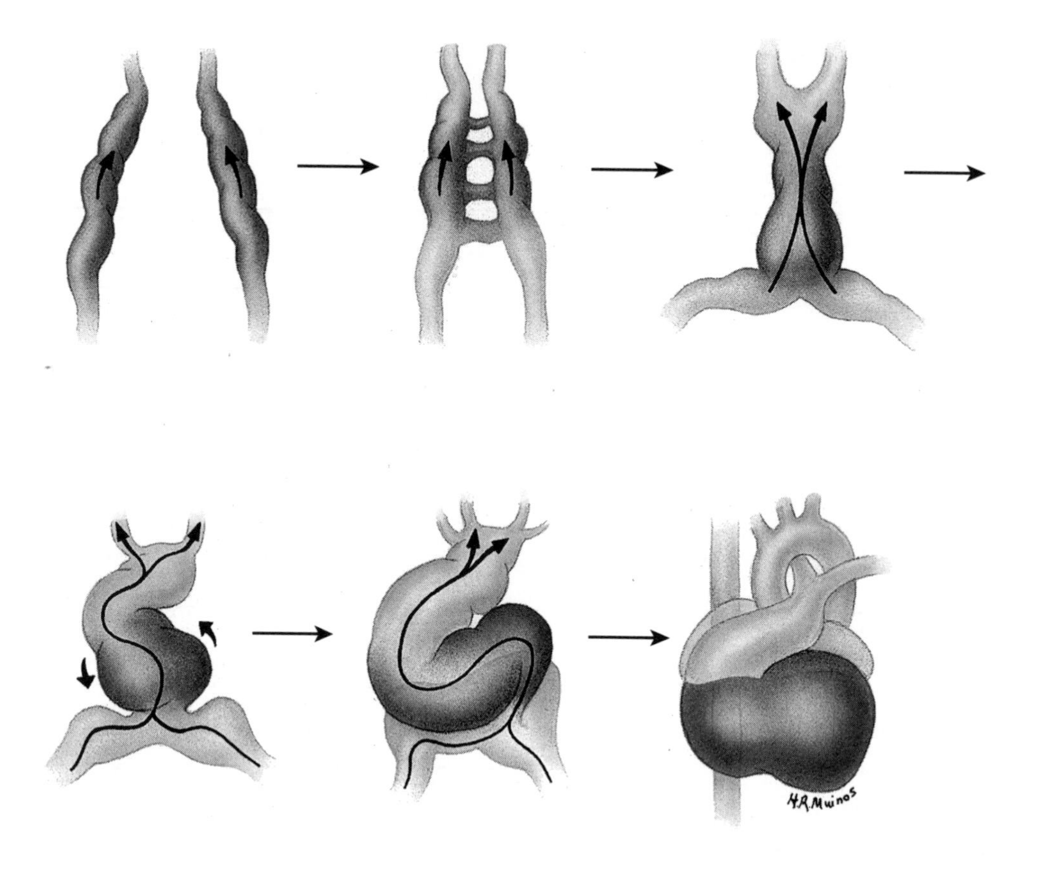

## NOTES

13

**Figure 13.15** Three procedures for reestablishing blood flow in occluded coronary arteries (page 409).

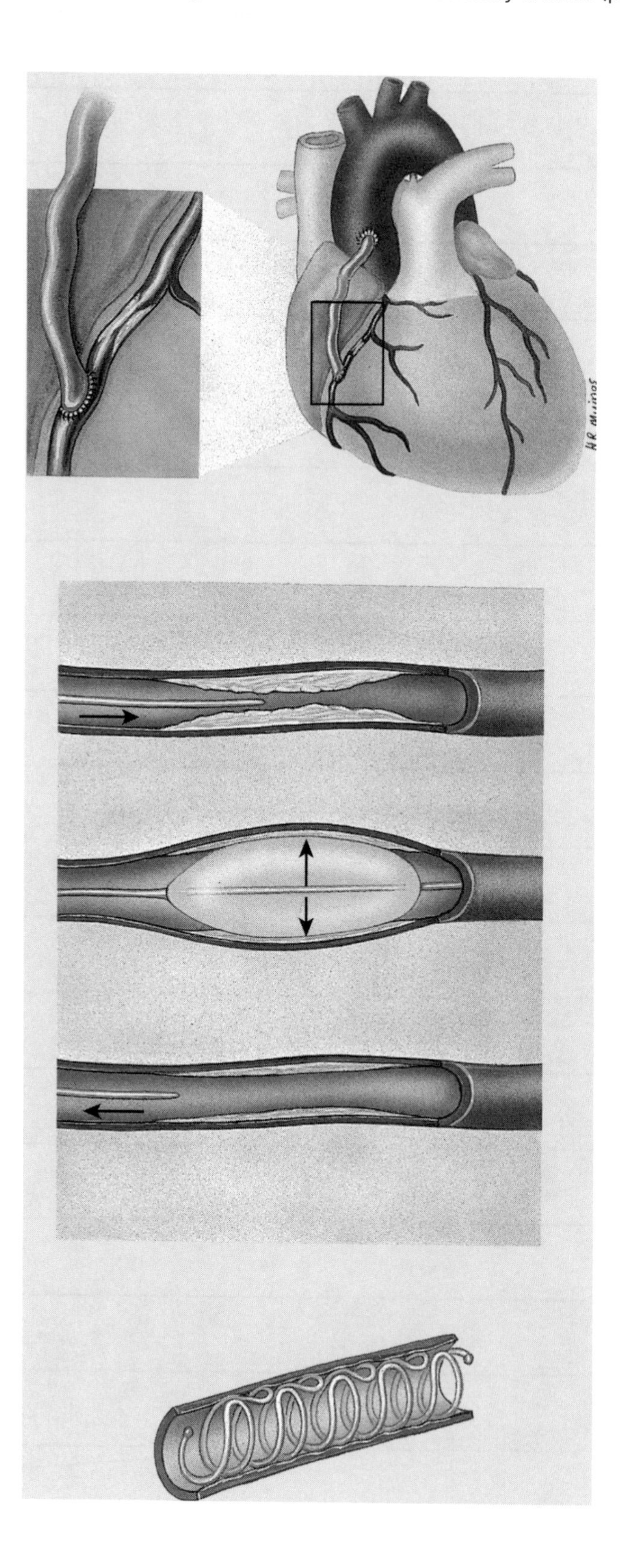

## NOTES

13

**Figure 14.1** Comparative structure of blood vessels (page 416).

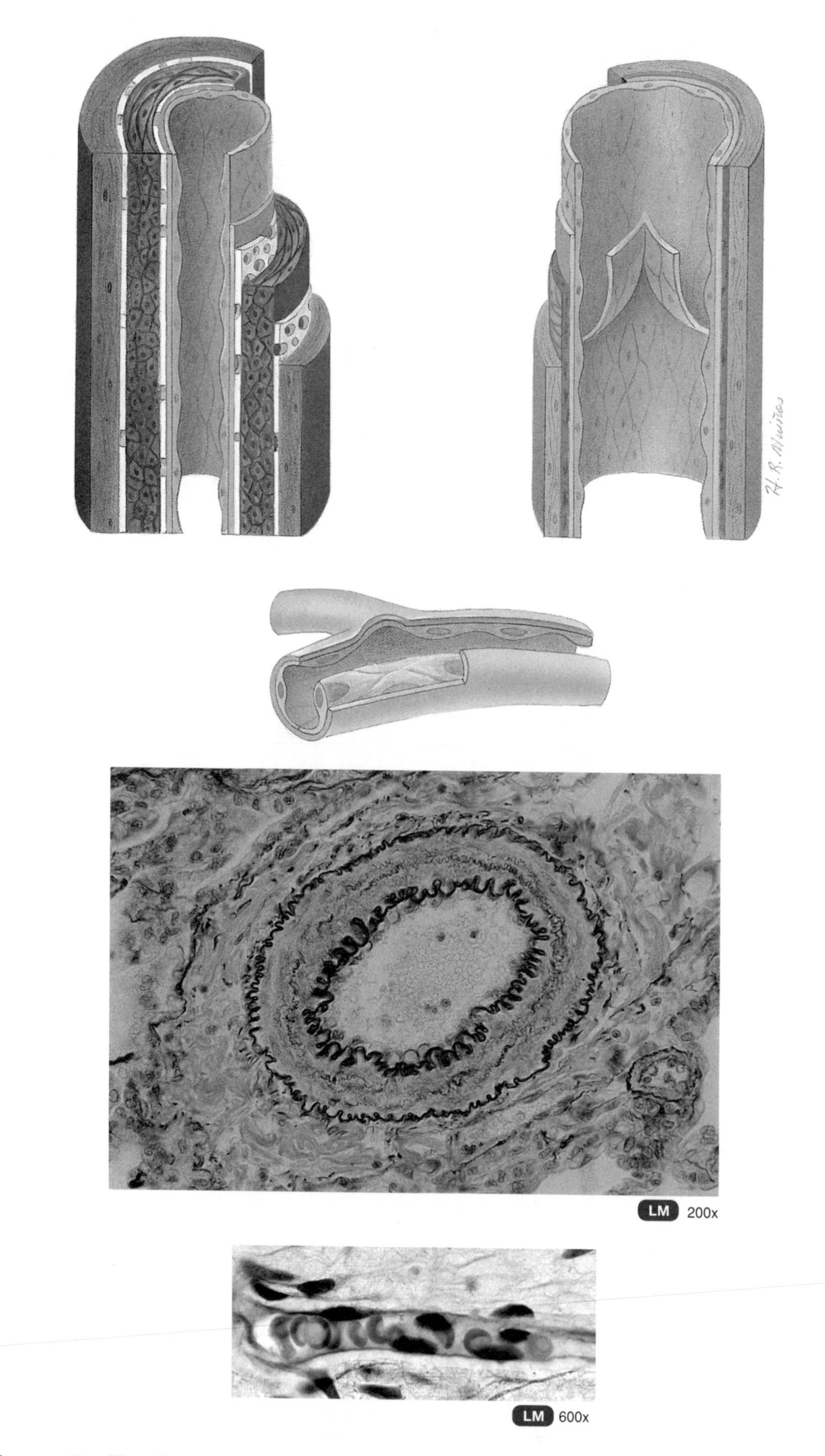

NOTES

14

**Figure 14.2** Arteriole, capillaries, and venule (page 417).

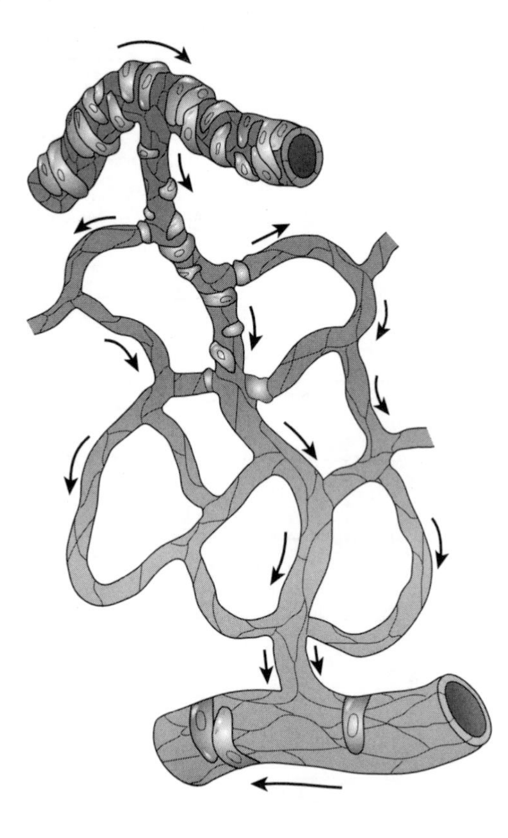

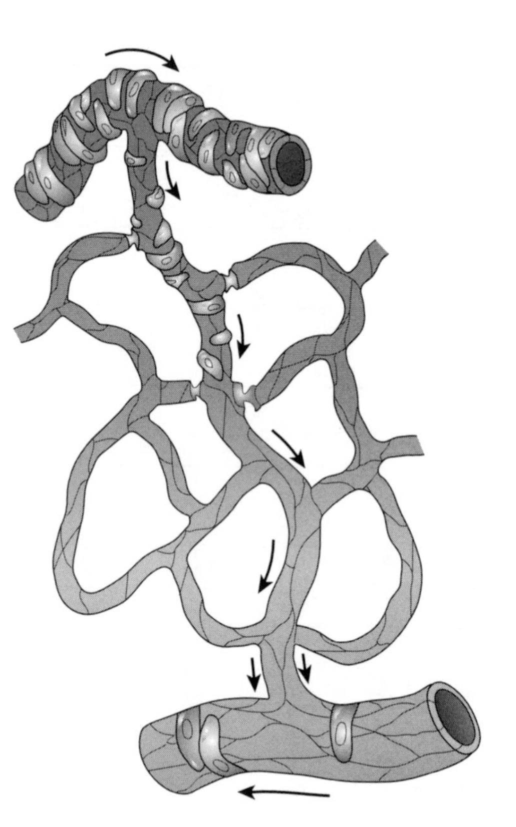

## NOTES

14

**Figure 14.3** Types of capillaries (shown in transverse sections) (page 418).

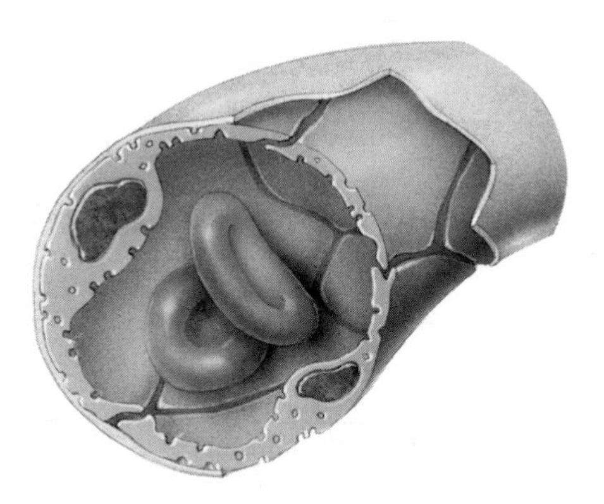

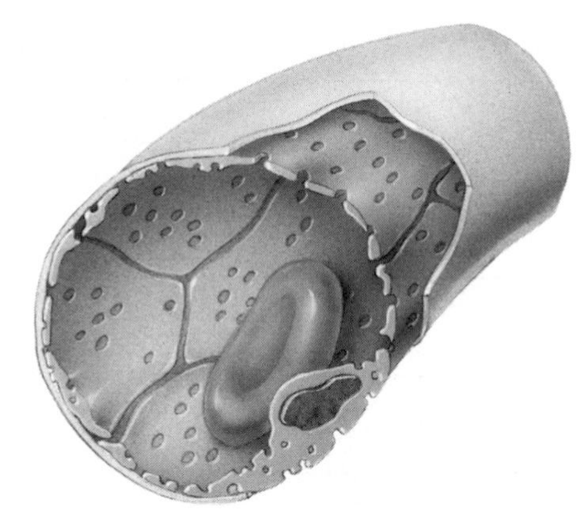

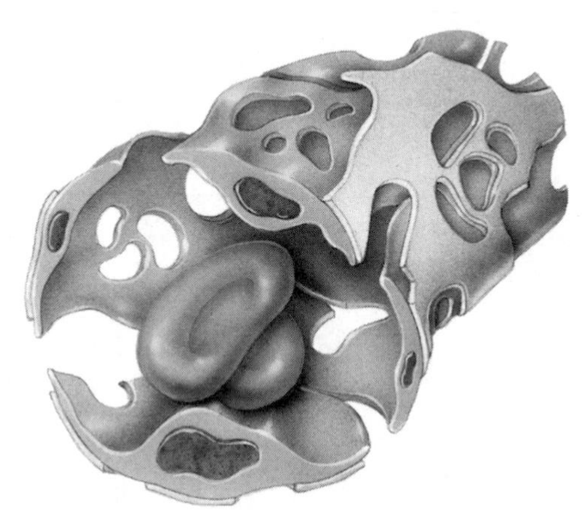

**Figure 14.4** Role of skeletal muscle contractions and venous valves in returning blood to the heart (page 419).

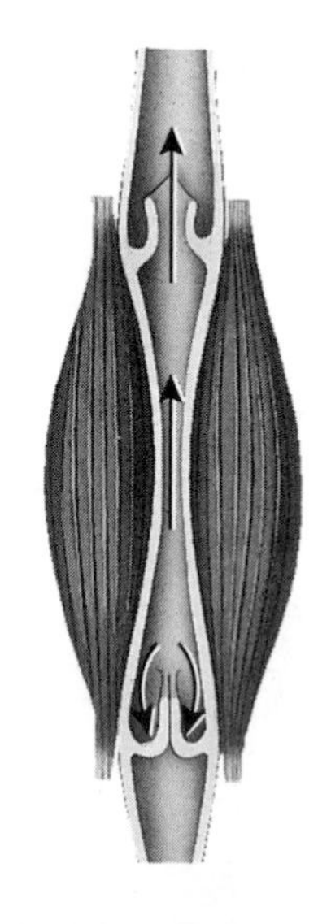

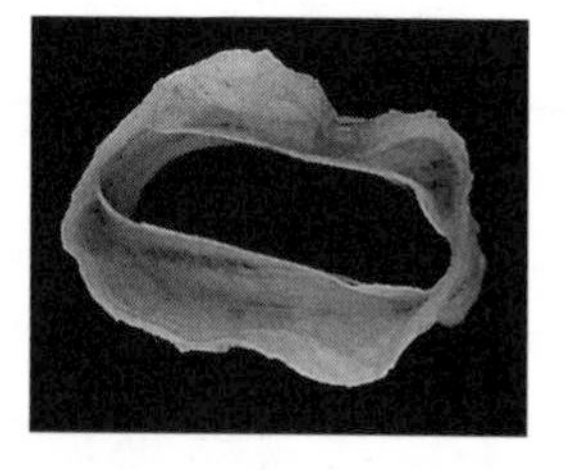

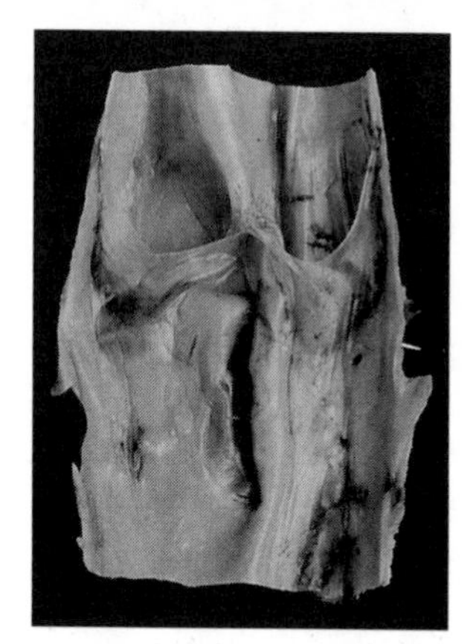

NOTES

14

**Figure 14.5** Generalized view of circulatory routes (page 421).

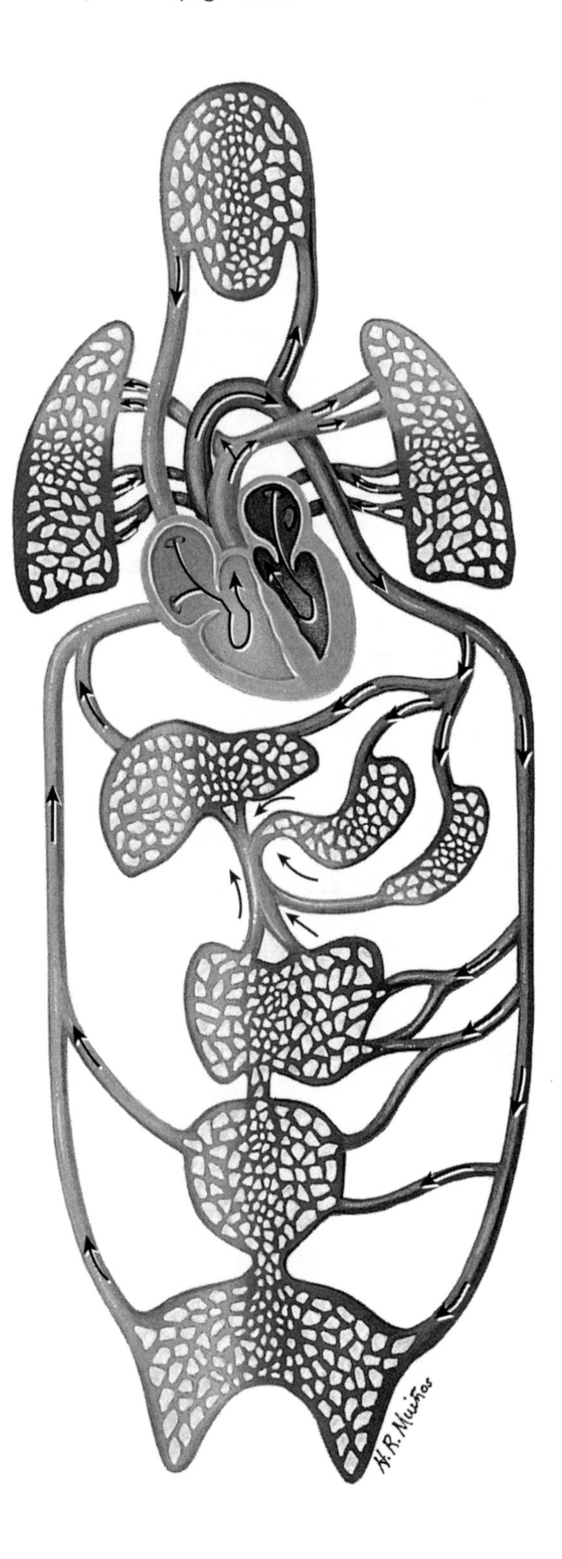

NOTES

14

**Figure 14.6a** Aorta and its principal branches (page 424).

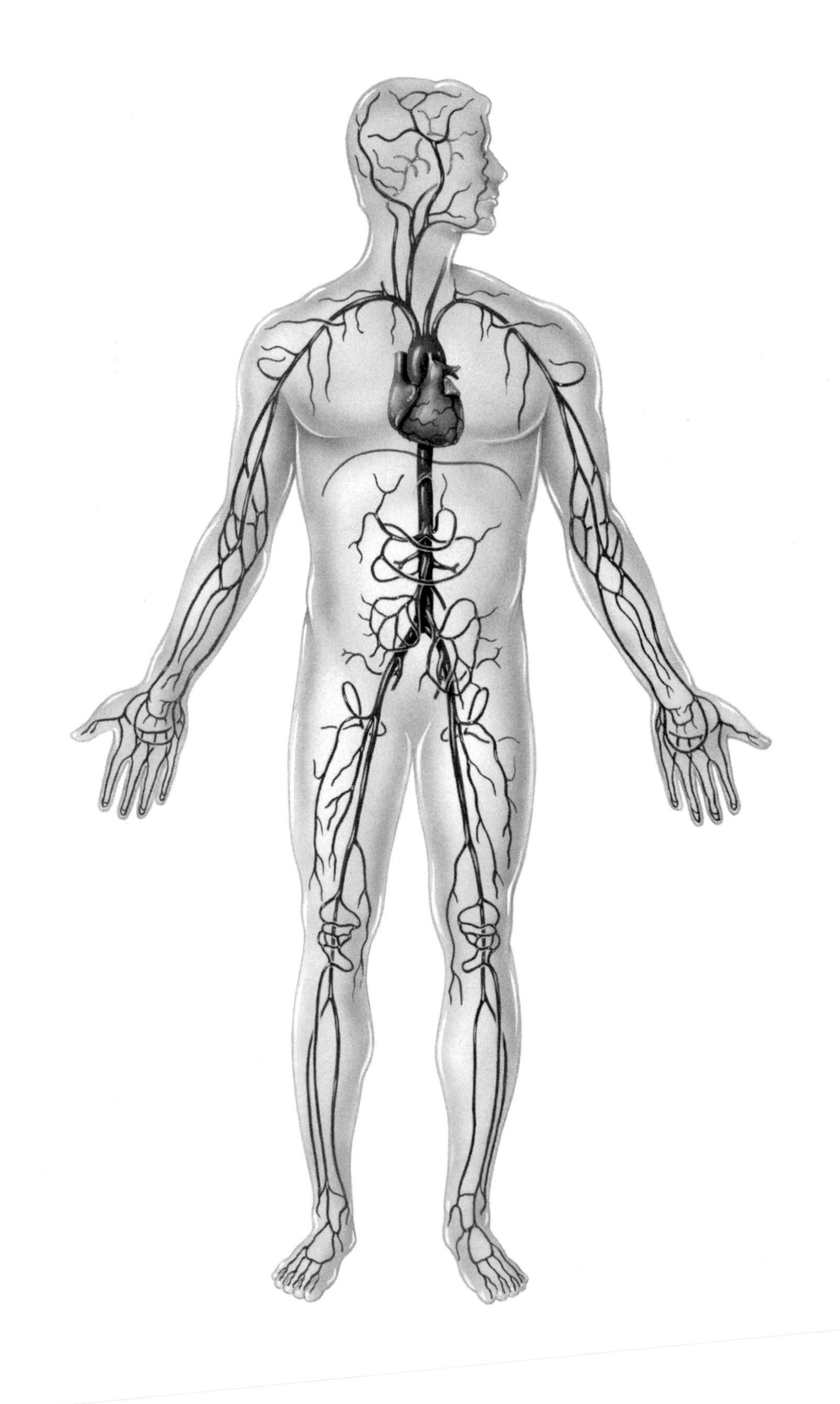

## NOTES

14

**Figure 14.6b** Aorta and its principal branches (page 425).

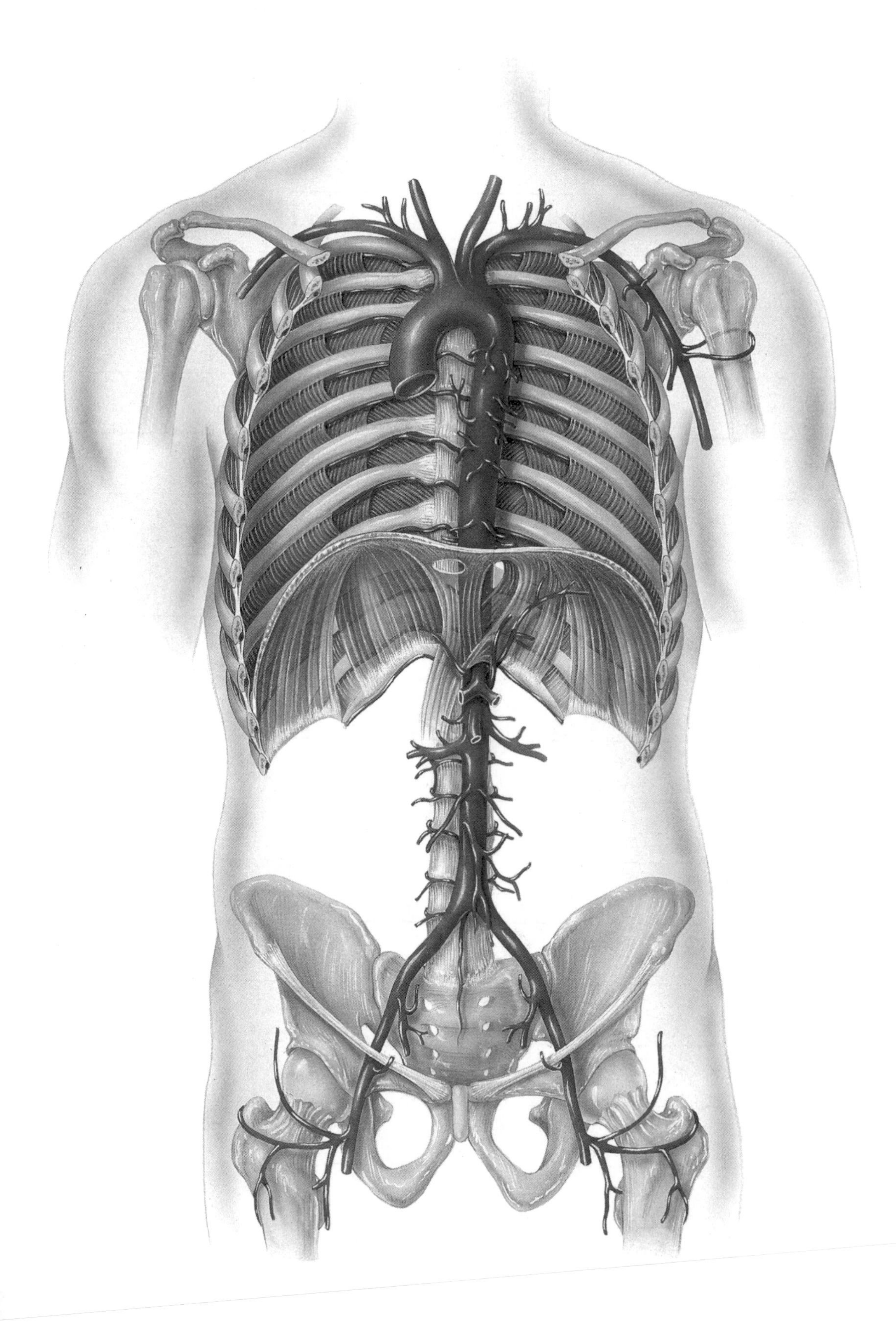

## NOTES

14

**Figure 14.7** Ascending aorta and its branches (page 427).

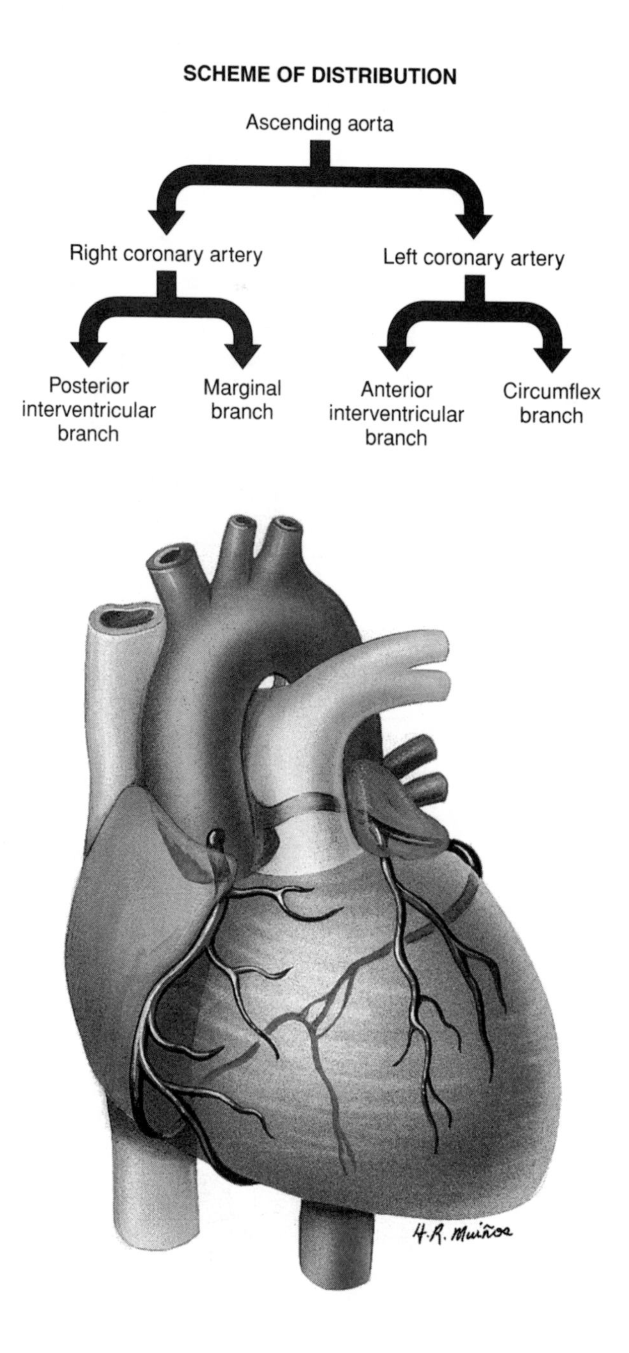

NOTES

14

**Figure 14.8a–c** Arch of the aorta and its branches (page 431).

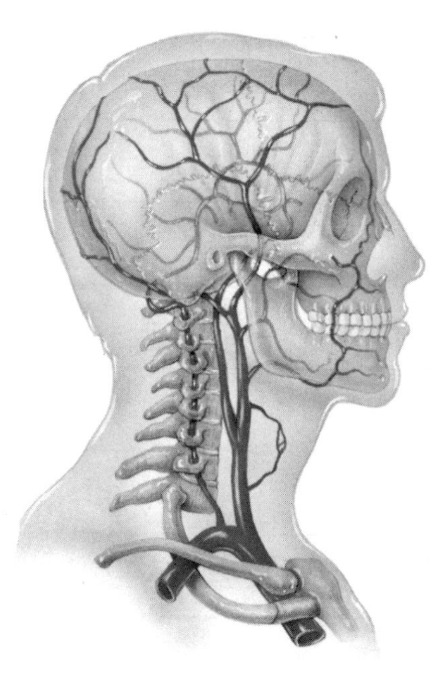

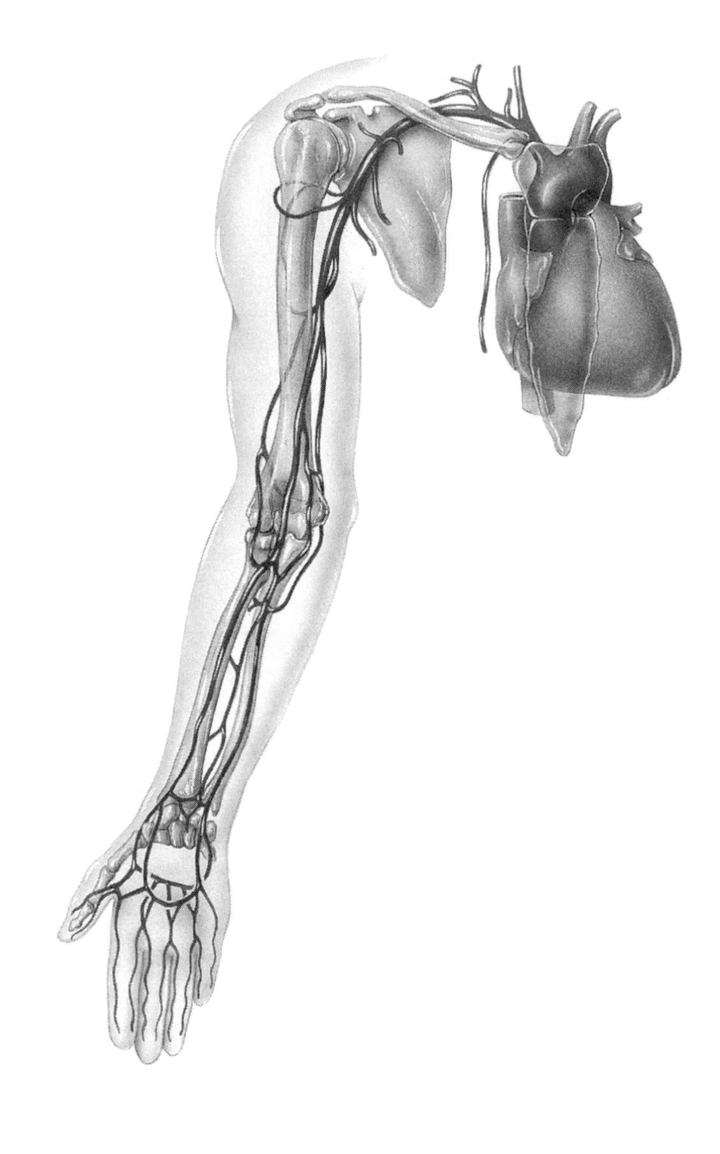

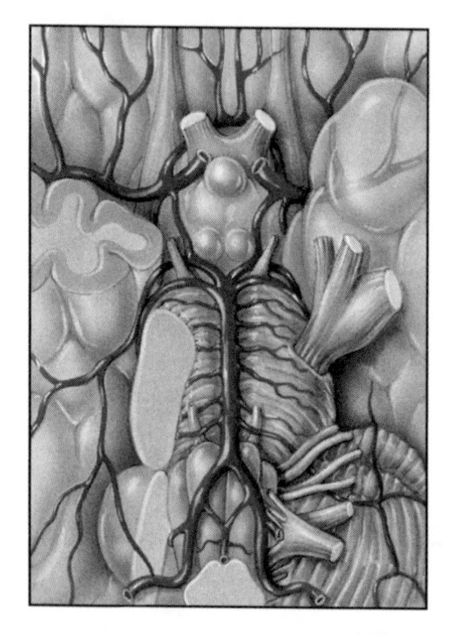

NOTES

14

**Figure 14.9a,b,c** Abdominal aorta and its principal branches (pages 437, 438).

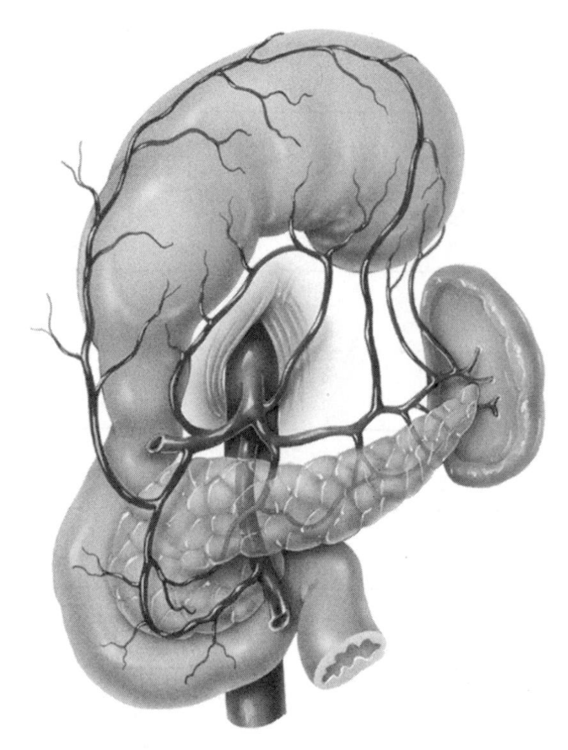

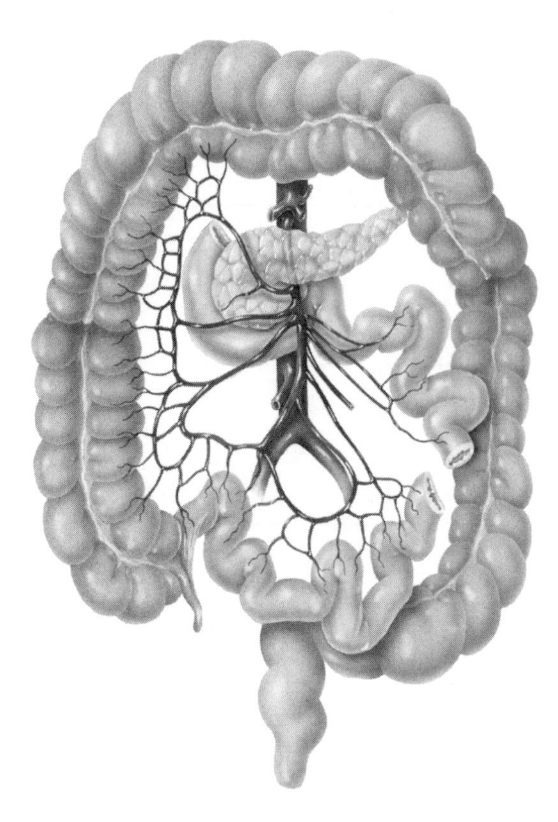

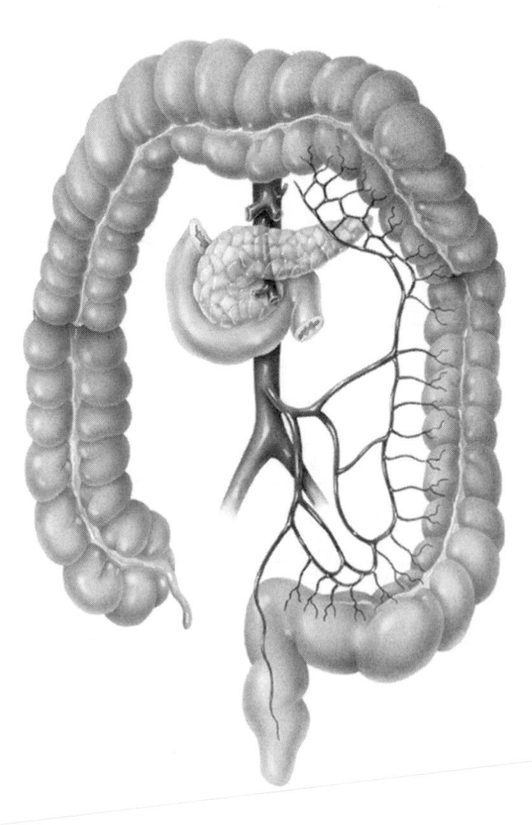

NOTES

14

**Figure 14.10a,b** Arteries of the pelvis and right lower limb (page 442).

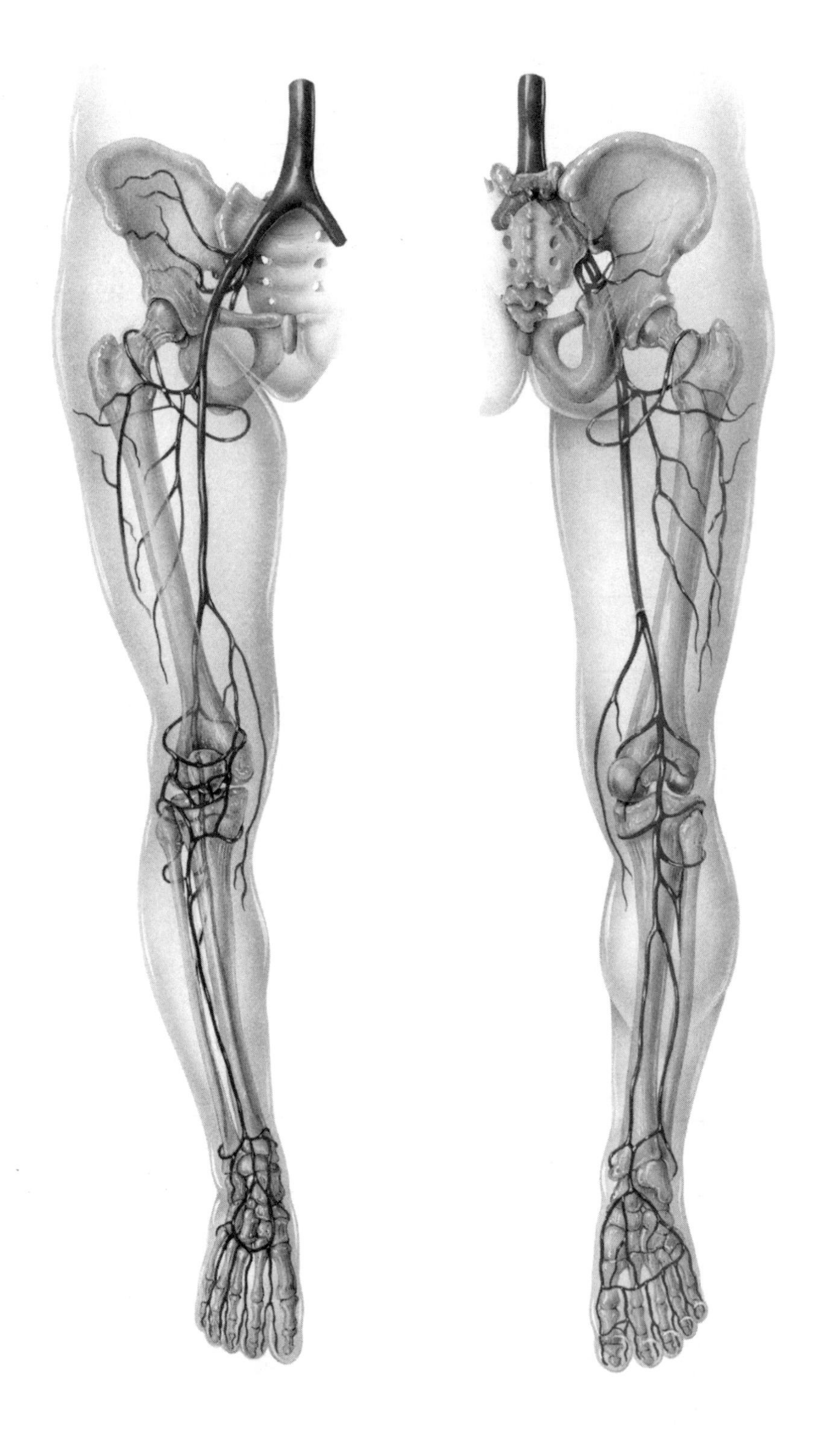

## NOTES

14

**Figure 14.11** Principal veins (page 445).

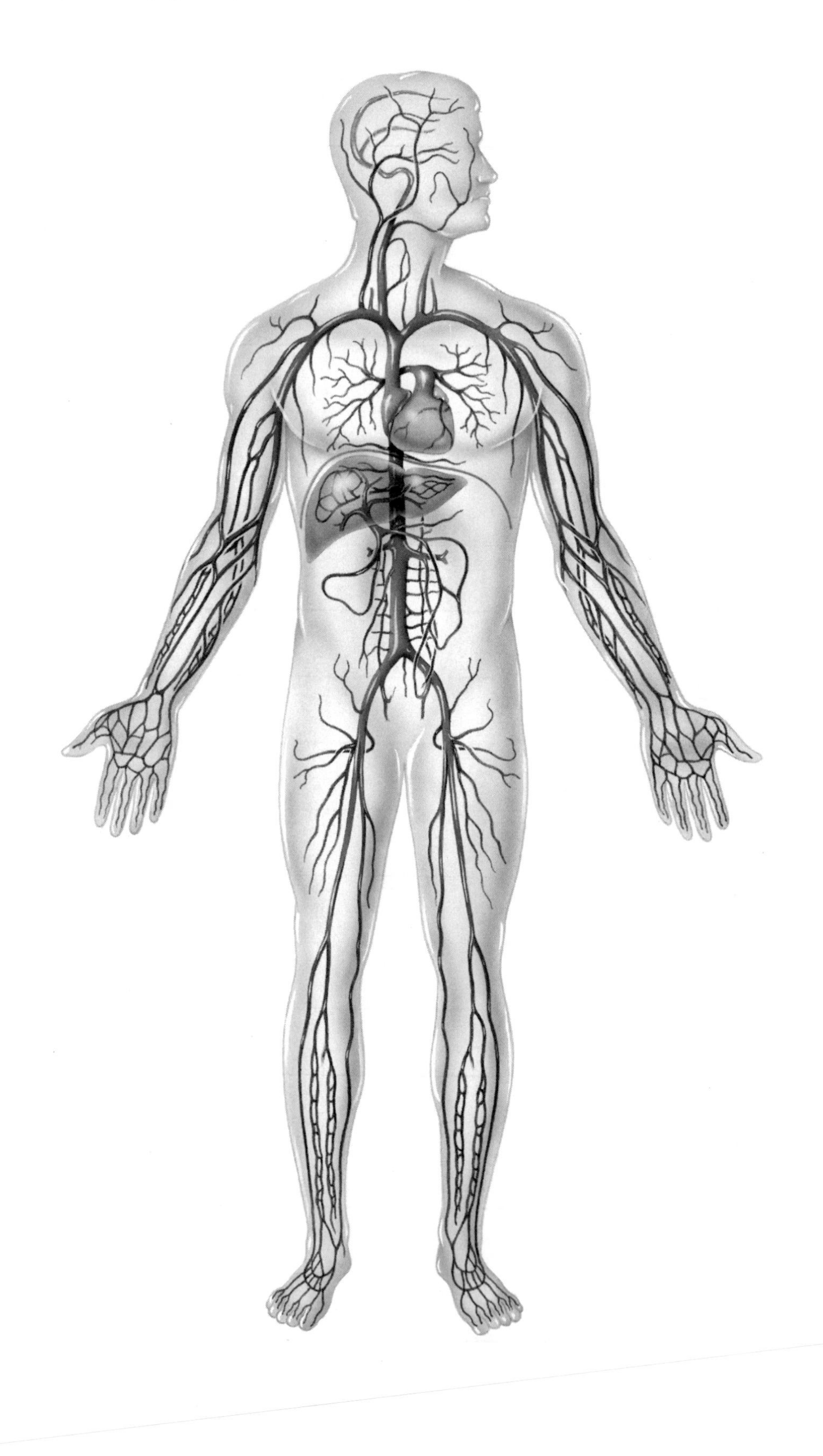

NOTES

14

**Figure 14.12a** Principal veins of the head and neck (page 448).

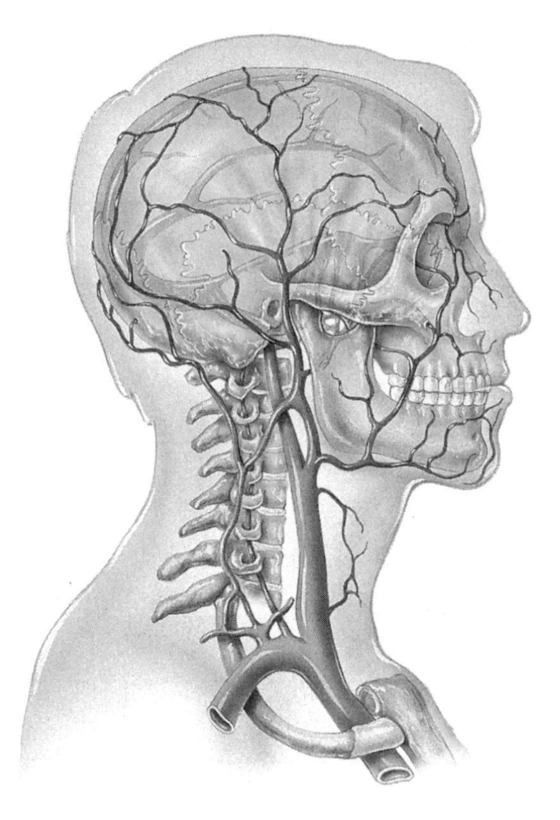

NOTES

14

**Figure 14 13a–c** Principal veins of the right upper limb (page 451).

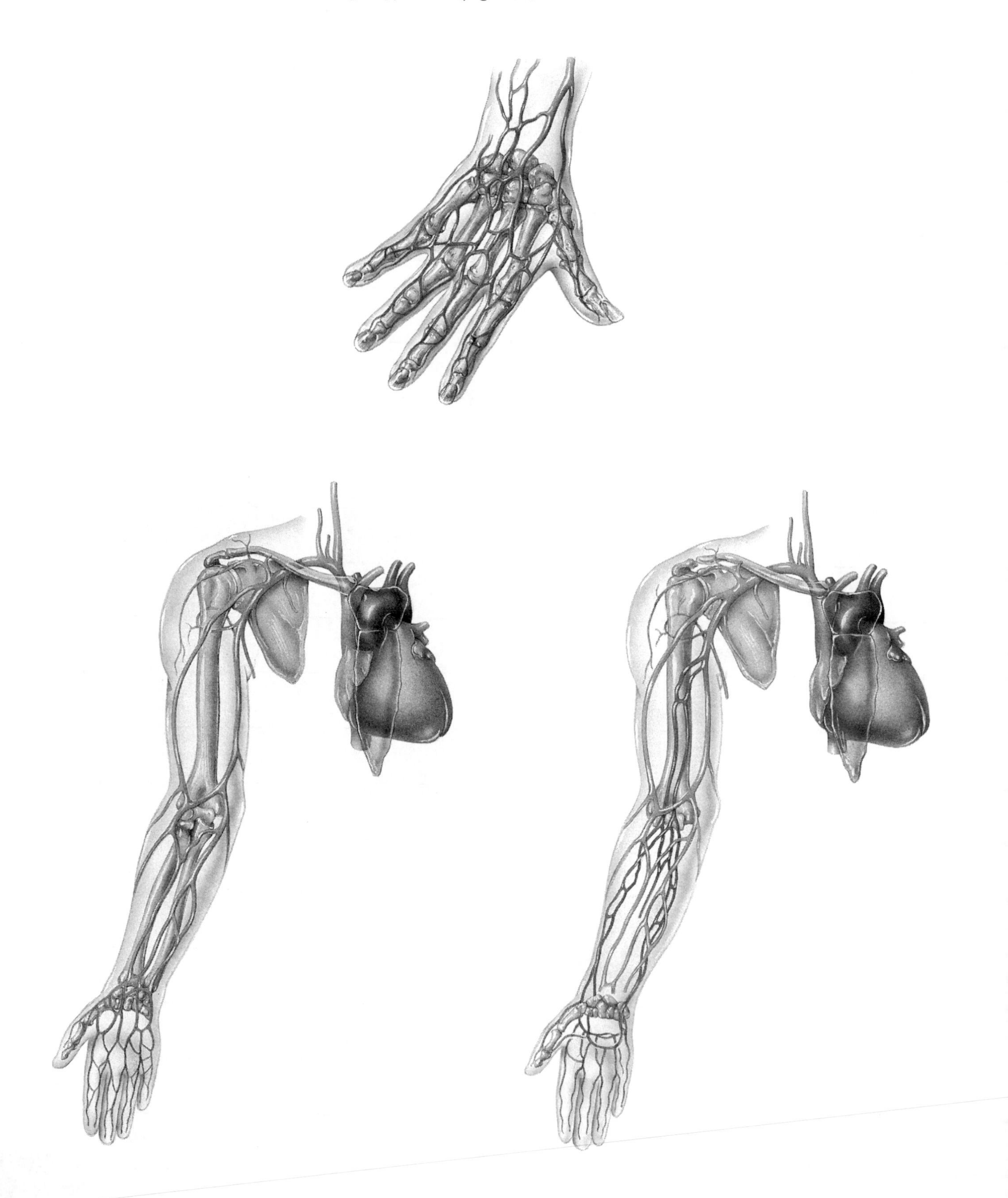

NOTES

14

**Figure 14.14a** Principal veins of the thorax, abdomen, and pelvis (page 454).

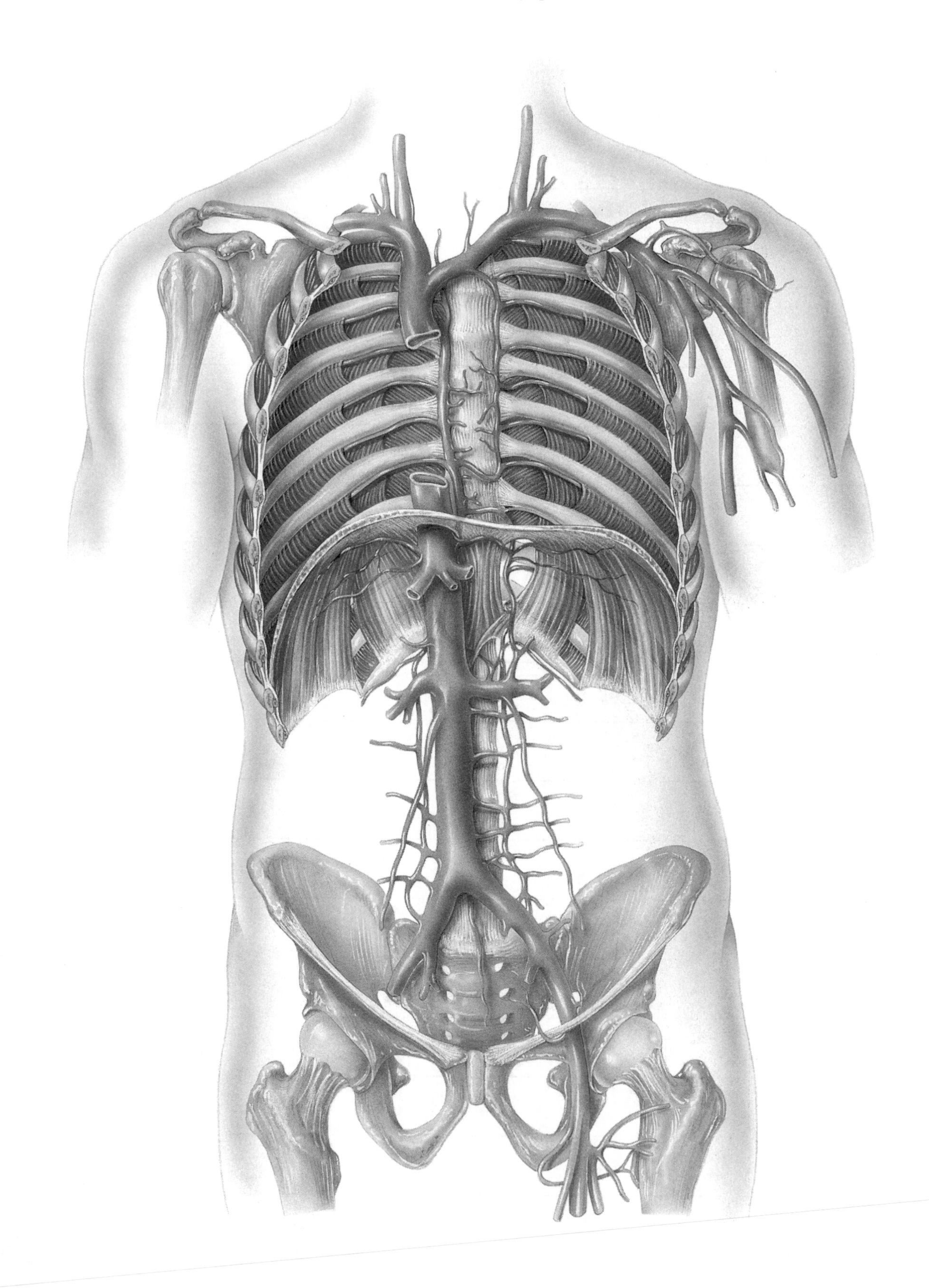

NOTES

14

**Figure 14.15** Principal veins of the pelvis and lower limbs (page 461).

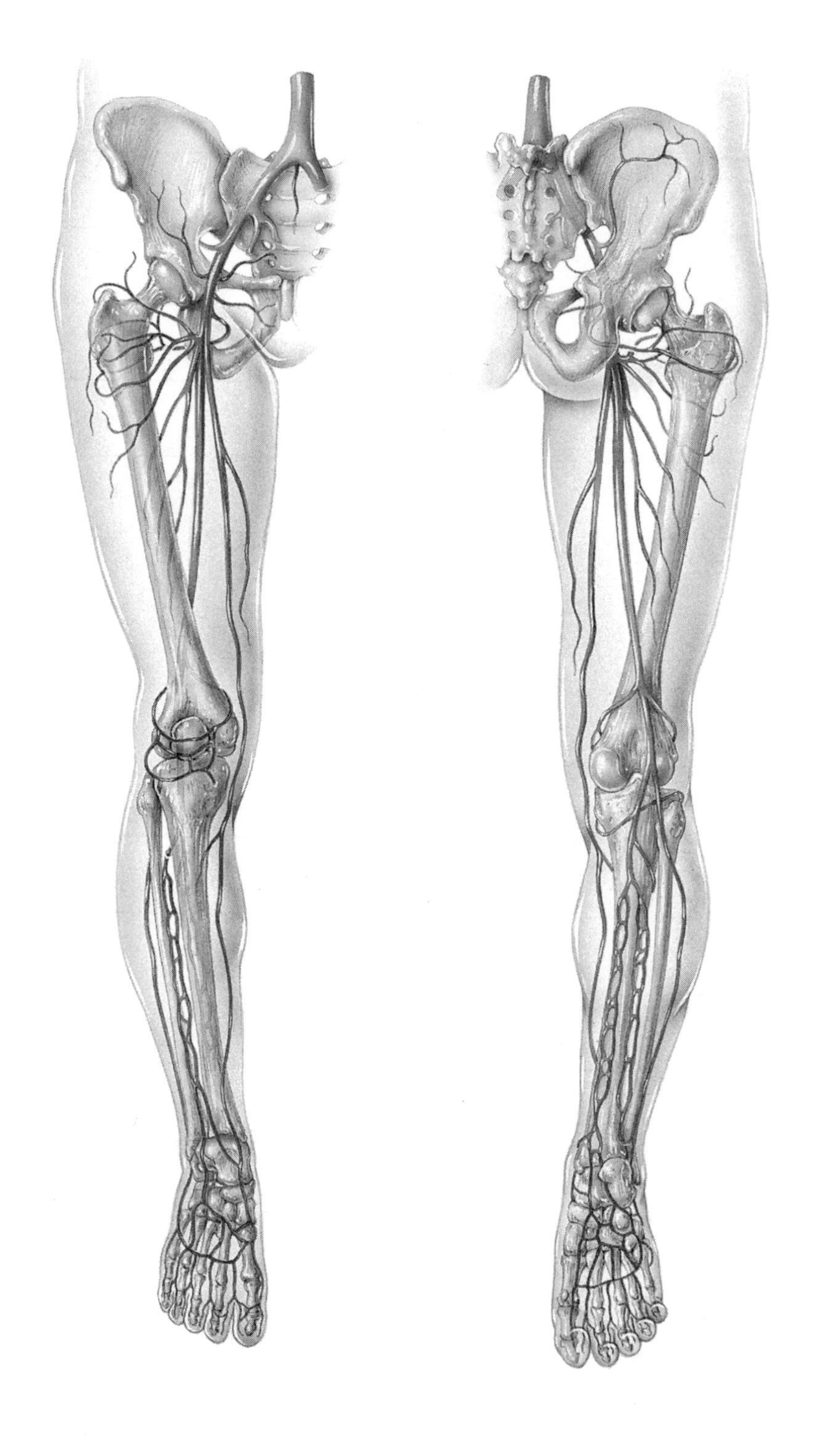

NOTES

14

**Figure 14.16a** Hepatic portal circulation (page 462).

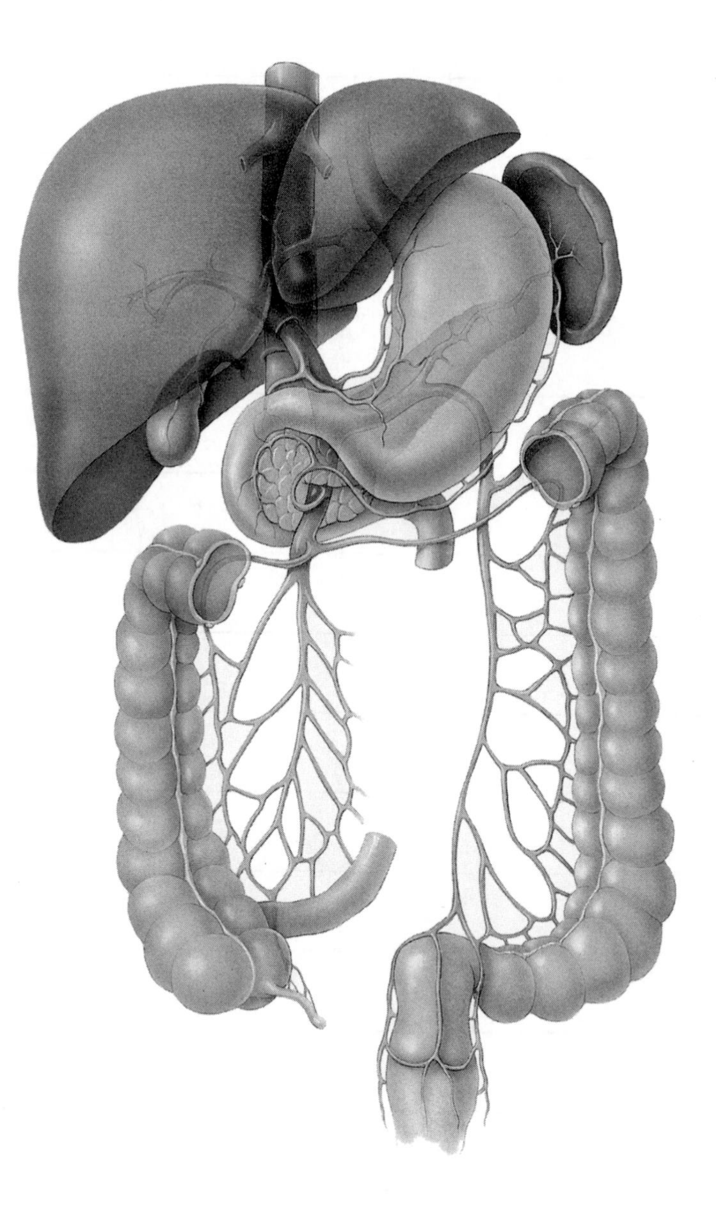

NOTES

14

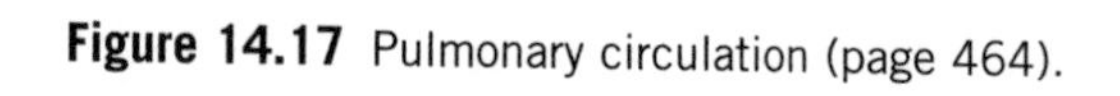

**Figure 14.17** Pulmonary circulation (page 464).

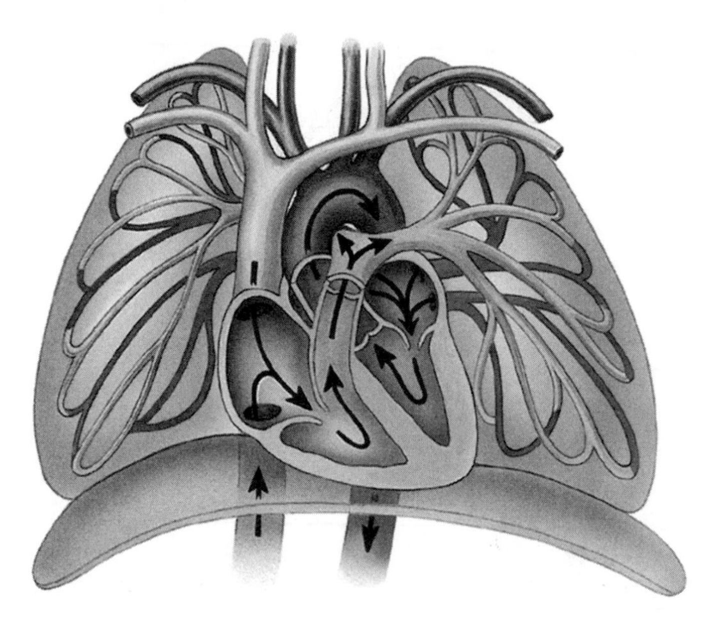

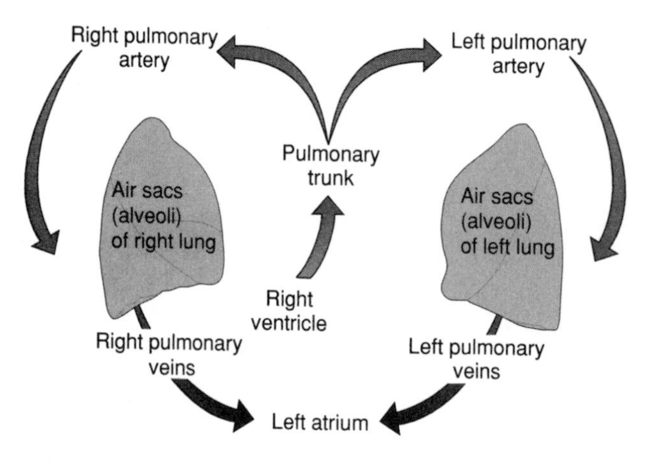

NOTES

14

**Figure 14.18a,b** Fetal circulation and changes at birth (page 466).

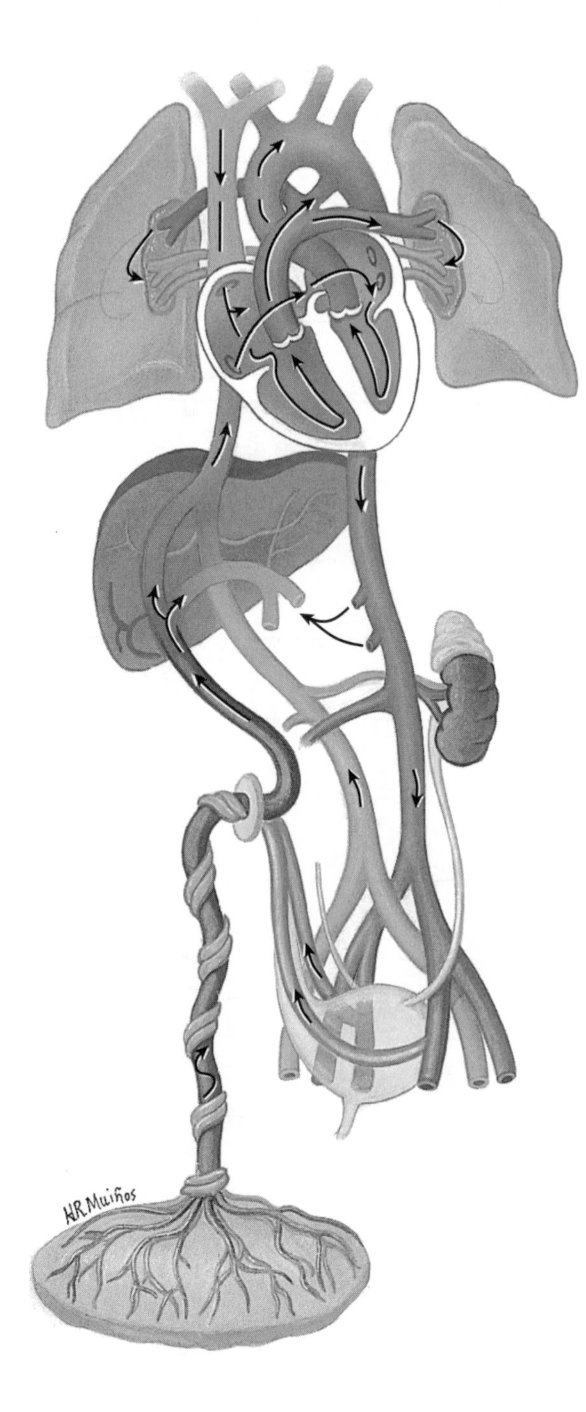


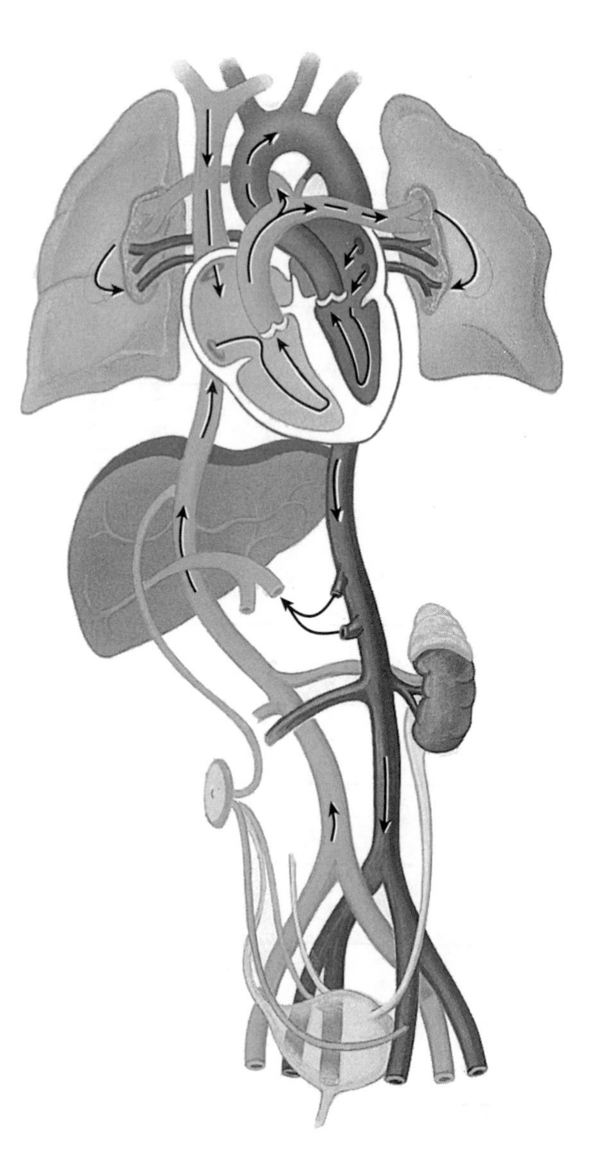

## NOTES

**Figure 14.19** Development of blood vessels and blood cells from blood islands (page 468).

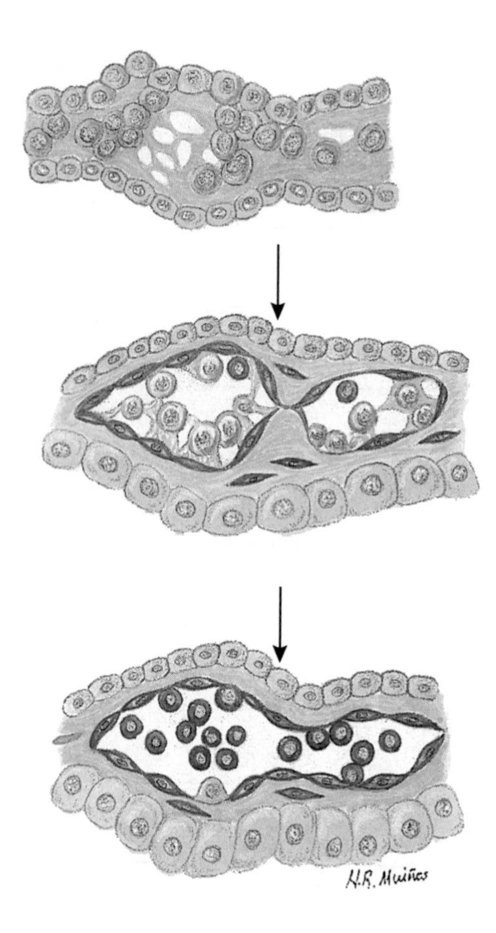

NOTES

14

**Figure 15.1** Components of the lymphatic system (page 475).

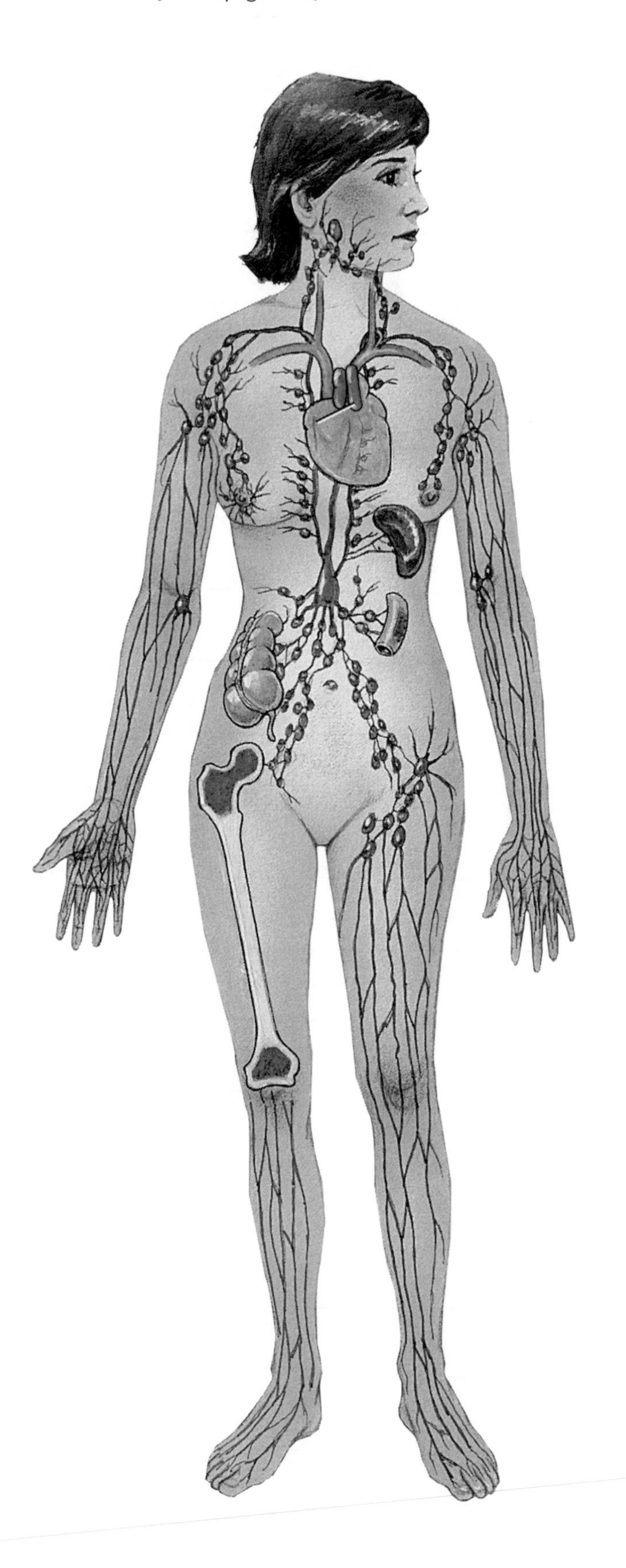

NOTES

**Figure 15.2** Lymphatic capillaries (page 476).

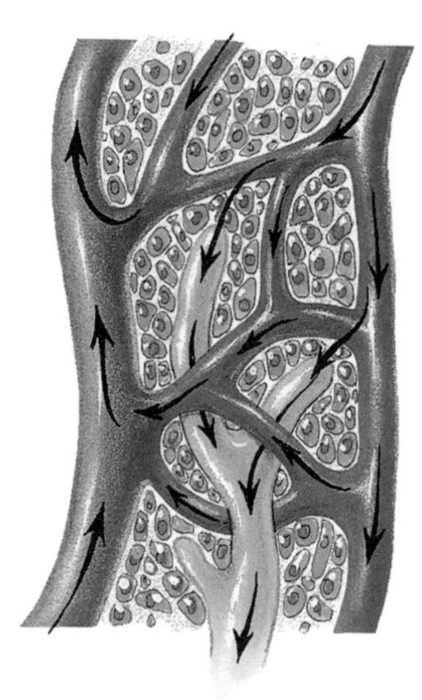

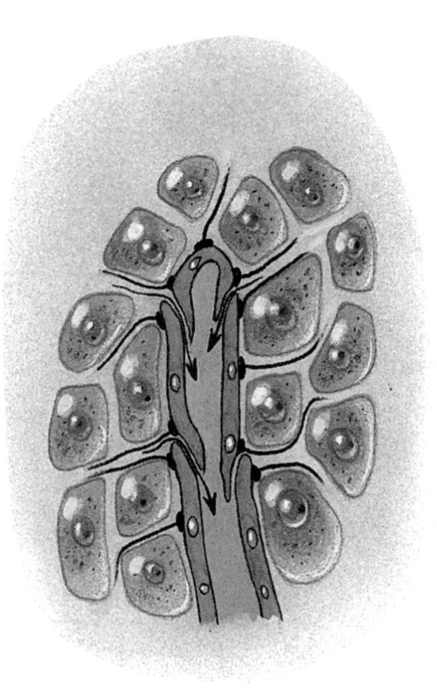

**Figure 15.3** Routes for drainage of lymph (page 477).

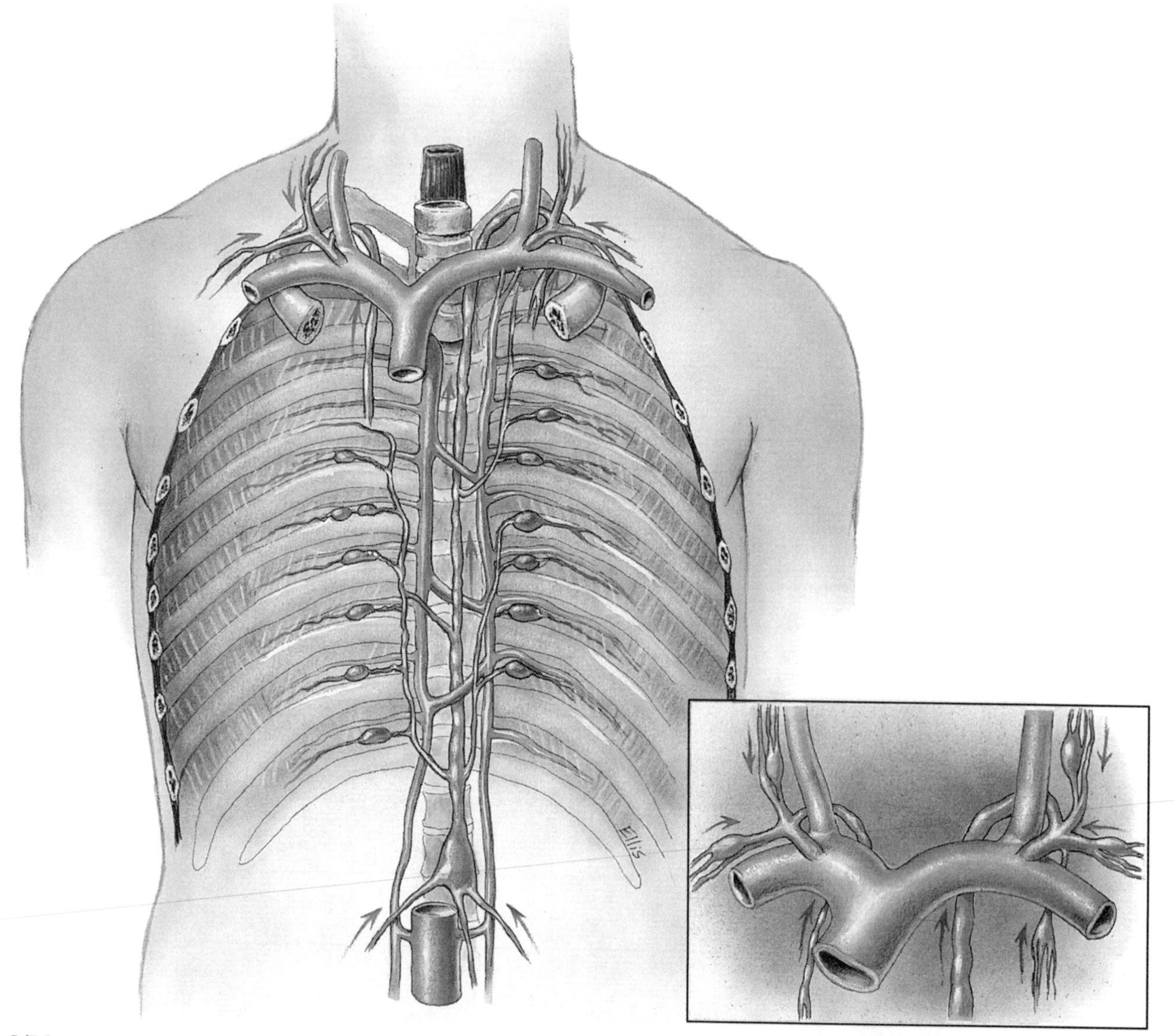

NOTES

**Figure 15.4** Relationship of the lymphatic system to the cardiovascular system (page 478).

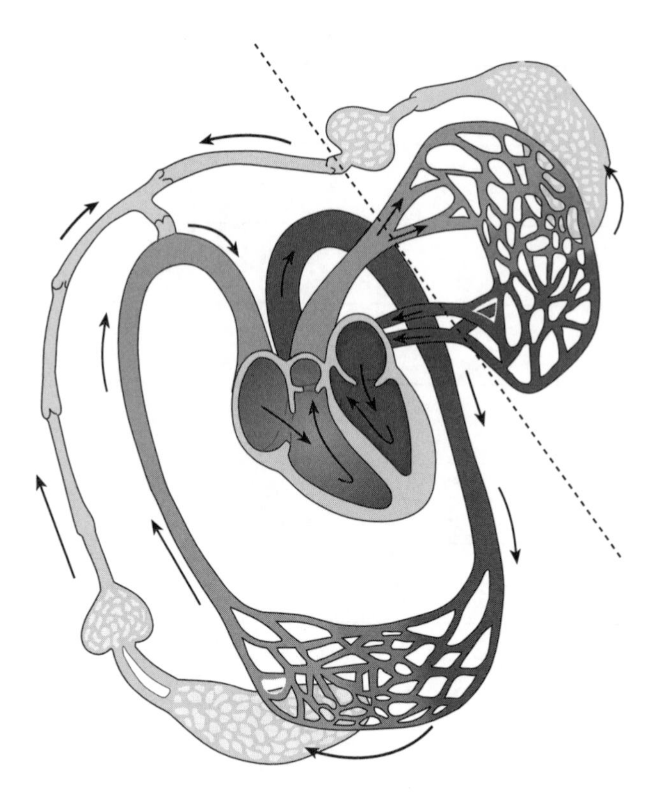

**Figure 15.5** Thymus (page 480).

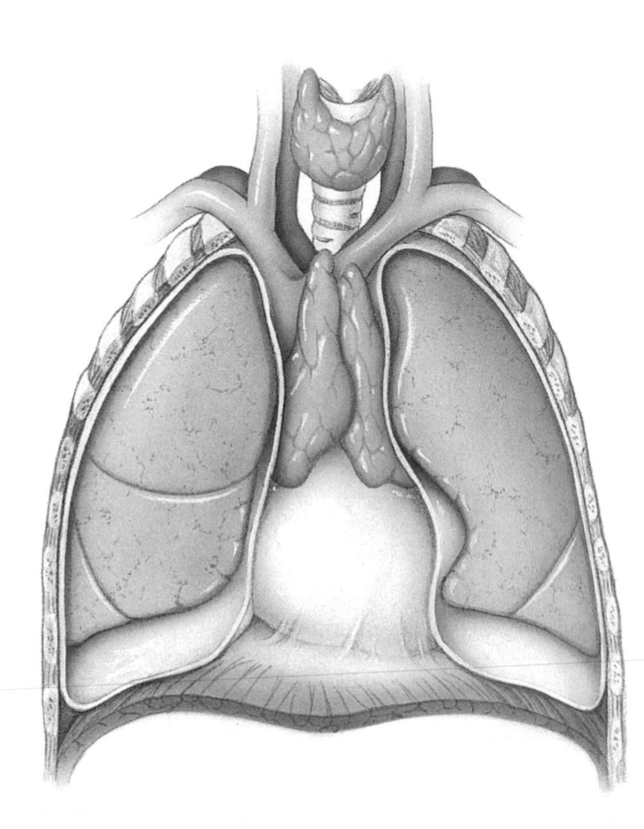

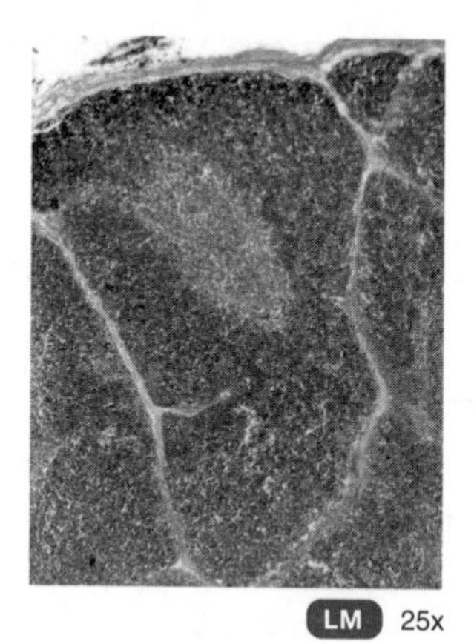



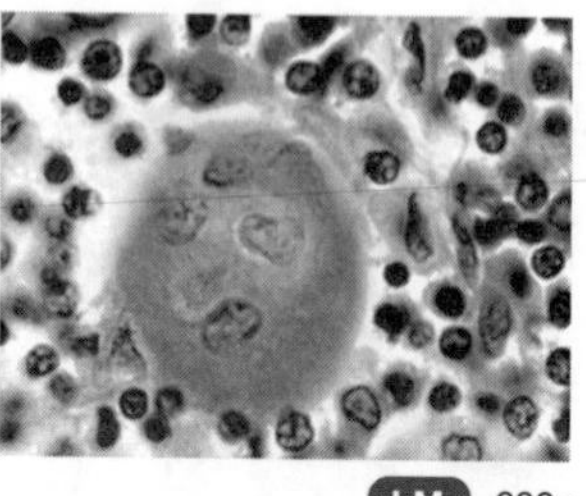

NOTES

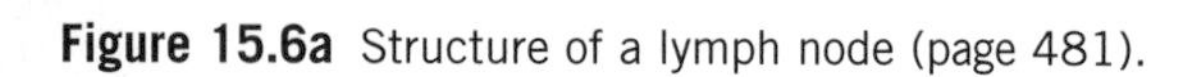

**Figure 15.6a** Structure of a lymph node (page 481).

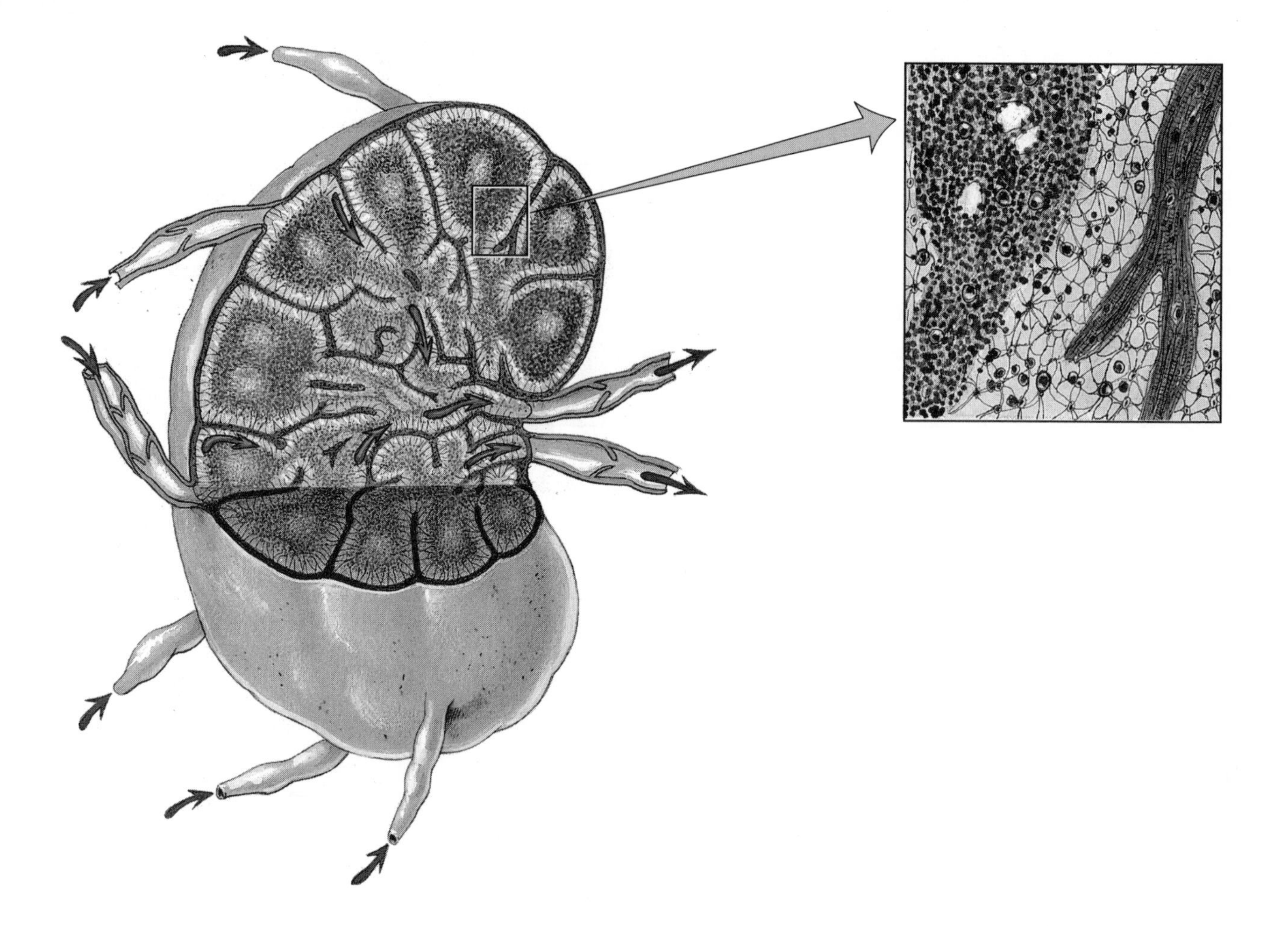

NOTES

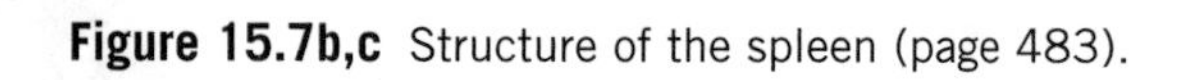

**Figure 15.7b,c** Structure of the spleen (page 483).

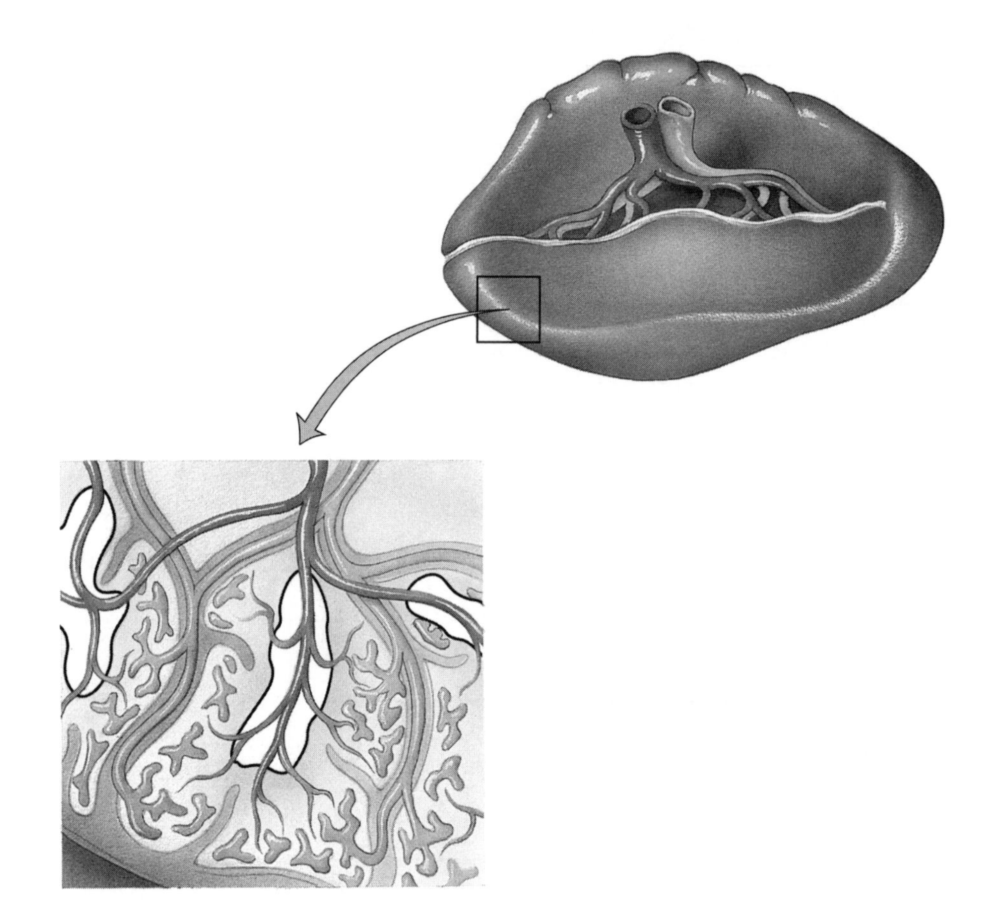

**Figure 15.8** Principal lymph nodes of the head and neck (page 485).

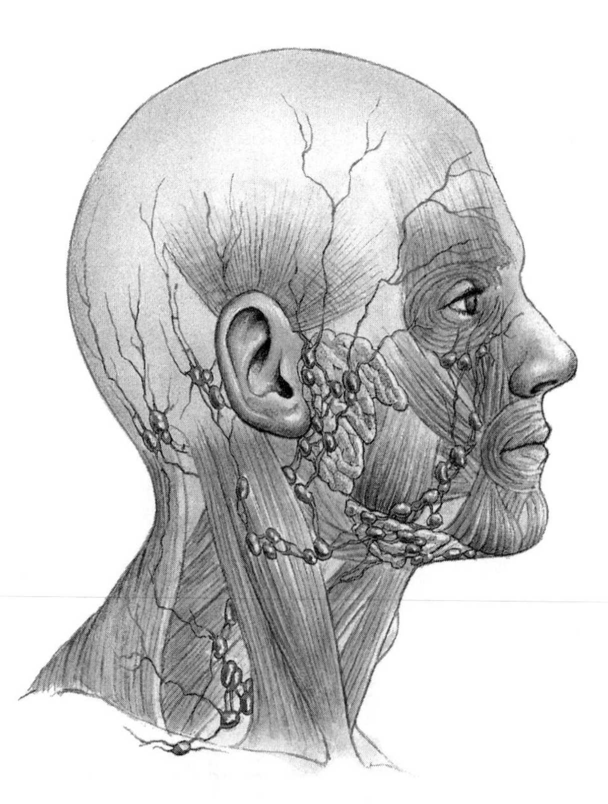

NOTES

**Figure 15.9** Principal lymph nodes of the upper limbs (page 486).

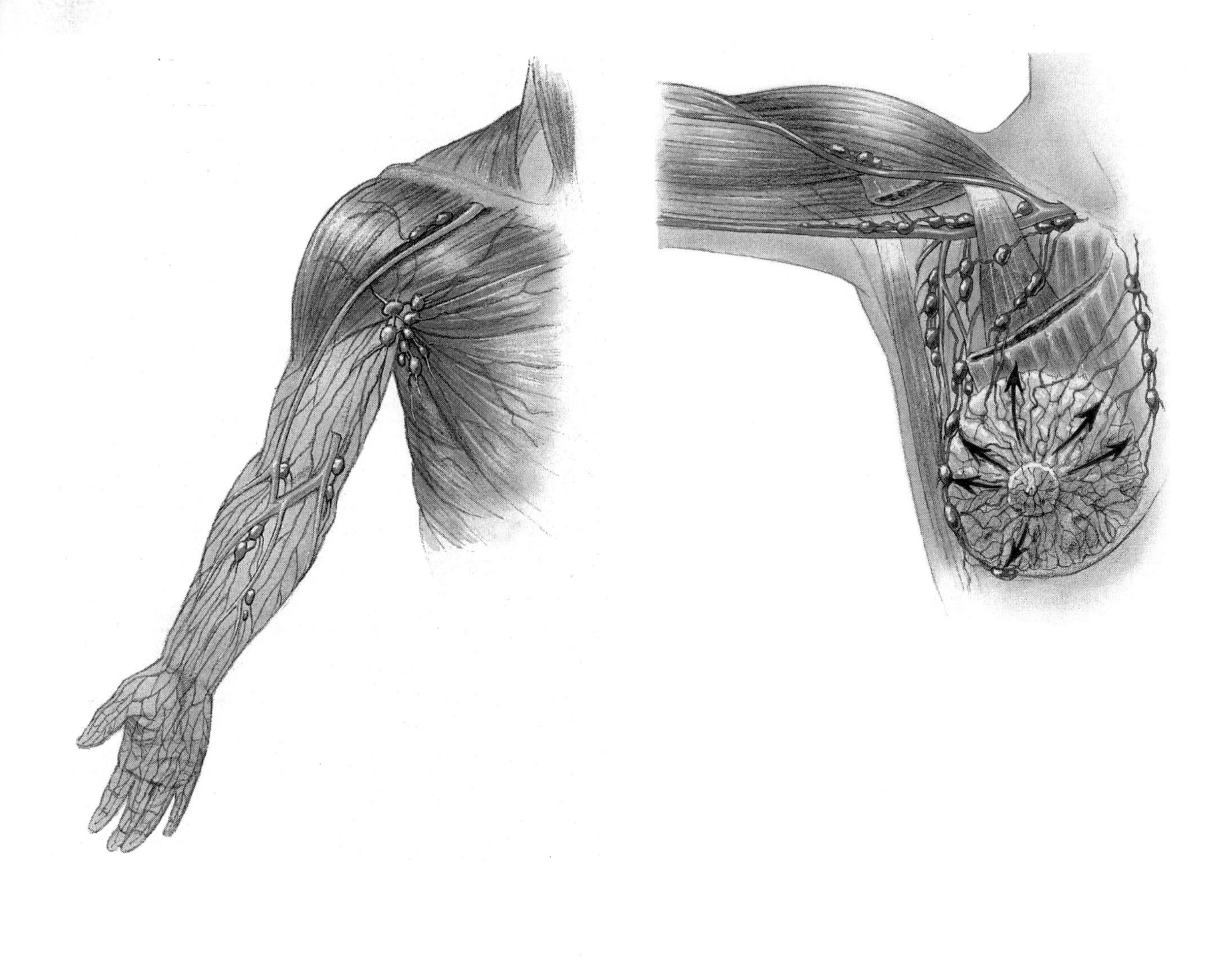

**NOTES**

**Figure 15.10** Principal lymph nodes of the lower limbs (page 487).

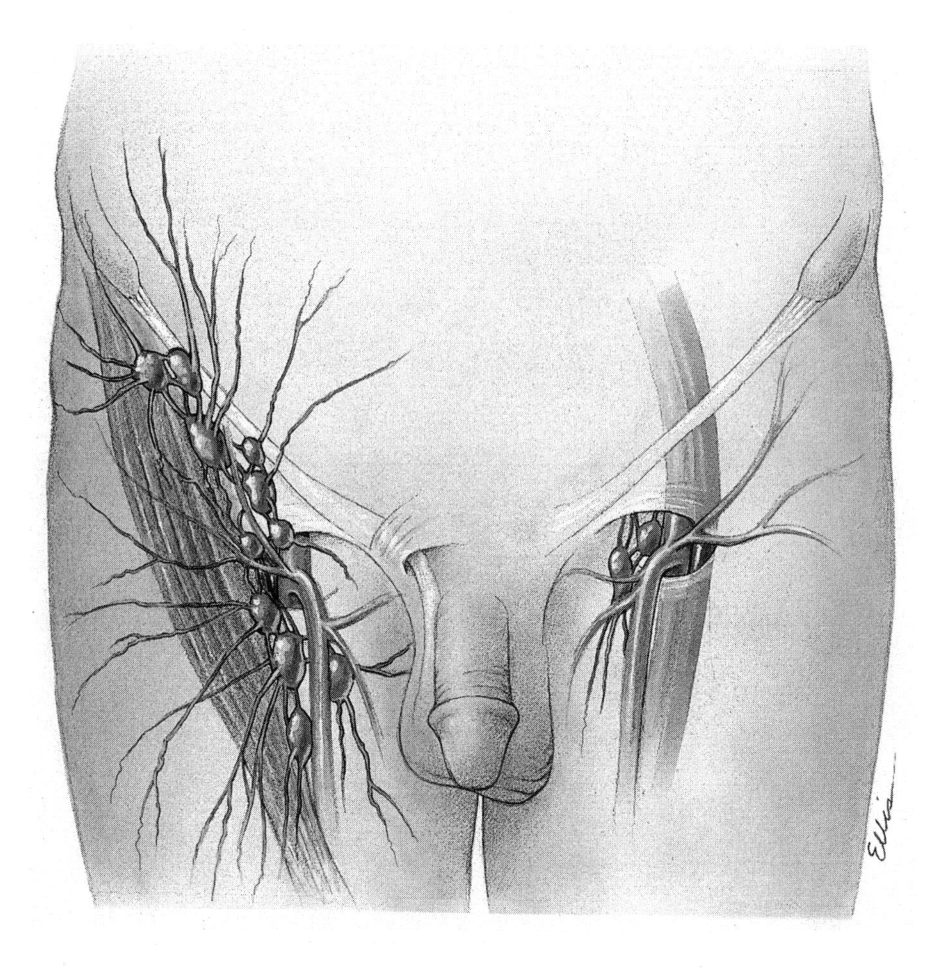

NOTES

**Figure 15.11a** Principal lymph nodes of the abdomen and pelvis (page 488).

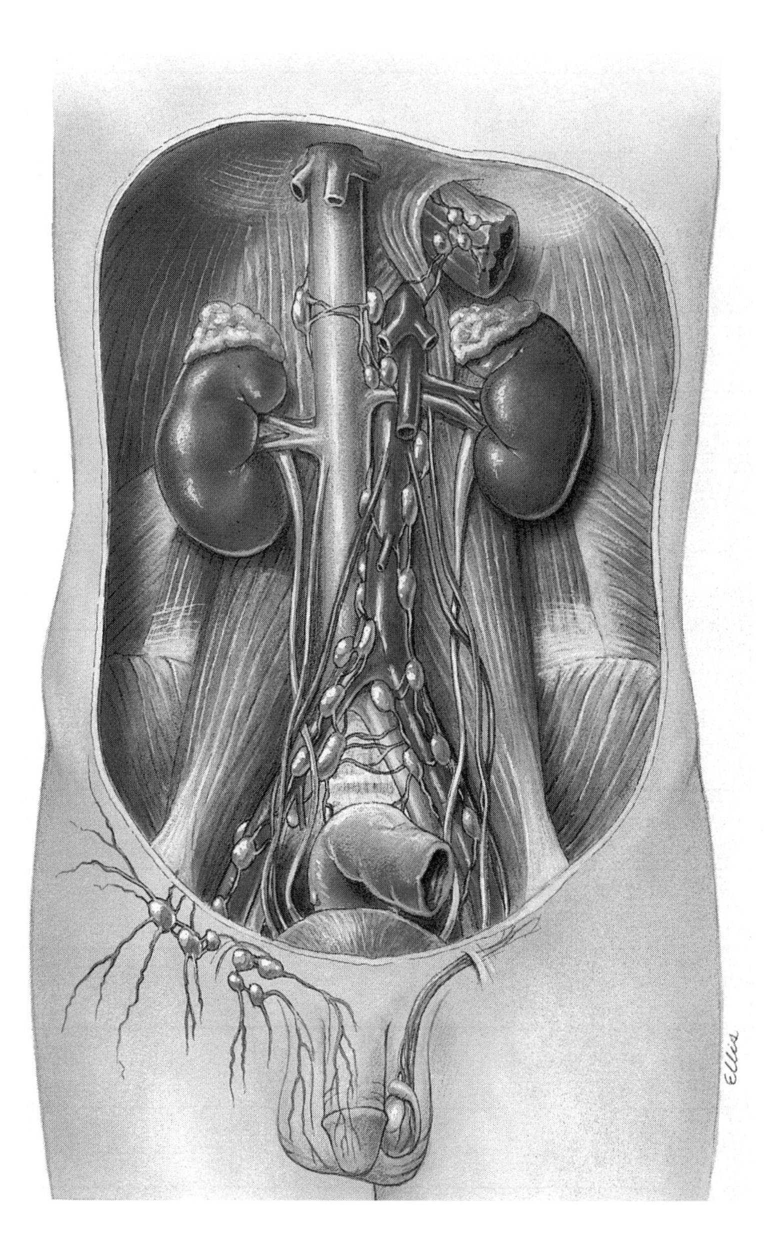

NOTES

**Figure 15.11b** Principal lymph nodes of the abdomen and pelvis (page 489).

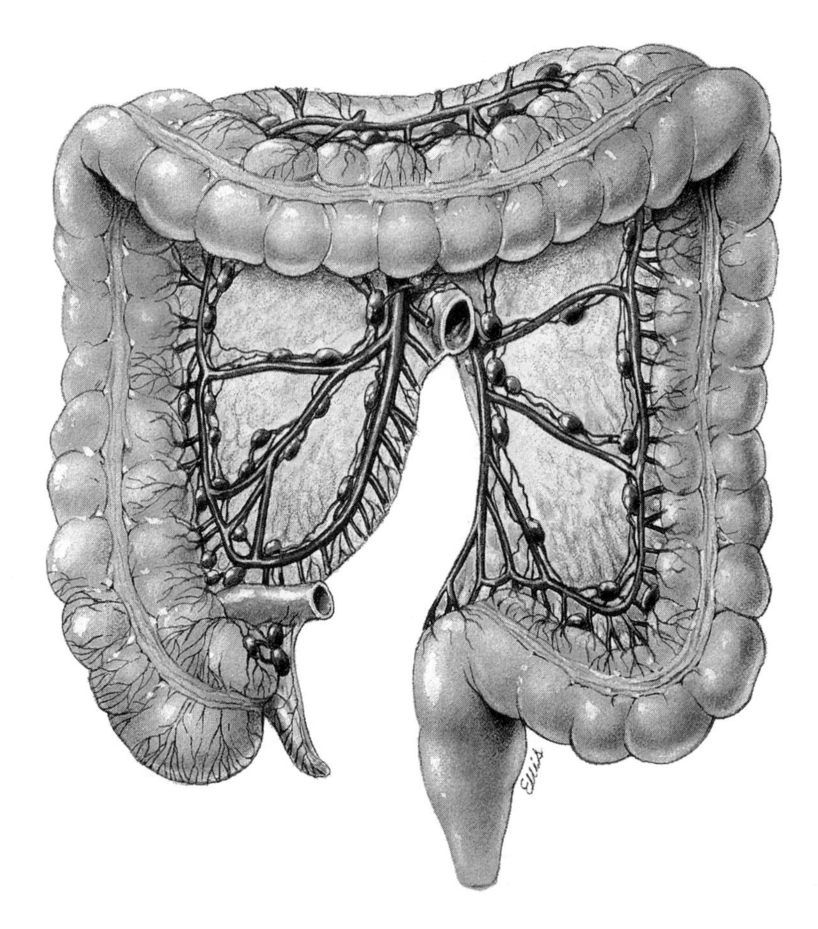

NOTES

**Figure 15.12** Principal lymph nodes of the thorax (page 491).

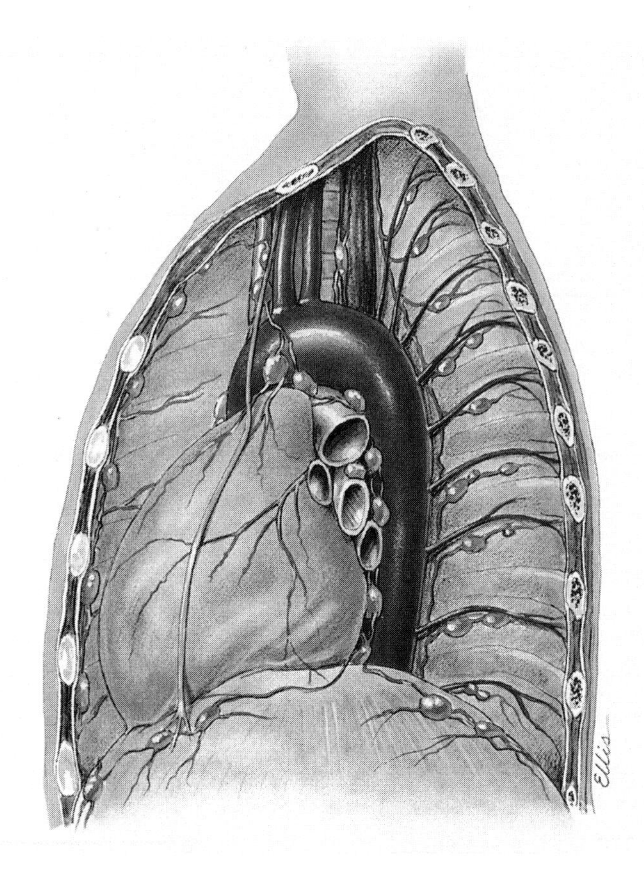


**Figure 15.13** Development of the lymphatic system (page 492).

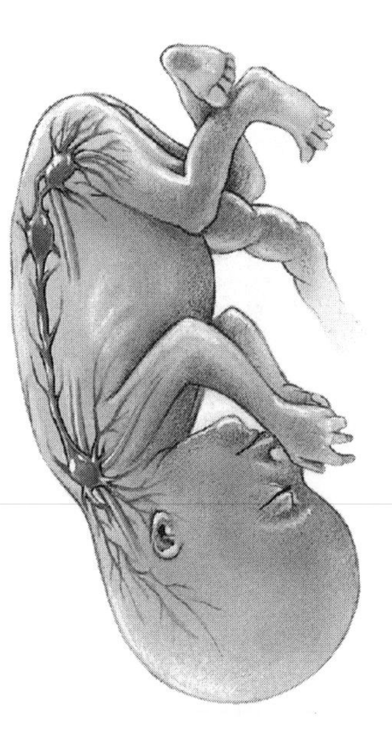

NOTES

**Figure 15.14** HIV, the causative agent of AIDS (page 493).

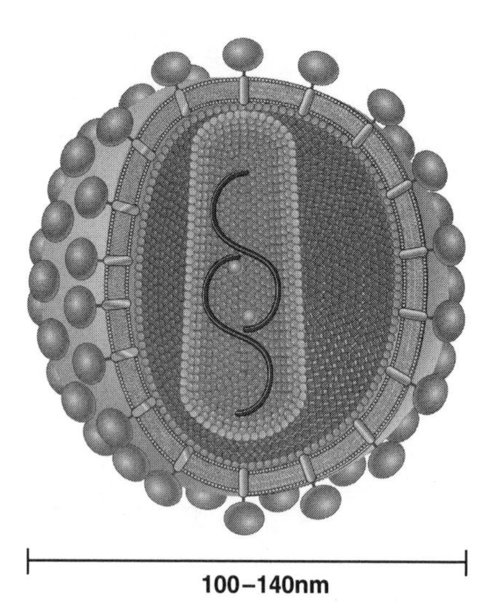

NOTES

**Figure 16.1** Major structure of the nervous system (page 500).

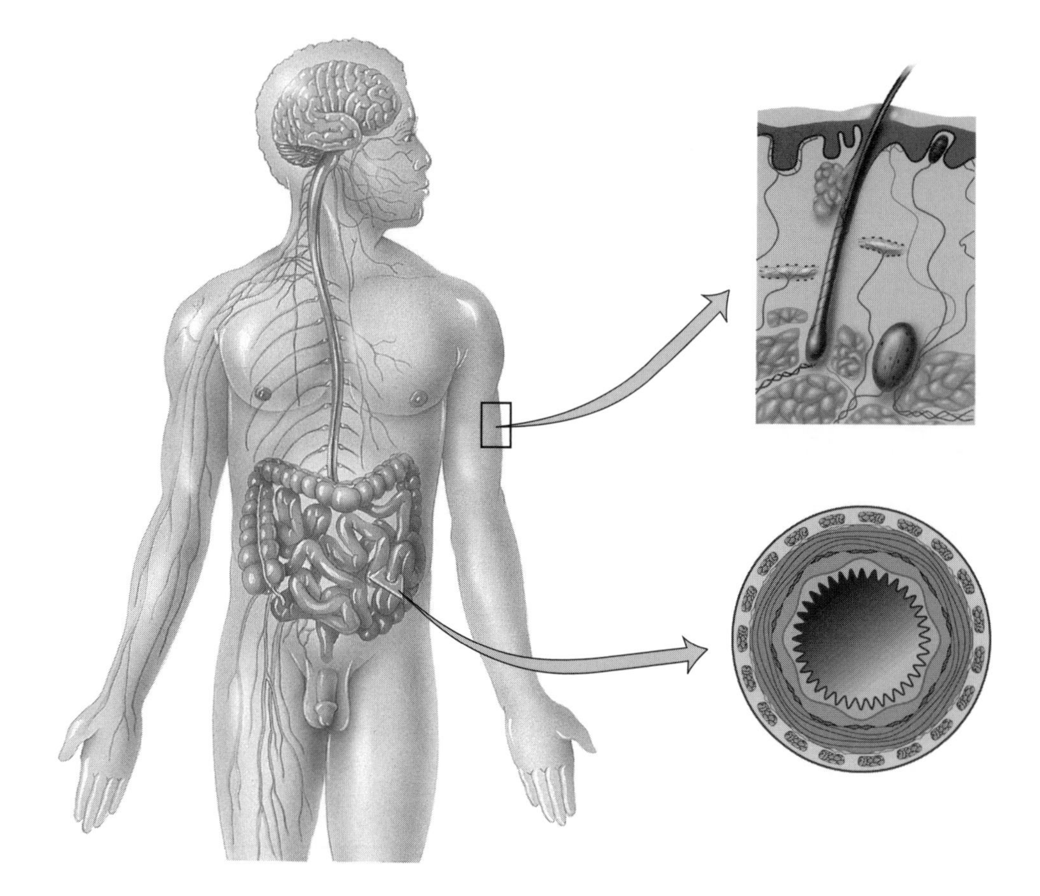

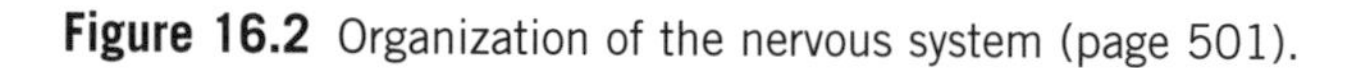

**Figure 16.2** Organization of the nervous system (page 501).

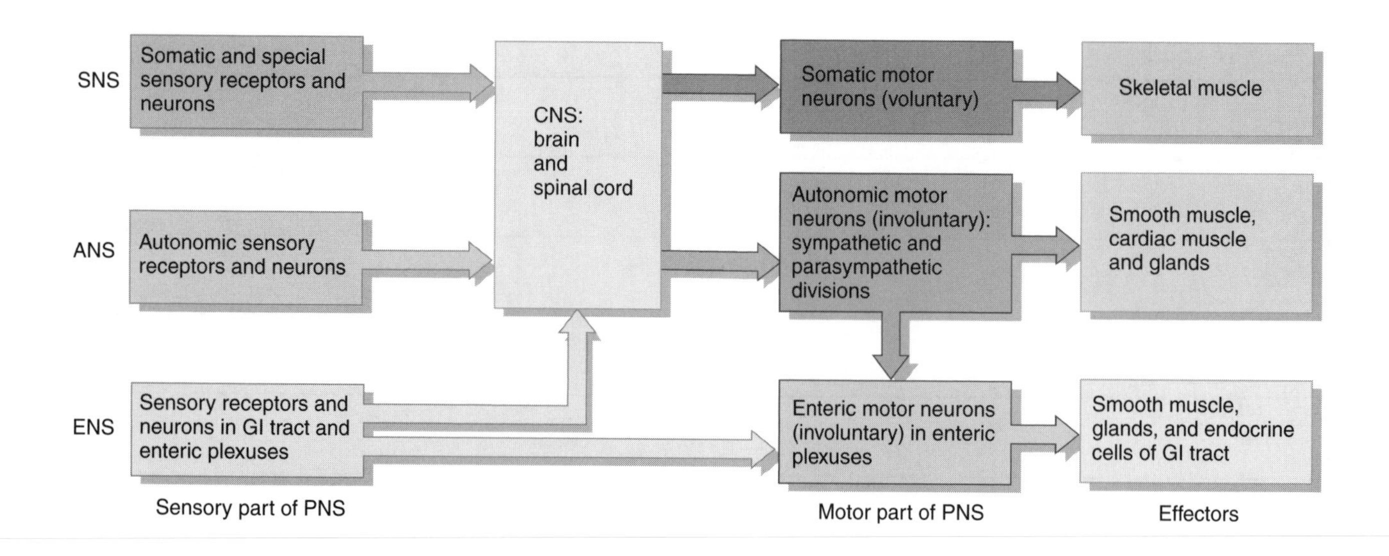

## NOTES

**Figure 16.3a,b** Structure of a typical neuron (page 505).

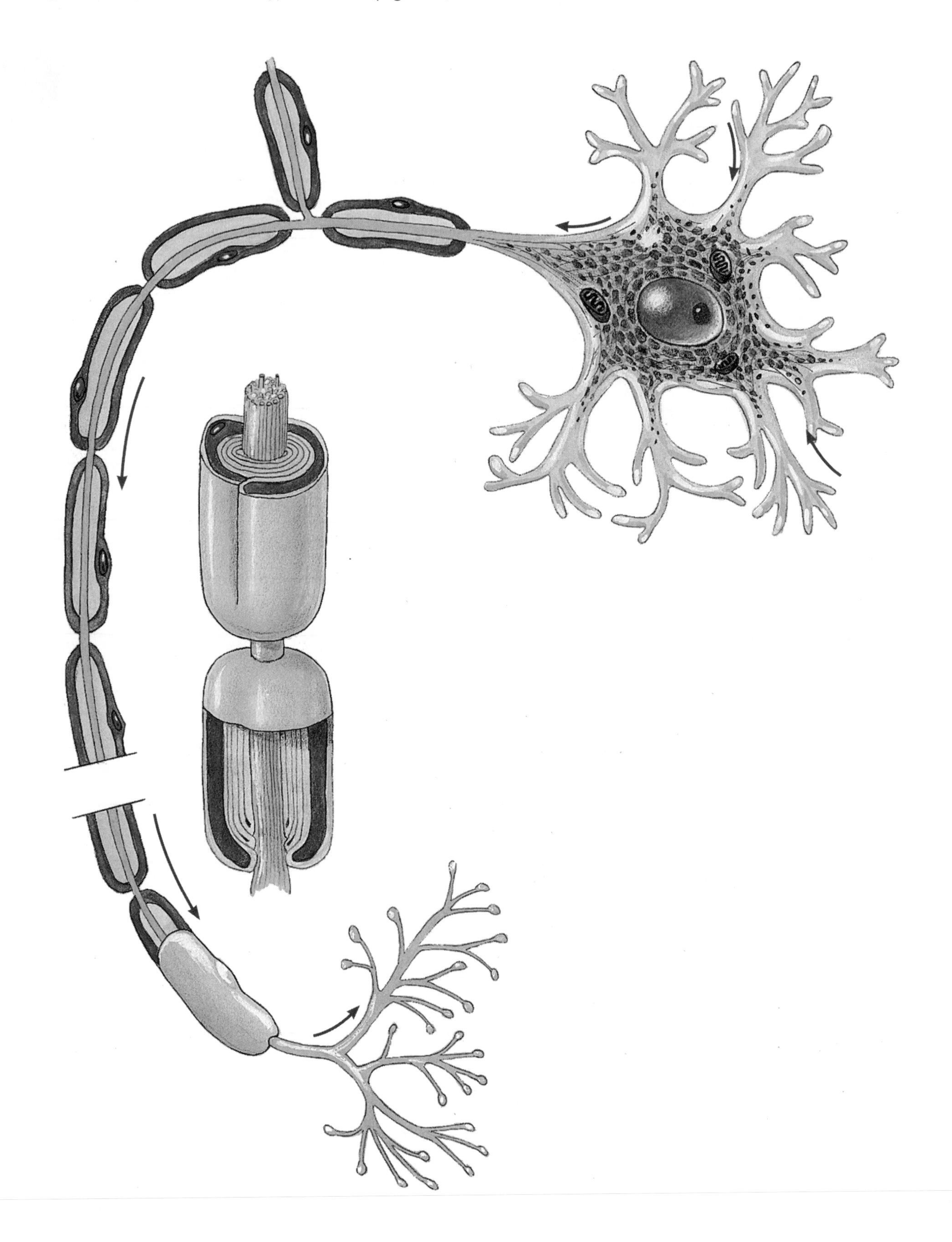

NOTES

16

**Figure 16.4a,b** Myelinated and unmyelinated axons (page 507).

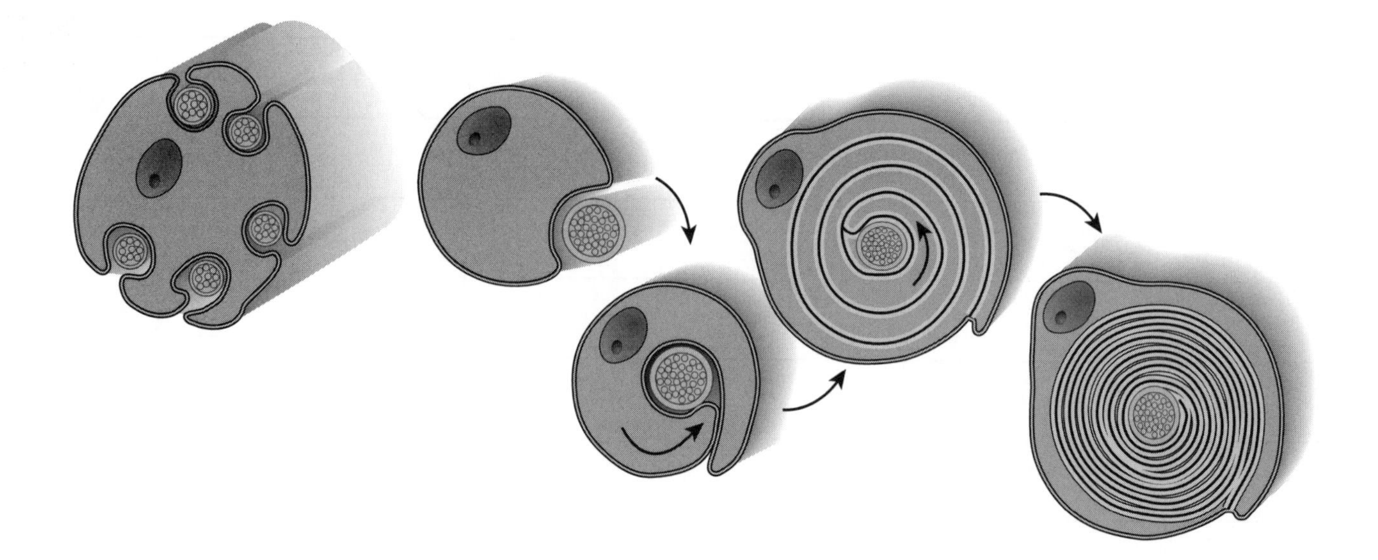

**Figure 16.5** Structural classification of neurons (page 508).

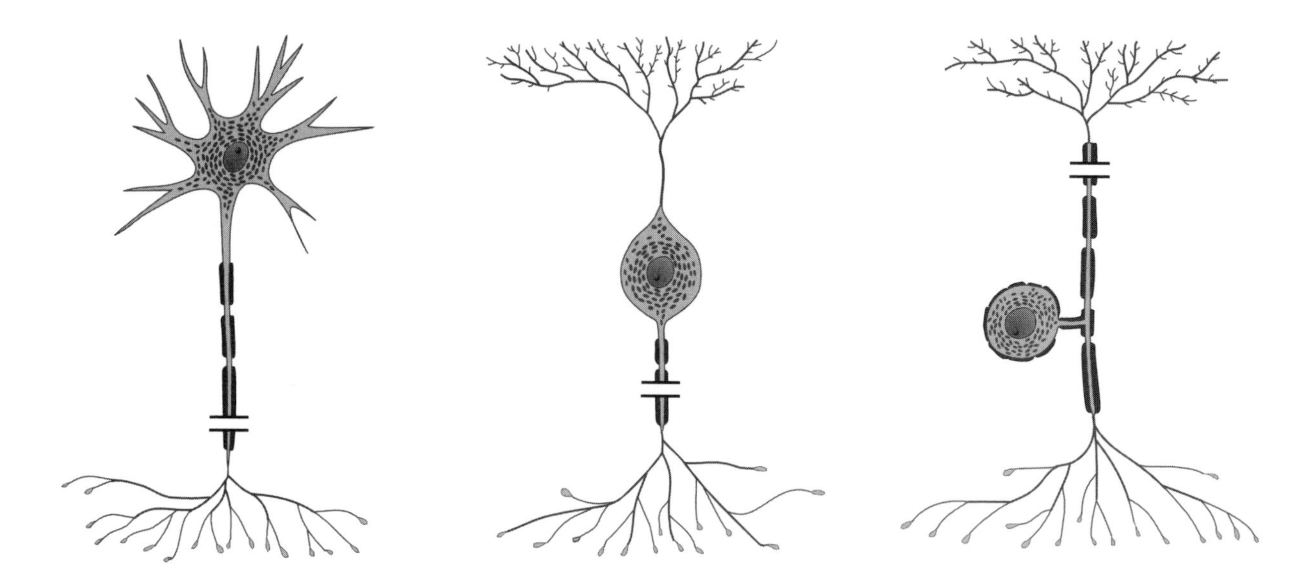

NOTES

16

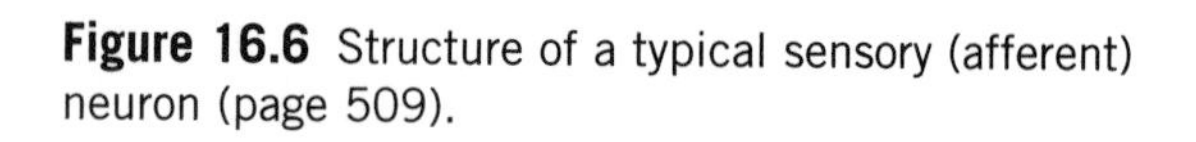

**Figure 16.6** Structure of a typical sensory (afferent) neuron (page 509).

**Figure 16.7** Two examples of interneurons (page 509).

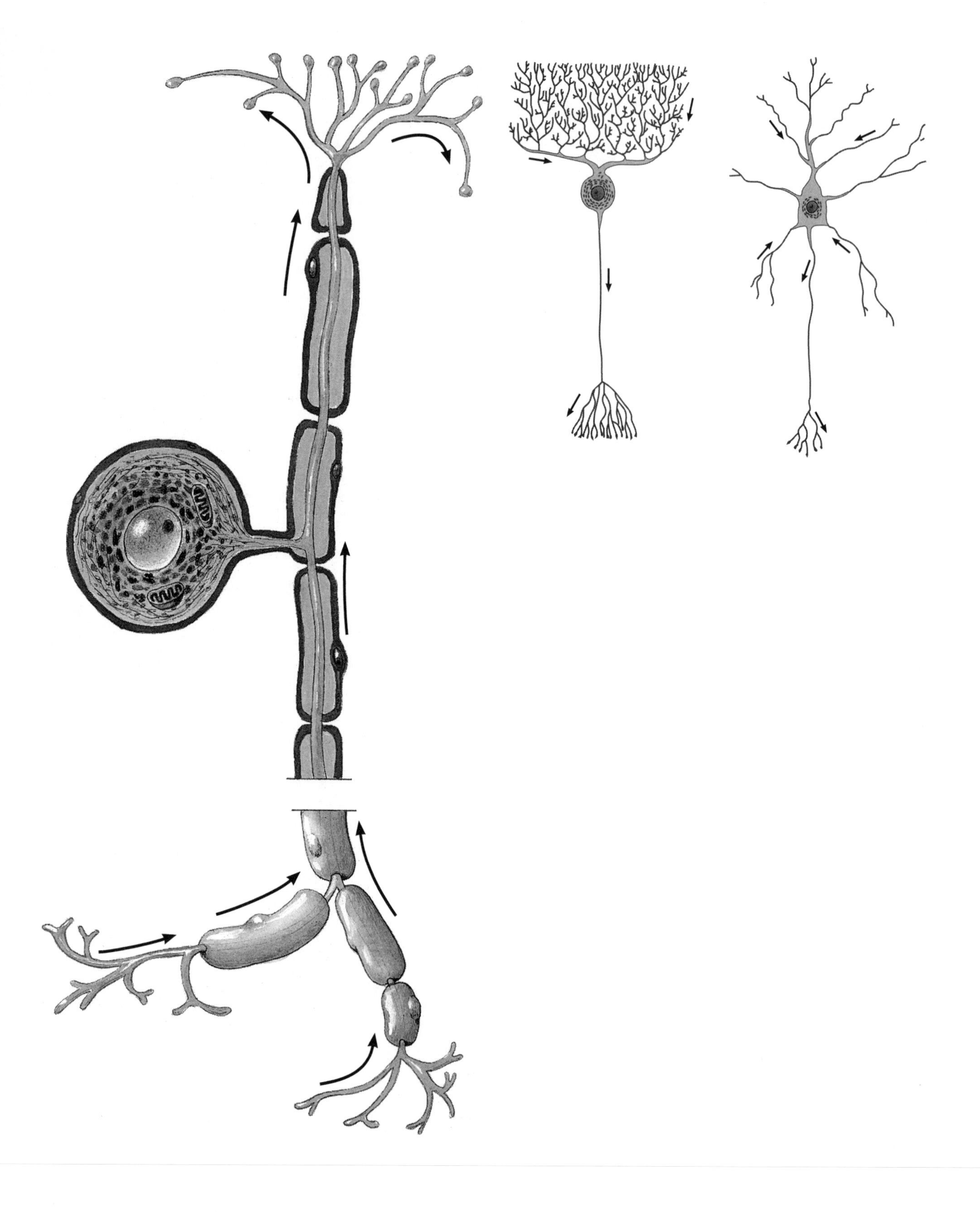

NOTES

16

**Figure 16.8** Distribution of gray and white matter in the spinal cord and brain (page 510).

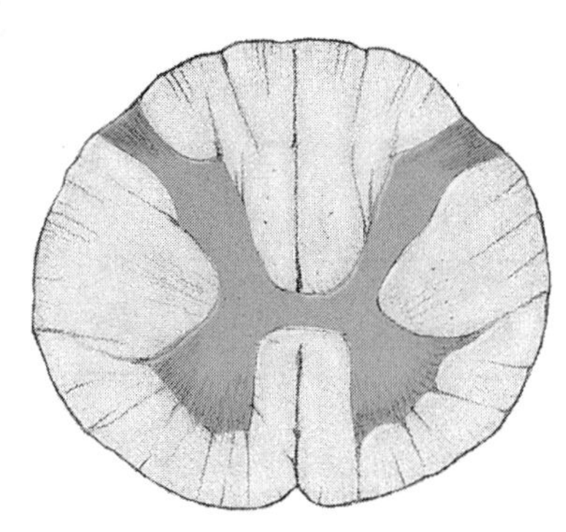

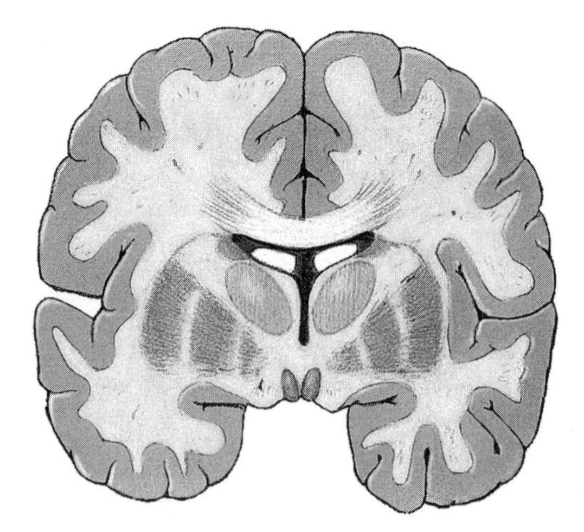

**Figure 16.9** Examples of neuronal circuits (page 511).

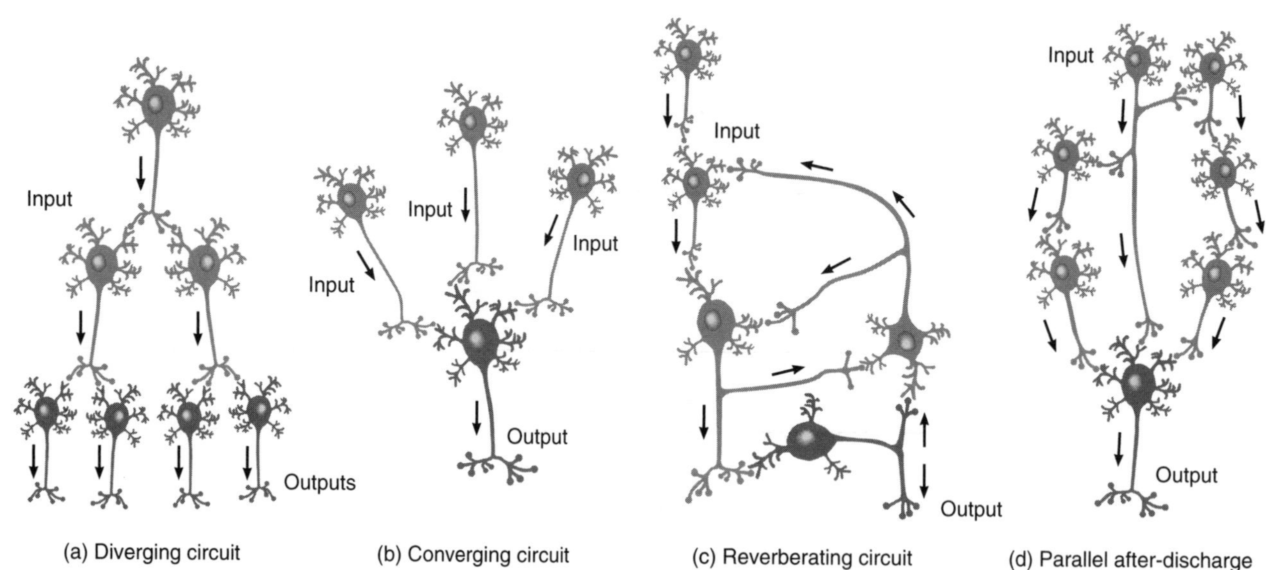

## NOTES

16

**Figure 17.1a** Gross anatomy of the spinal cord (page 517).

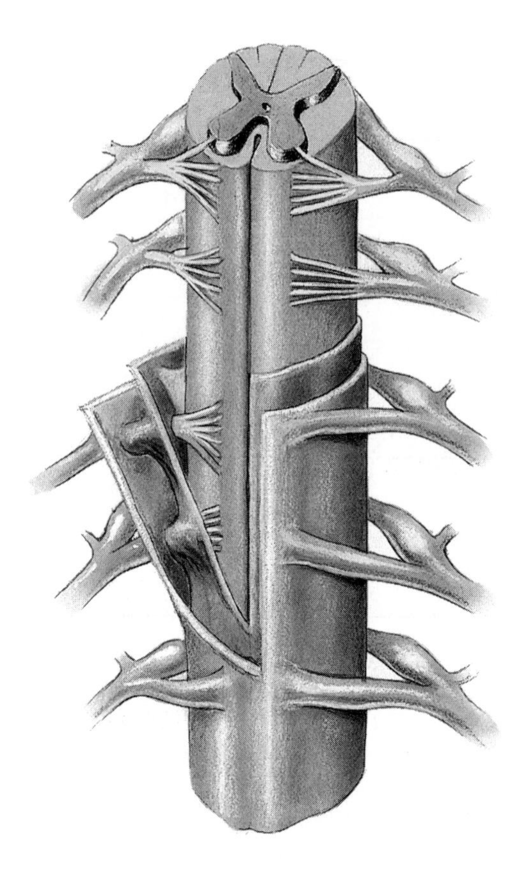

NOTES

17

**Figure 17.2a** External anatomy of the spinal cord and the spinal nerves (page 518).

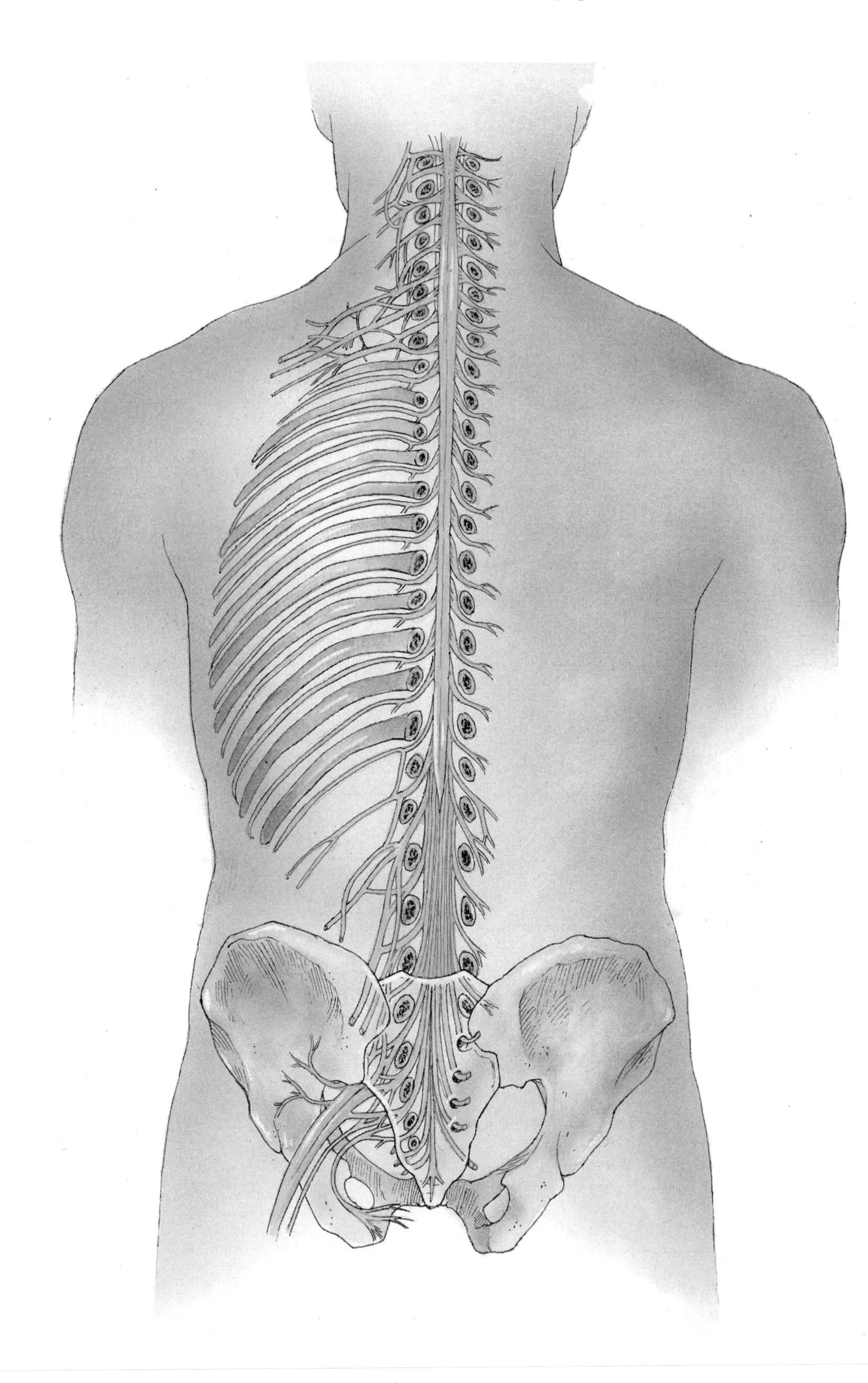

NOTES

**Figure 17.3a** Internal anatomy of the spinal cord: the organization of gray matter and white matter (page 521).

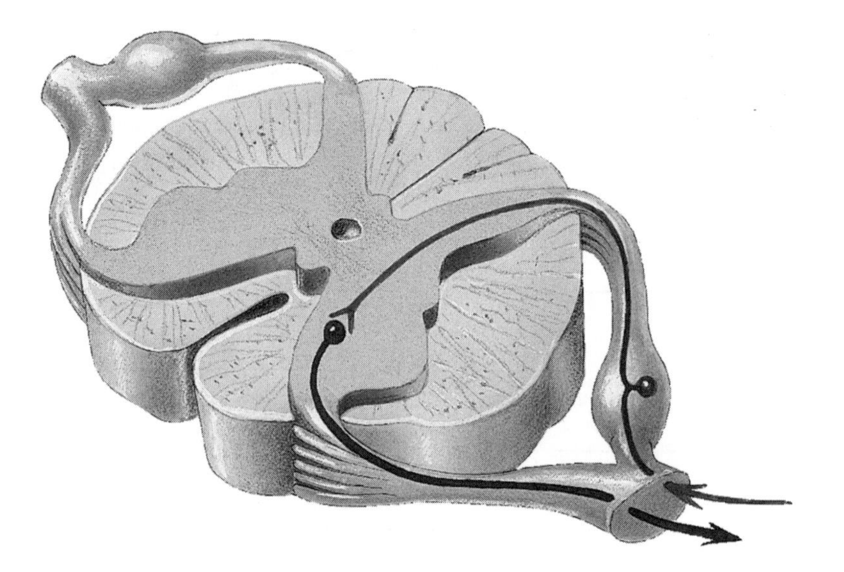

**Figure 17.4** The locations of selected sensory amd motor tracts (page 523).

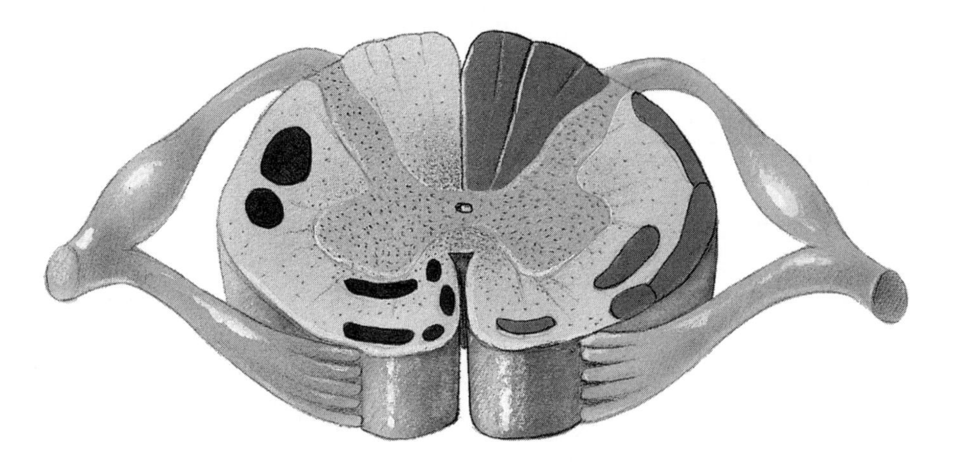

NOTES

17

**Figure 17.5** General components of a reflex arc (page 524).

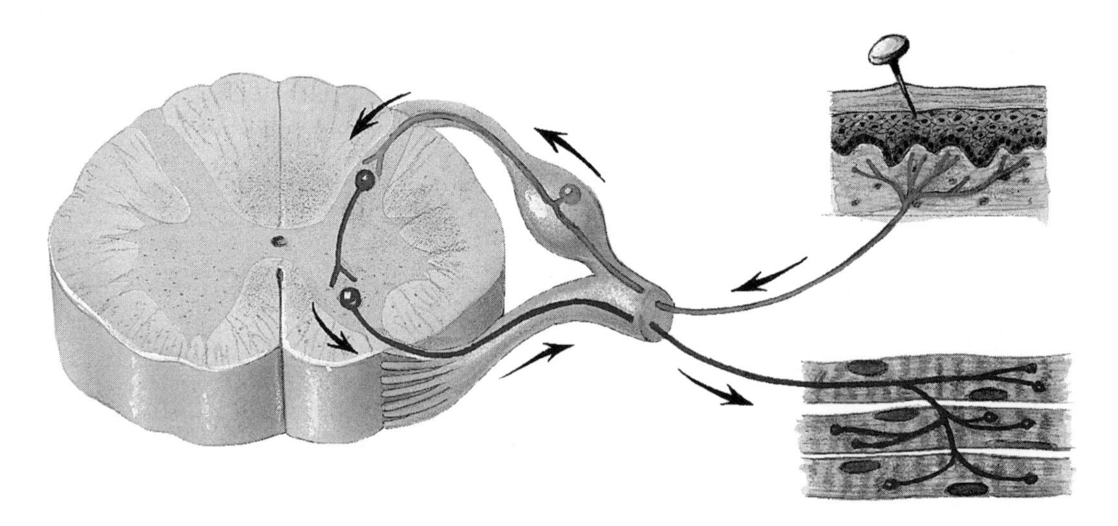

**Figure 17.6a** Organization and connective tissue coverings of a spinal nerve (page 526).

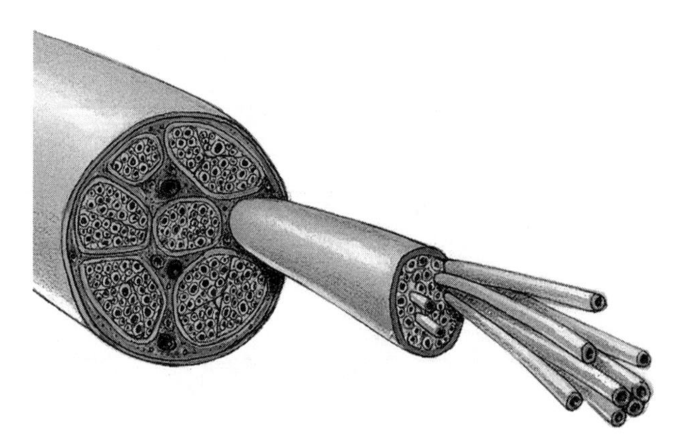

NOTES

**Figure 17.7** Branches of a typical spinal nerve (page 527).

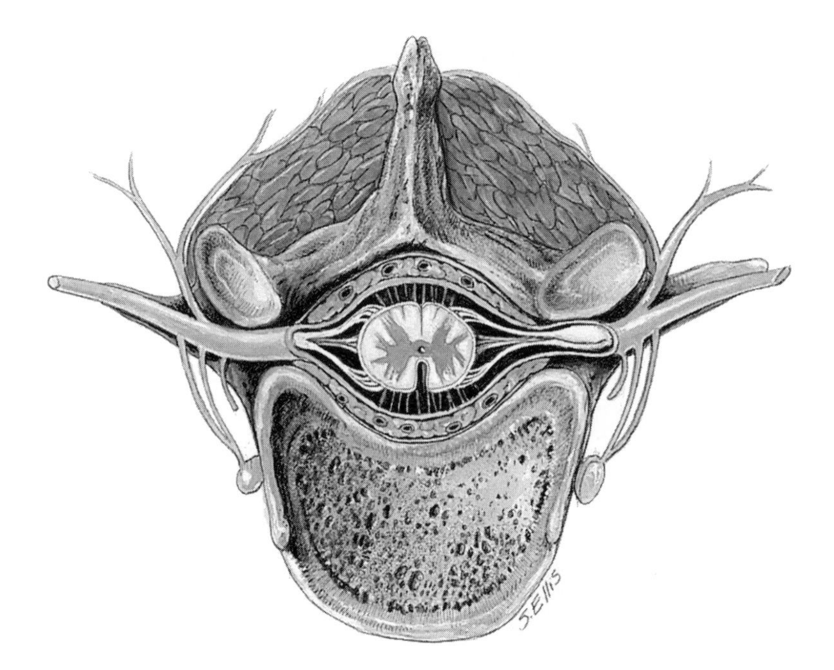

**Figure 17.8** Cervical plexus in anterior view (page 529).

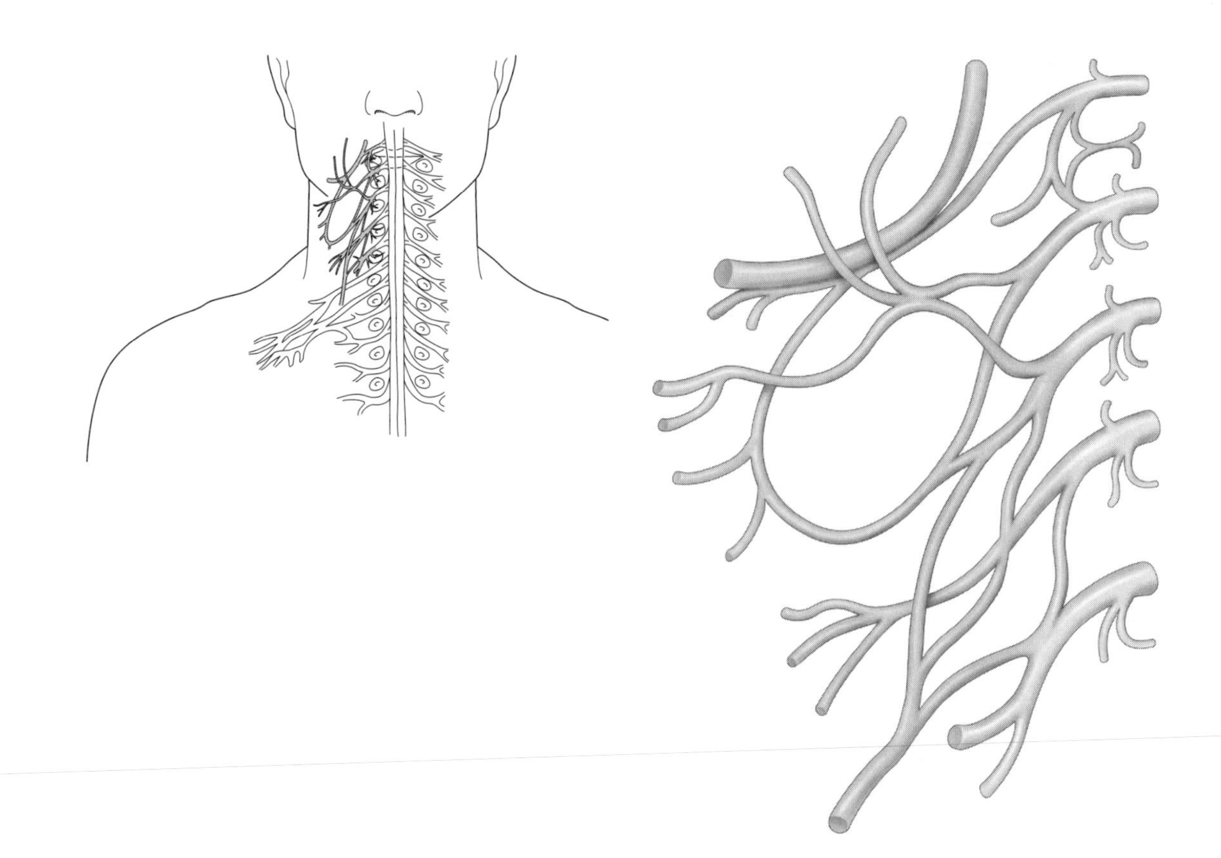

NOTES

**Figure 17.9a** Brachial plexus in anterior view (page 531).

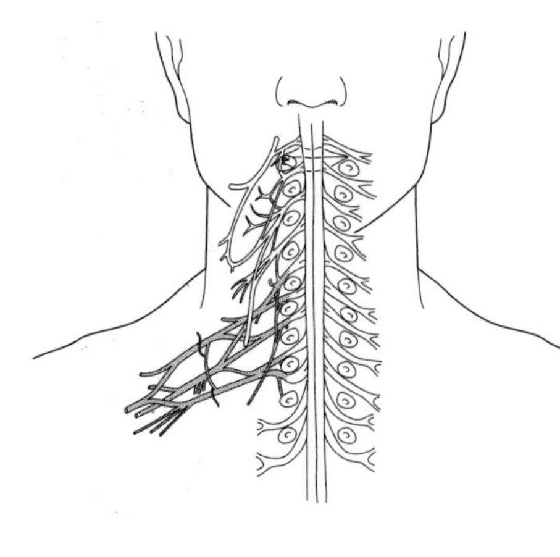

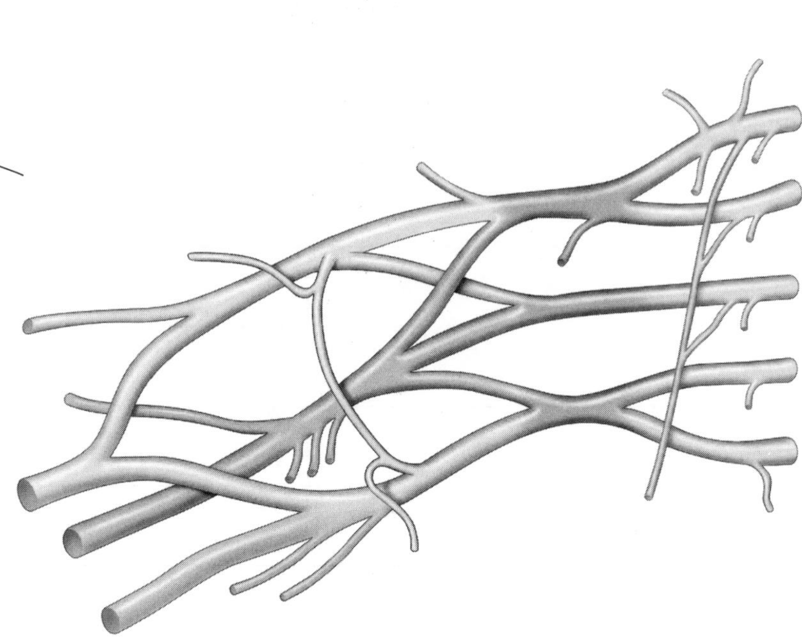

NOTES

17

**Figure 17.9b** Brachial plexus in anterior view (page 532).

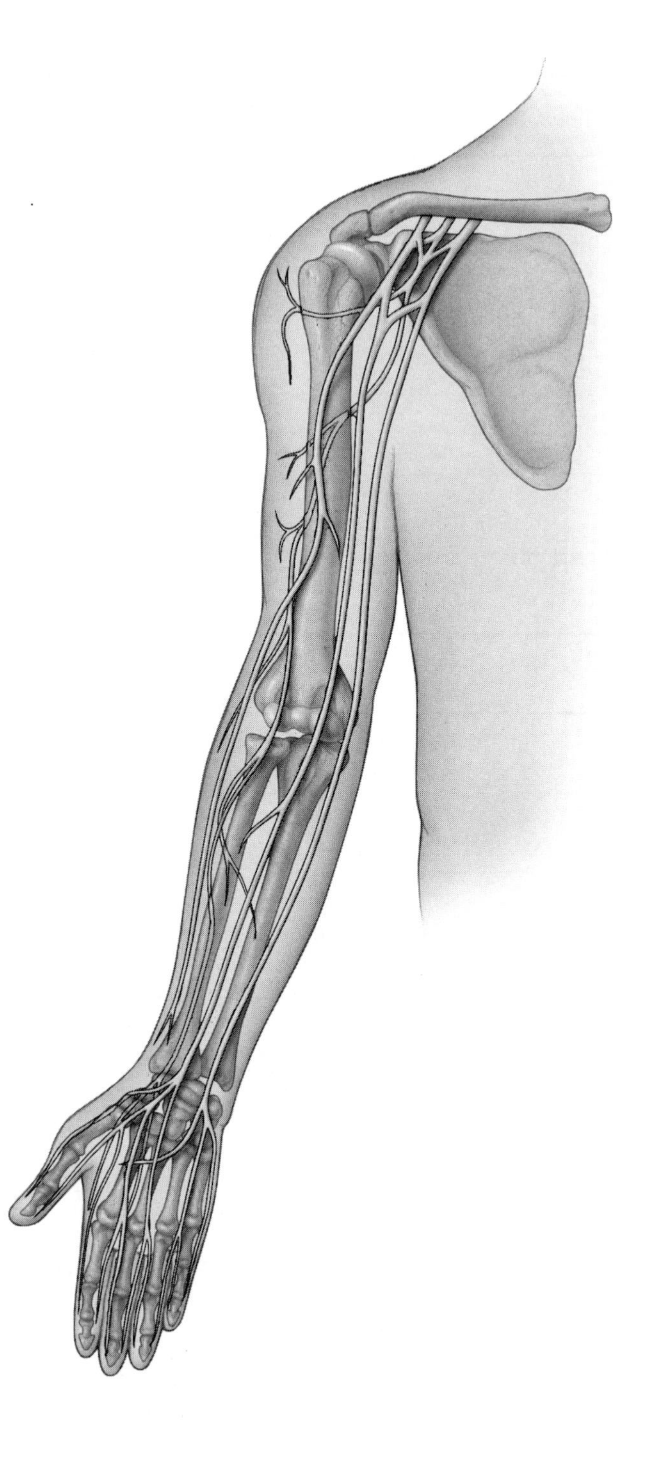

NOTES

17

**Figure 17.10** Injuries to the brachial plexus (page 533).

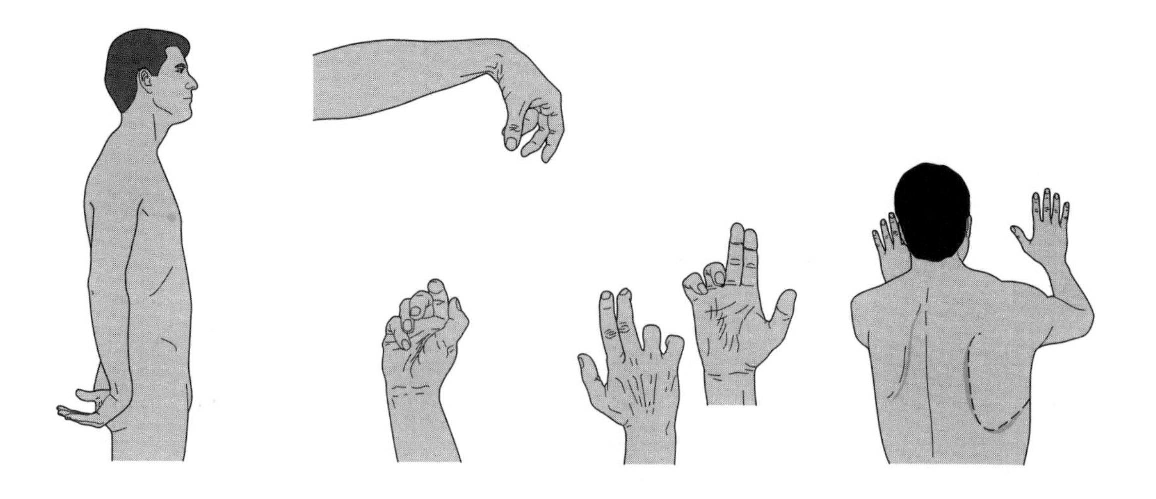

**Figure 17.11a** Lumbar plexus in anterior view (page 534).

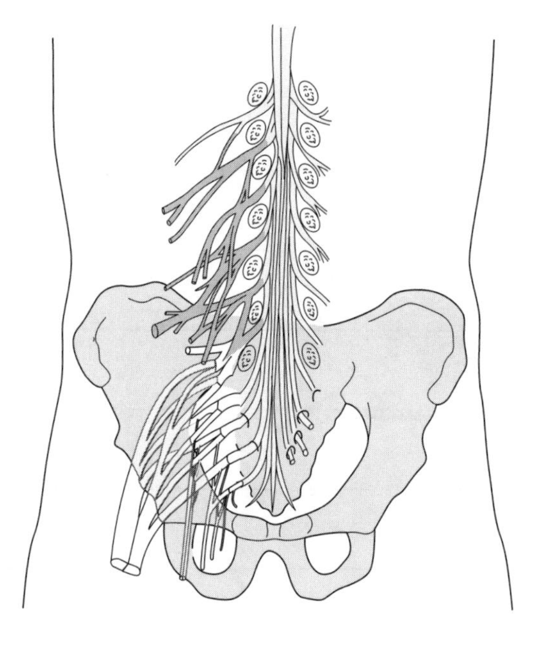

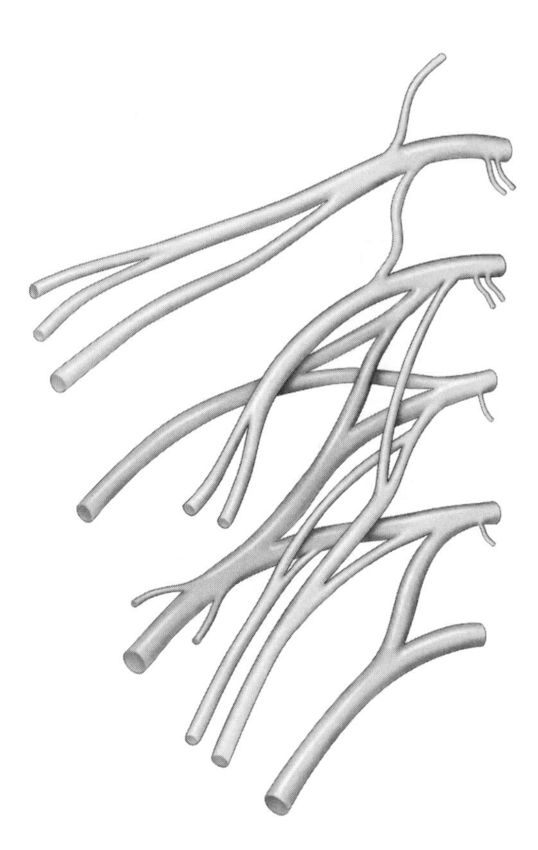

NOTES

**Figure 17.11b** Lumbar plexus in anterior view (page 535).

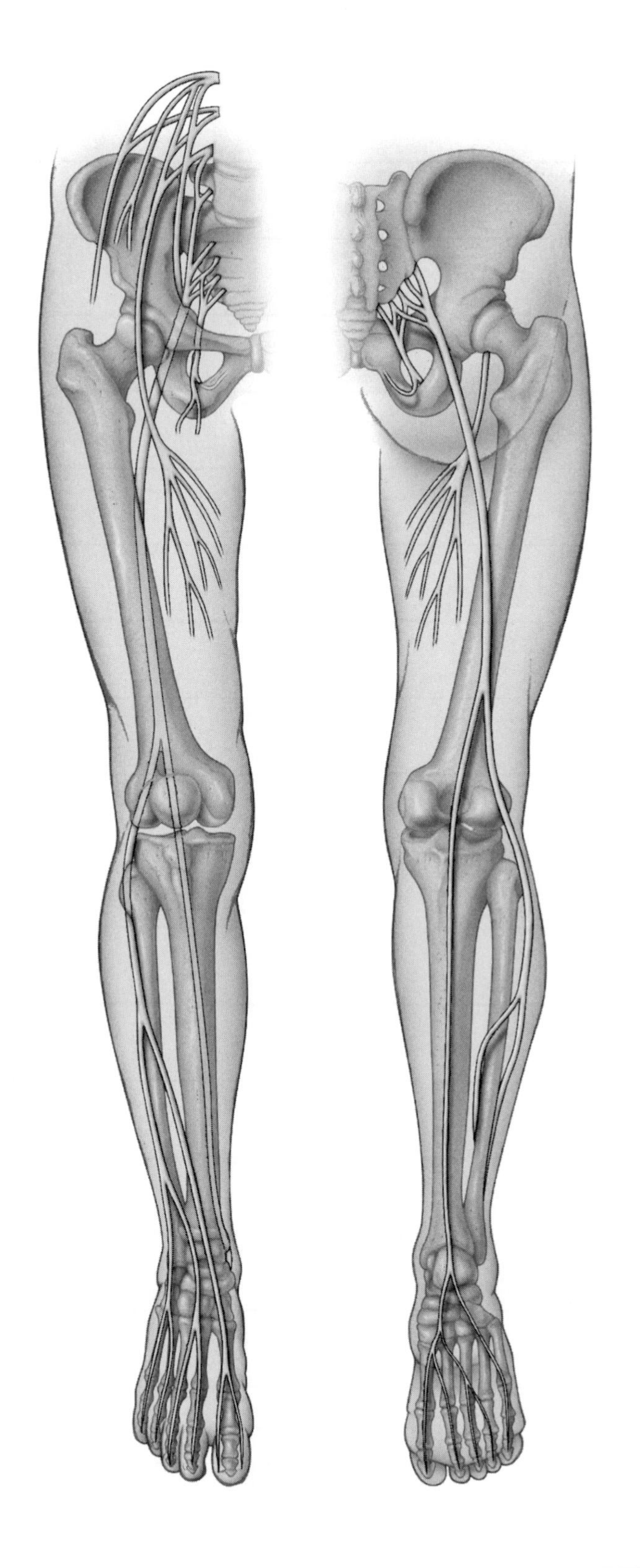

NOTES

17

**Figure 17.12** Sacral plexus in anterior view (page 538).

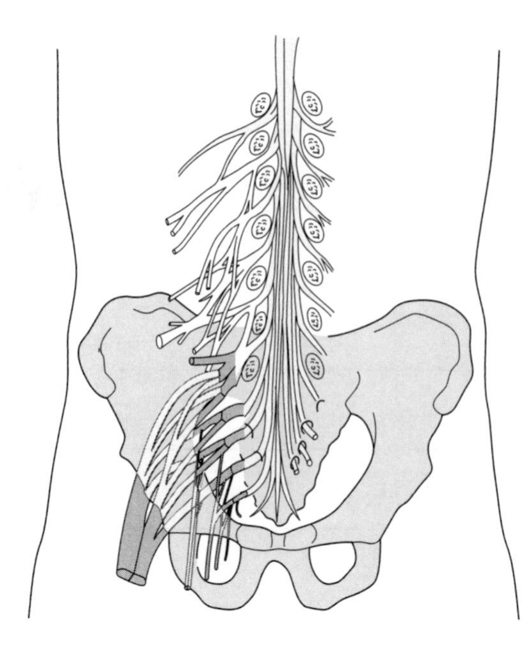

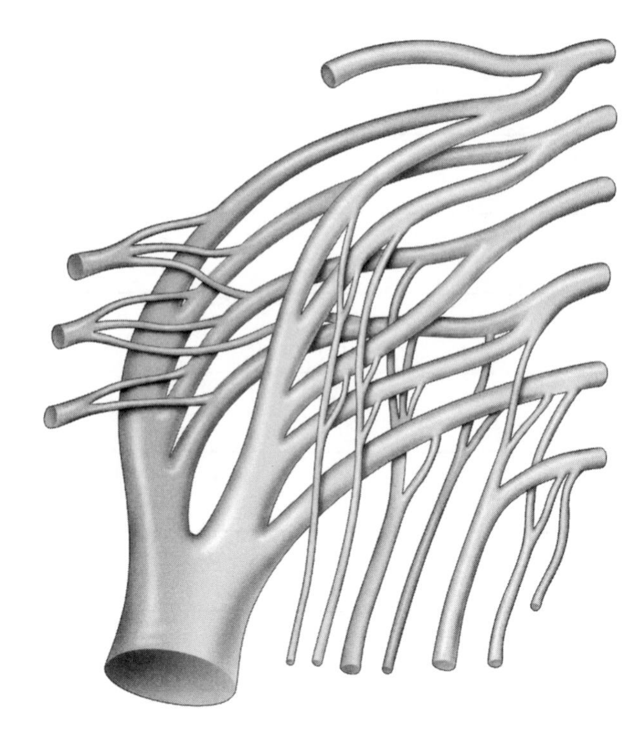

NOTES

17

**Figure 17.13** Distribution of dermatomes (page 539).

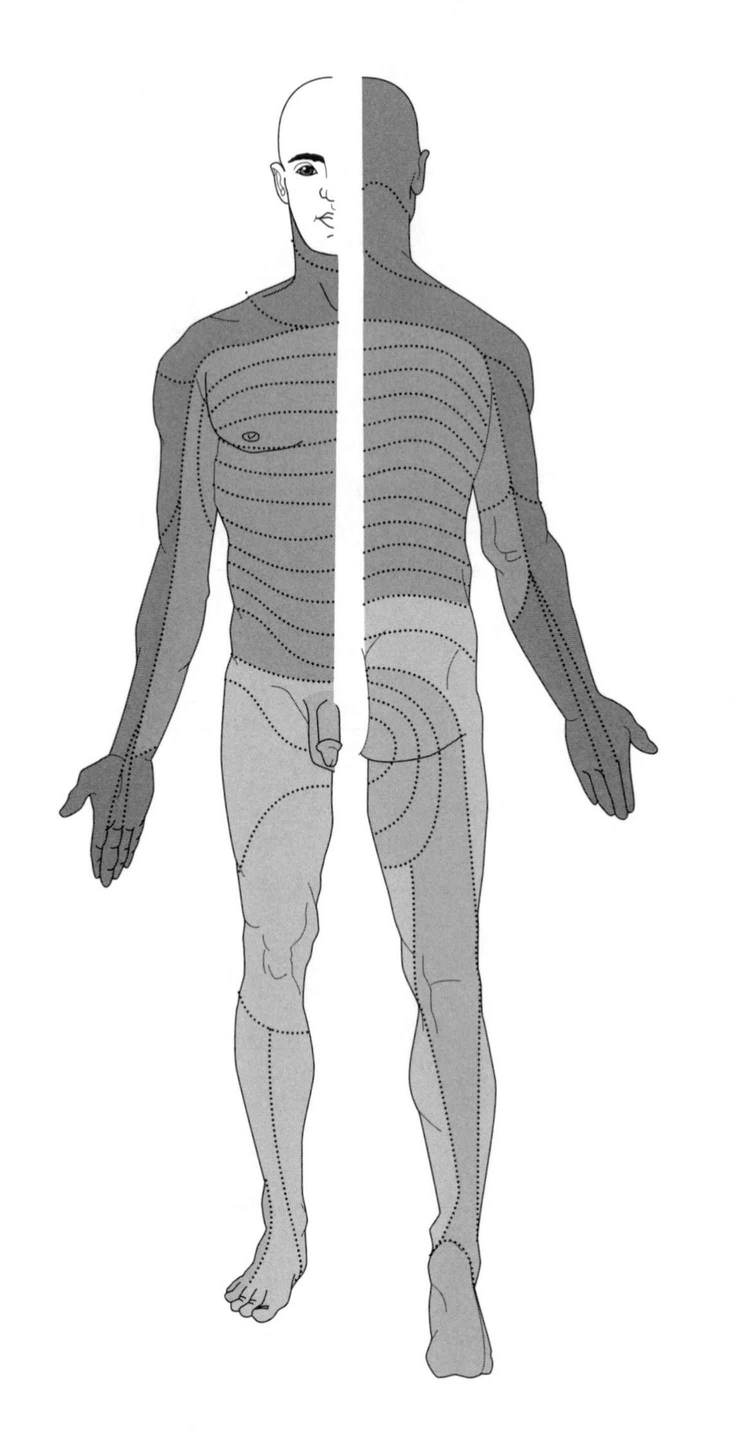

NOTES

17

**Figure 18.1a** The brain (page 545).

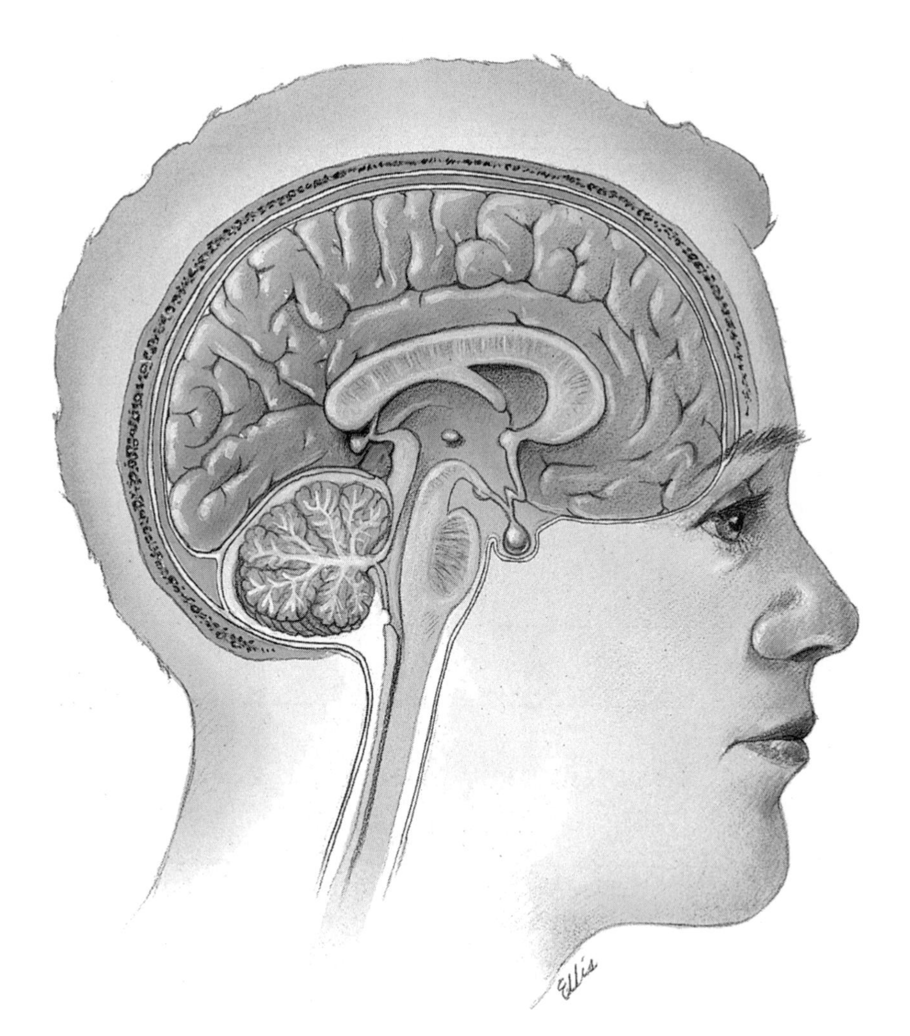

**Figure 18.2** The protective coverings of the brain (page 547).

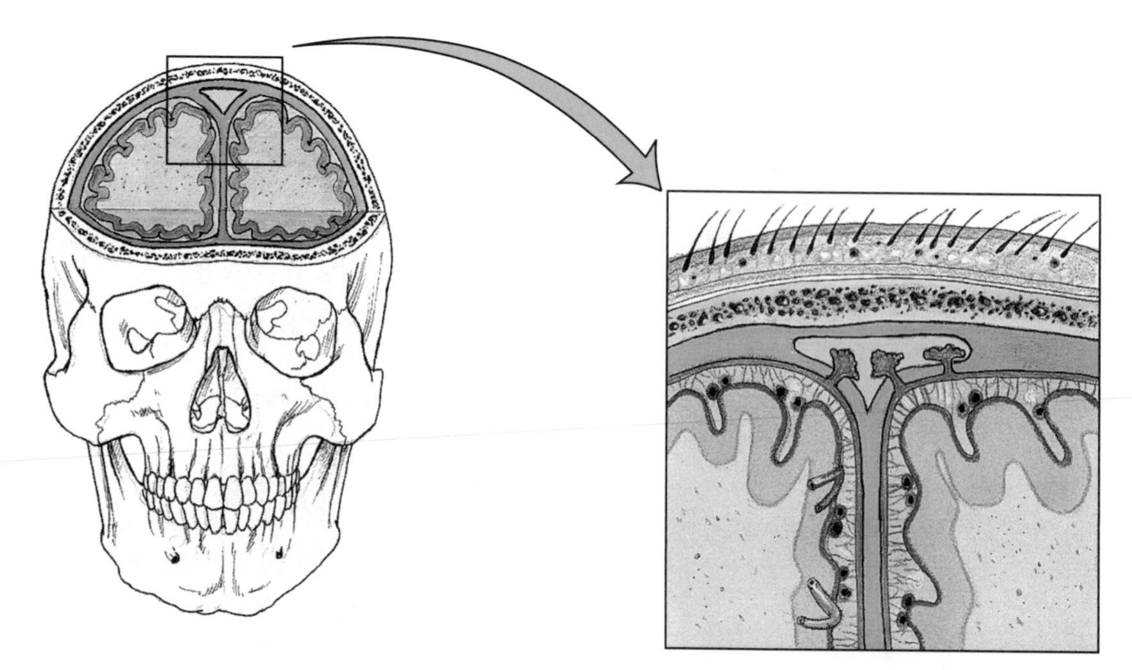

NOTES

18

**Figure 18.3** Locations of ventricles within a "transparent" brain (page 548).

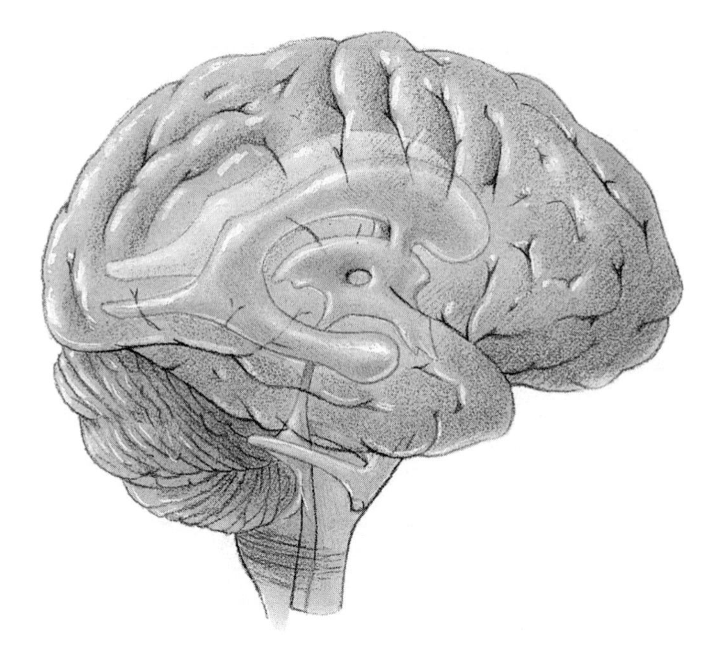

NOTES

18

**Figure 18.4a** Pathways of circulating cerebrospinal fluid (page 549).

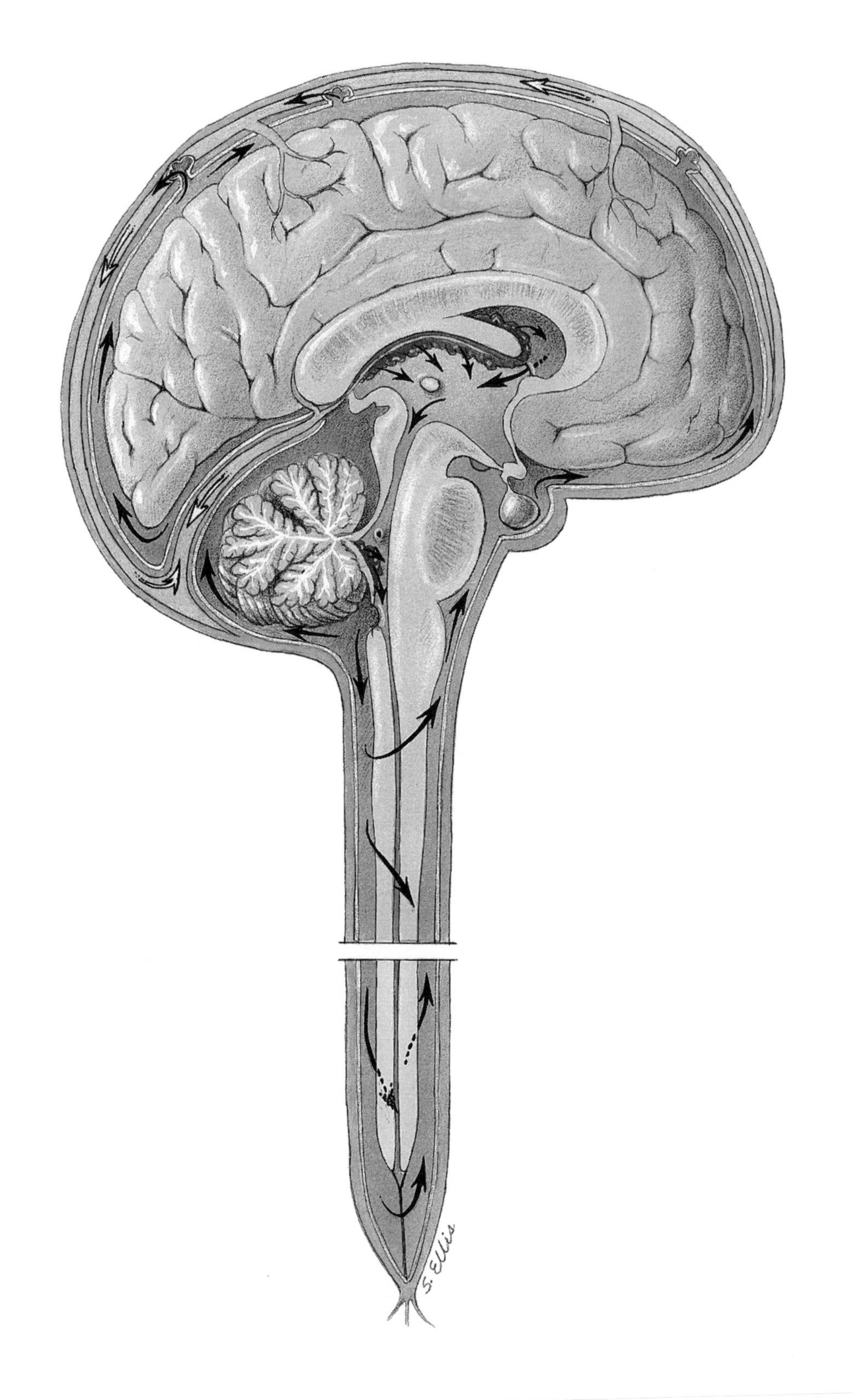

NOTES

18

**Figure 18.4b** Pathways of circulating cerebrospinal fluid (page 550).

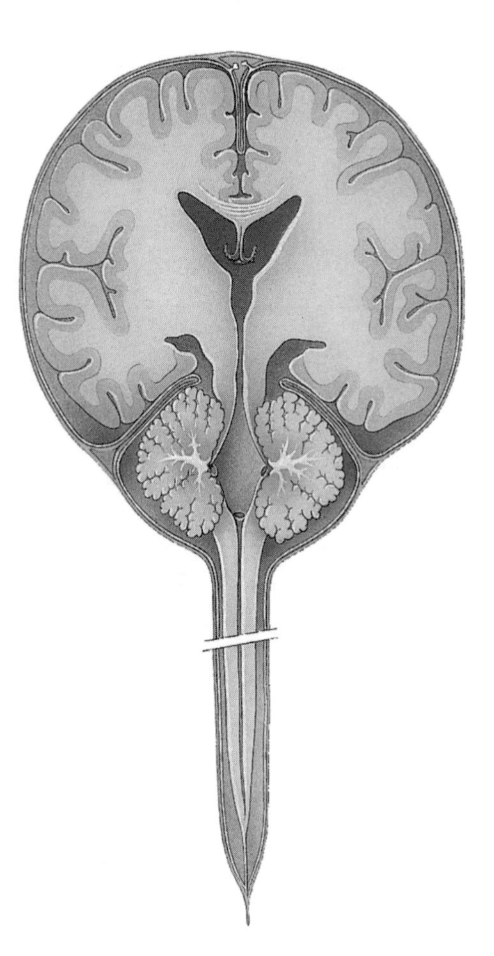

NOTES

**Figure 18.5** Medulla oblongata in relation to the rest of the brain stem (page 551).

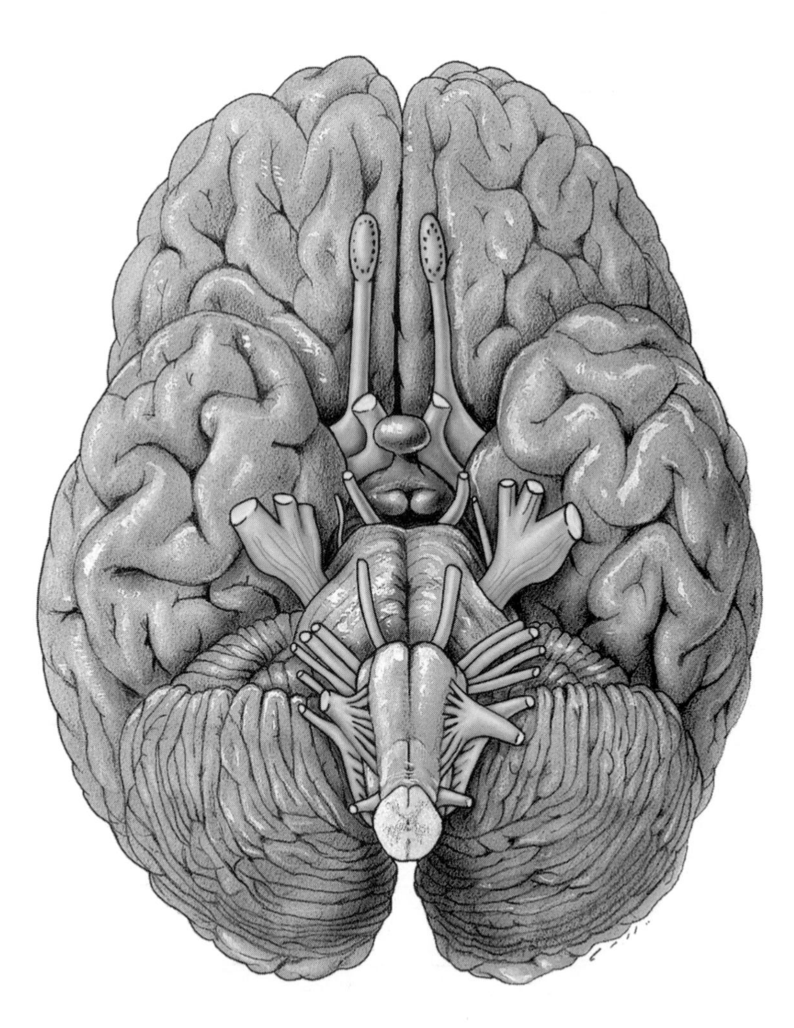

**Figure 18.6** Internal anatomy of the medulla oblongata (page 552).

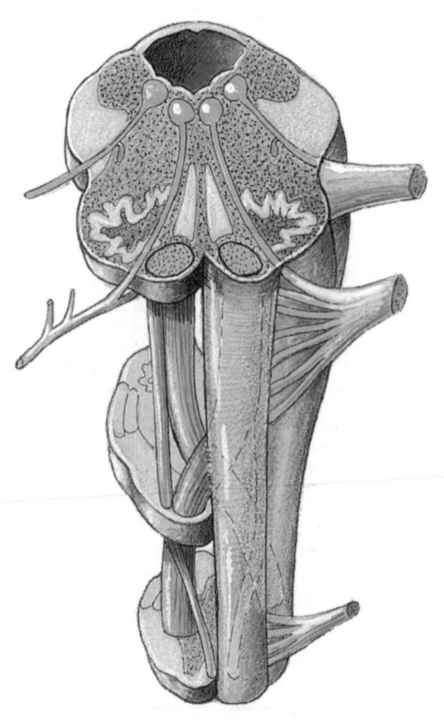

NOTES

18

**Figure 18.7a,b** Midbrain (page 554).

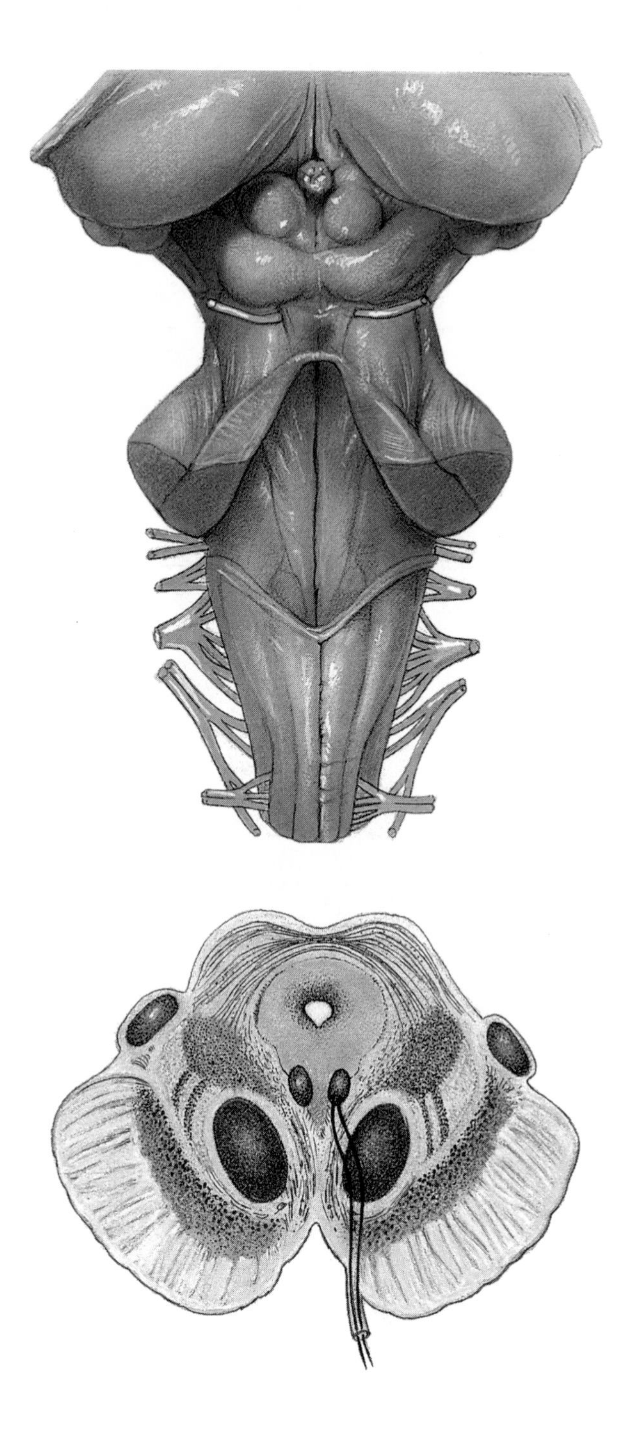

## NOTES

18

**Figure 18.8** Cerebellum (page 557).

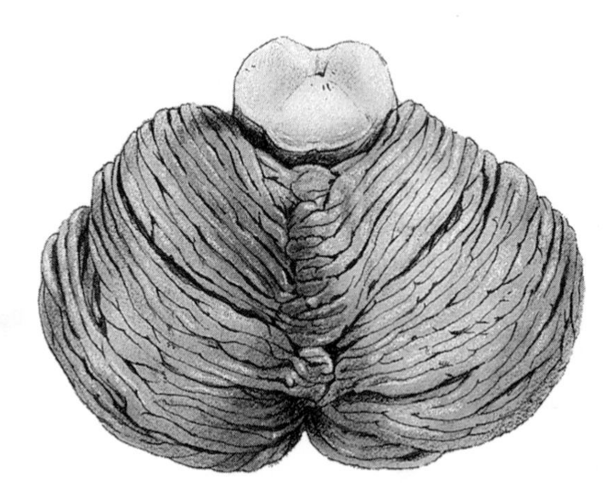

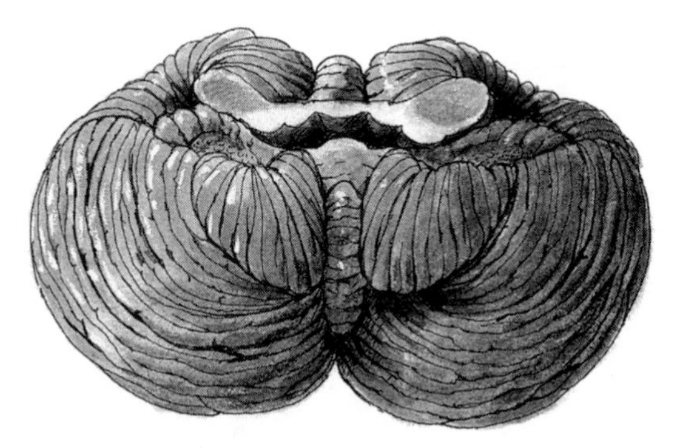

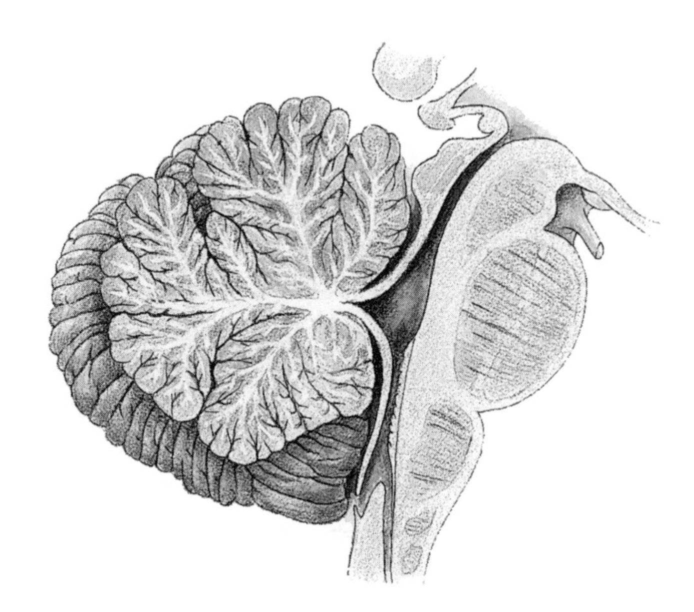

NOTES

18

**Figure 18.9** Thalamus (page 559).

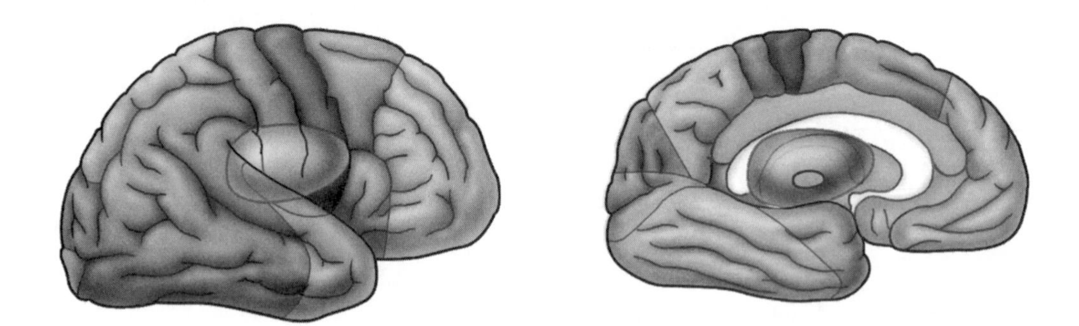

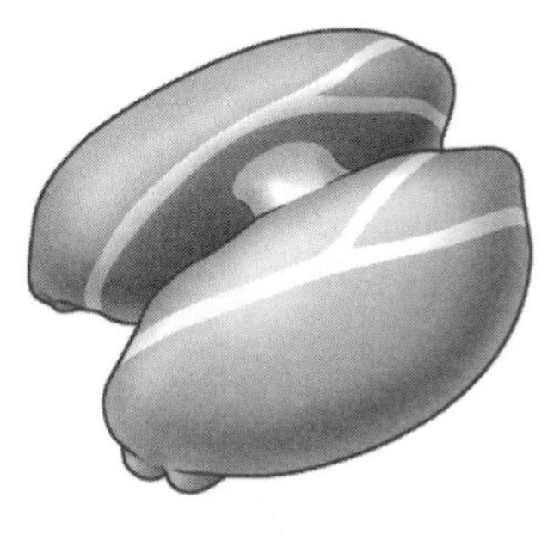

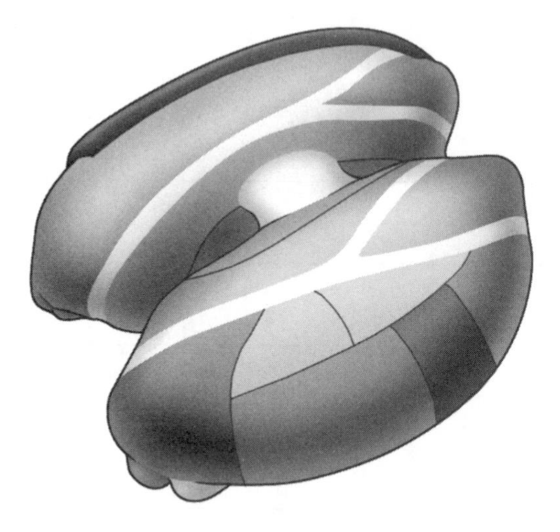

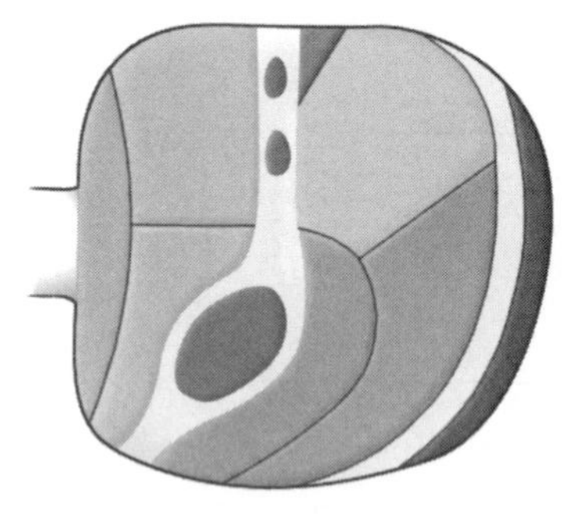

NOTES

18

**Figure 18.10** Hypothalamus (page 560).

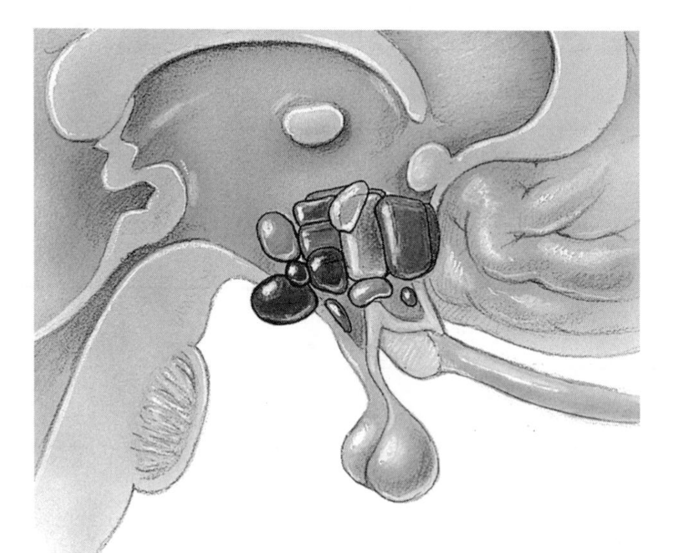

**Figure 18.11a,b** Cerebrum (page 562).

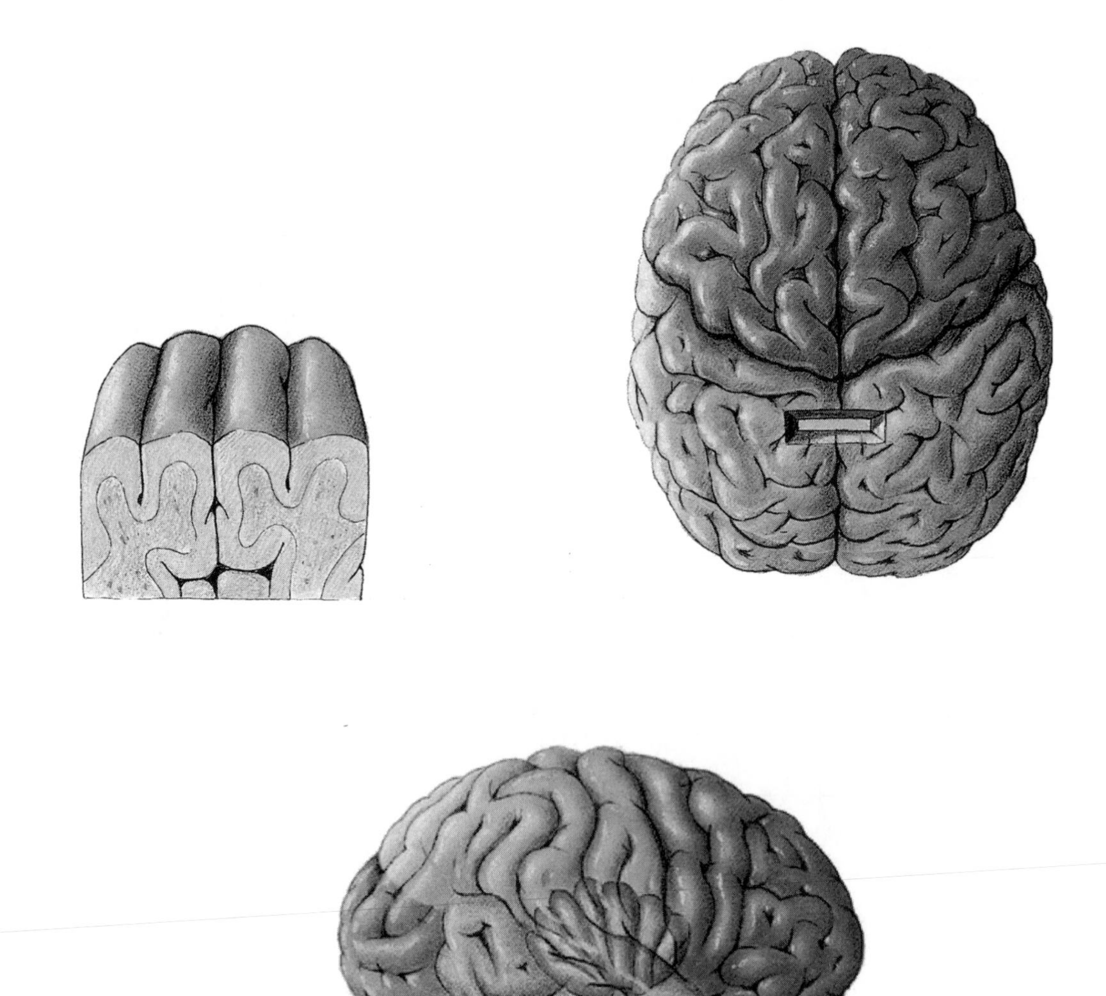

## NOTES

**Figure 18.13a** Basal ganglia (page 565).

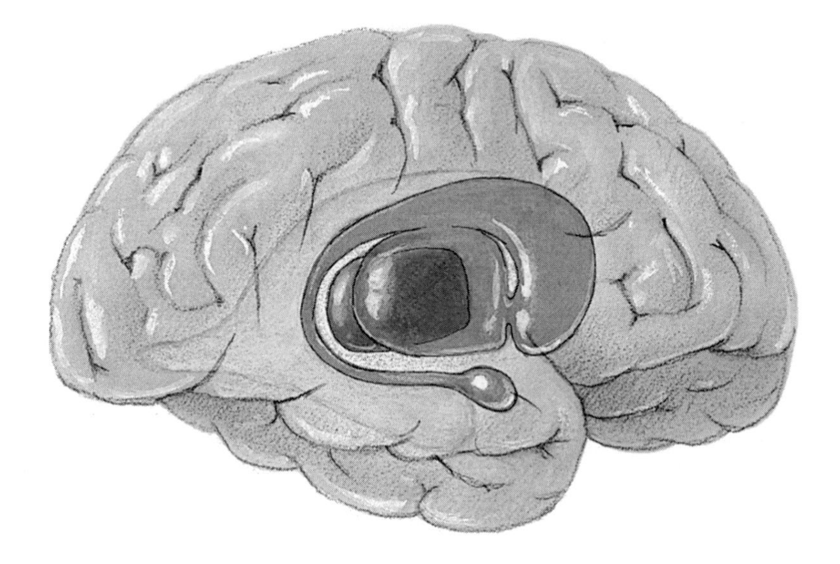

**Figure 18.13b** Basal ganglia (page 566).

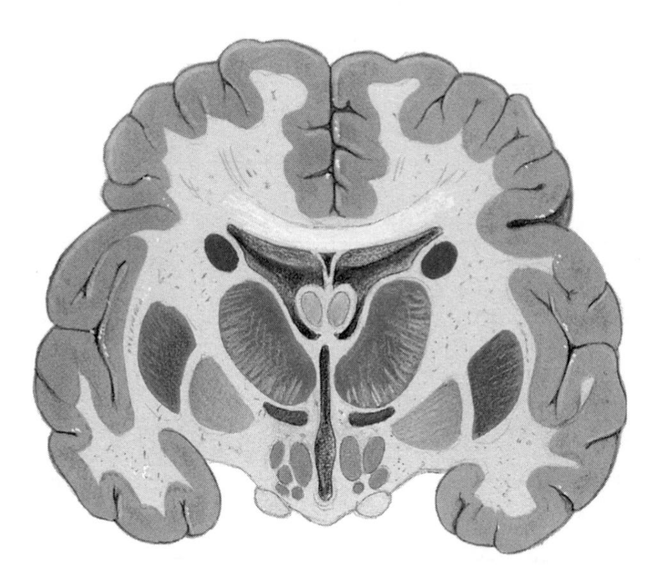

NOTES

**Figure 18.14** Components of the limbic system and surrounding structures (page 567).

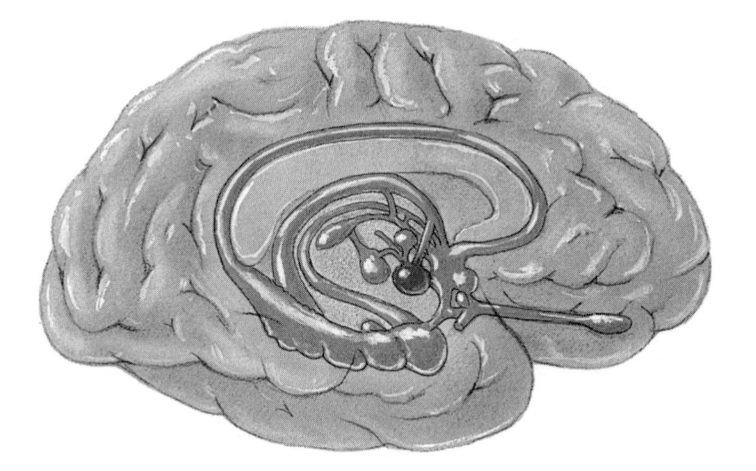

**Figure 18.15** Functional areas of the cerebrum (page 568).

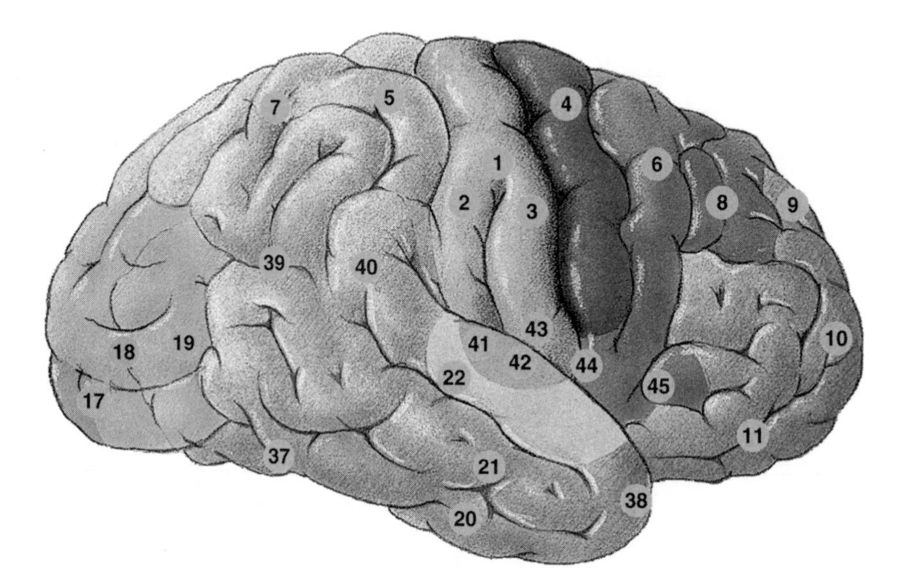

NOTES

**Figure 18.16** Summary of the principal functional differences between the left and right cerebral hemispheres (page 571).

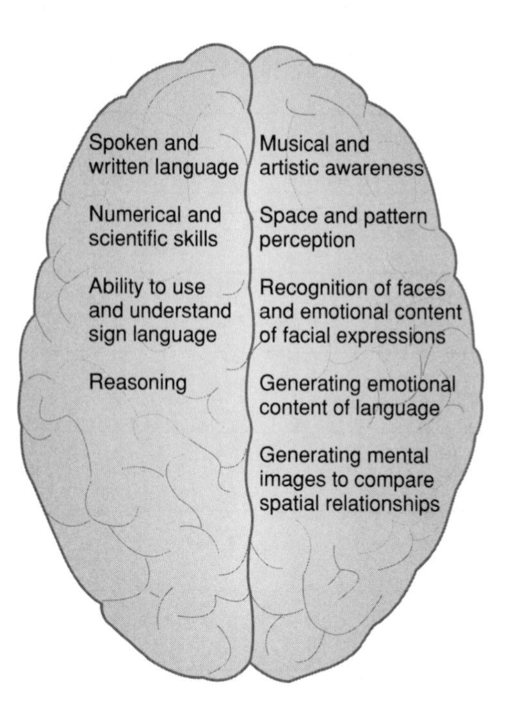

NOTES

18

**Figure 18.17** Oculomotor (III), trochlear (IV), and abducens (VI) nerves (page 573).

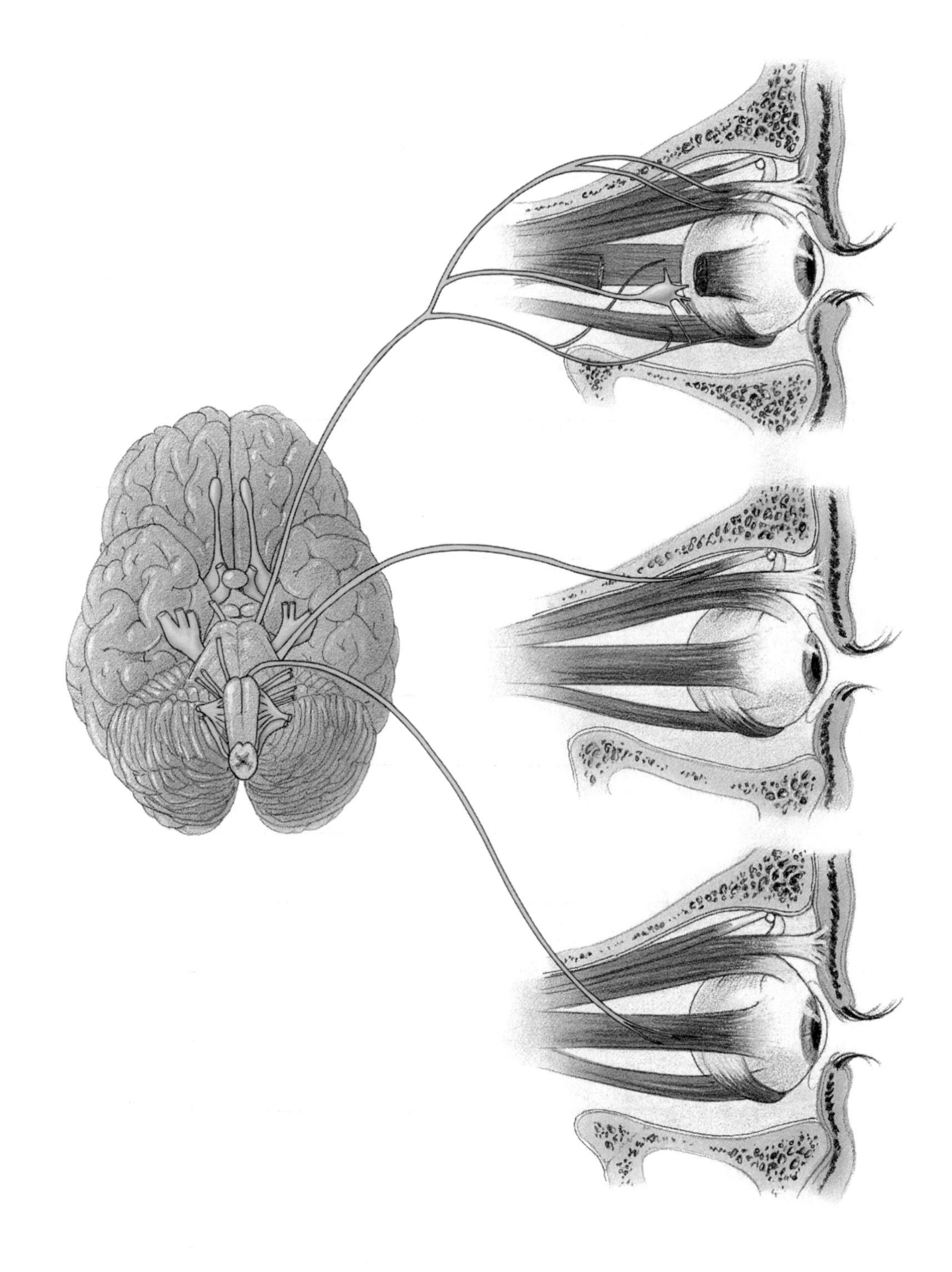

NOTES

18

**Figure 18.18** Trigeminal (V) nerve (page 574).

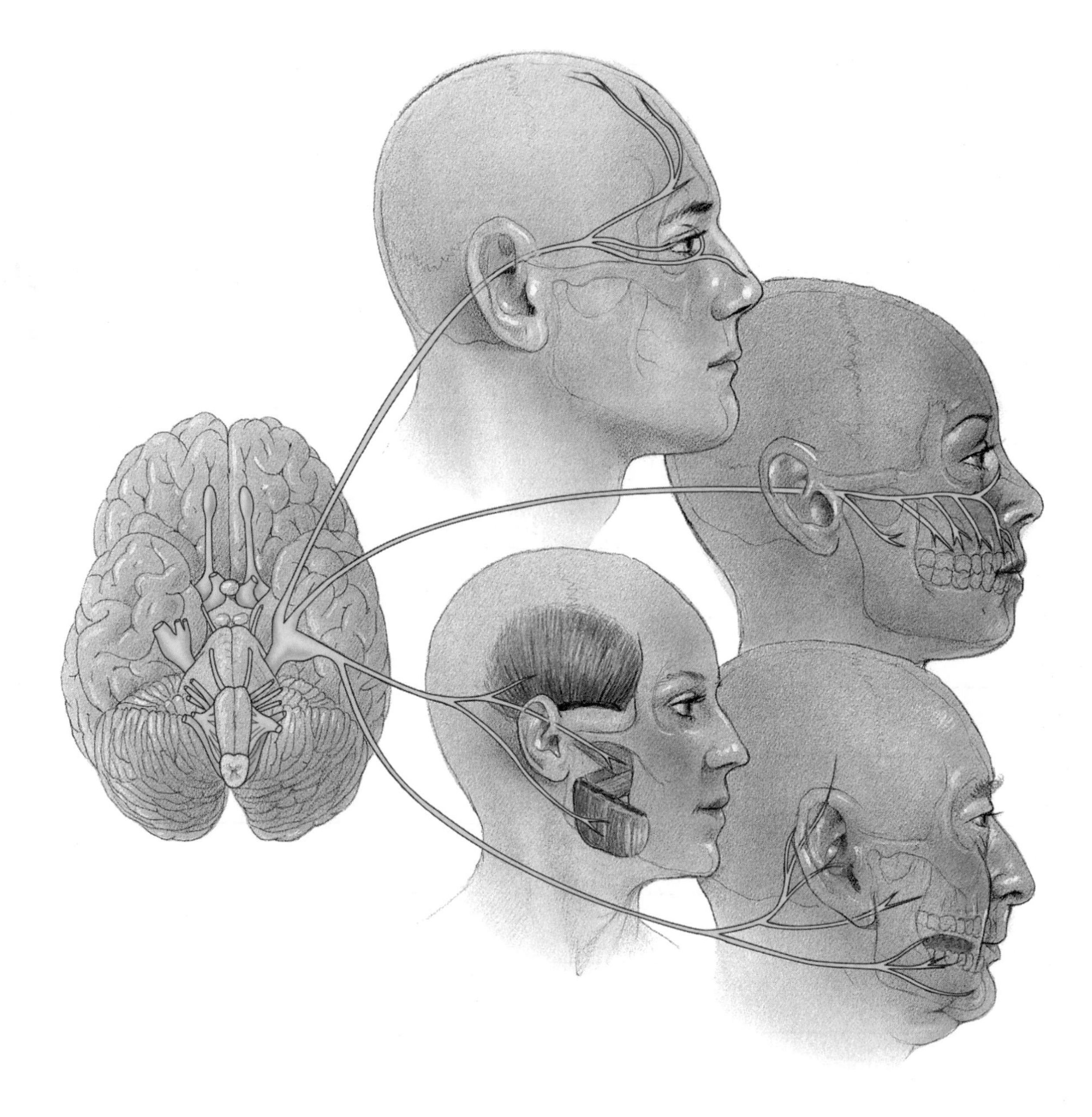

NOTES

18

**Figure 18.19** Facial (VII) nerve (page 575).

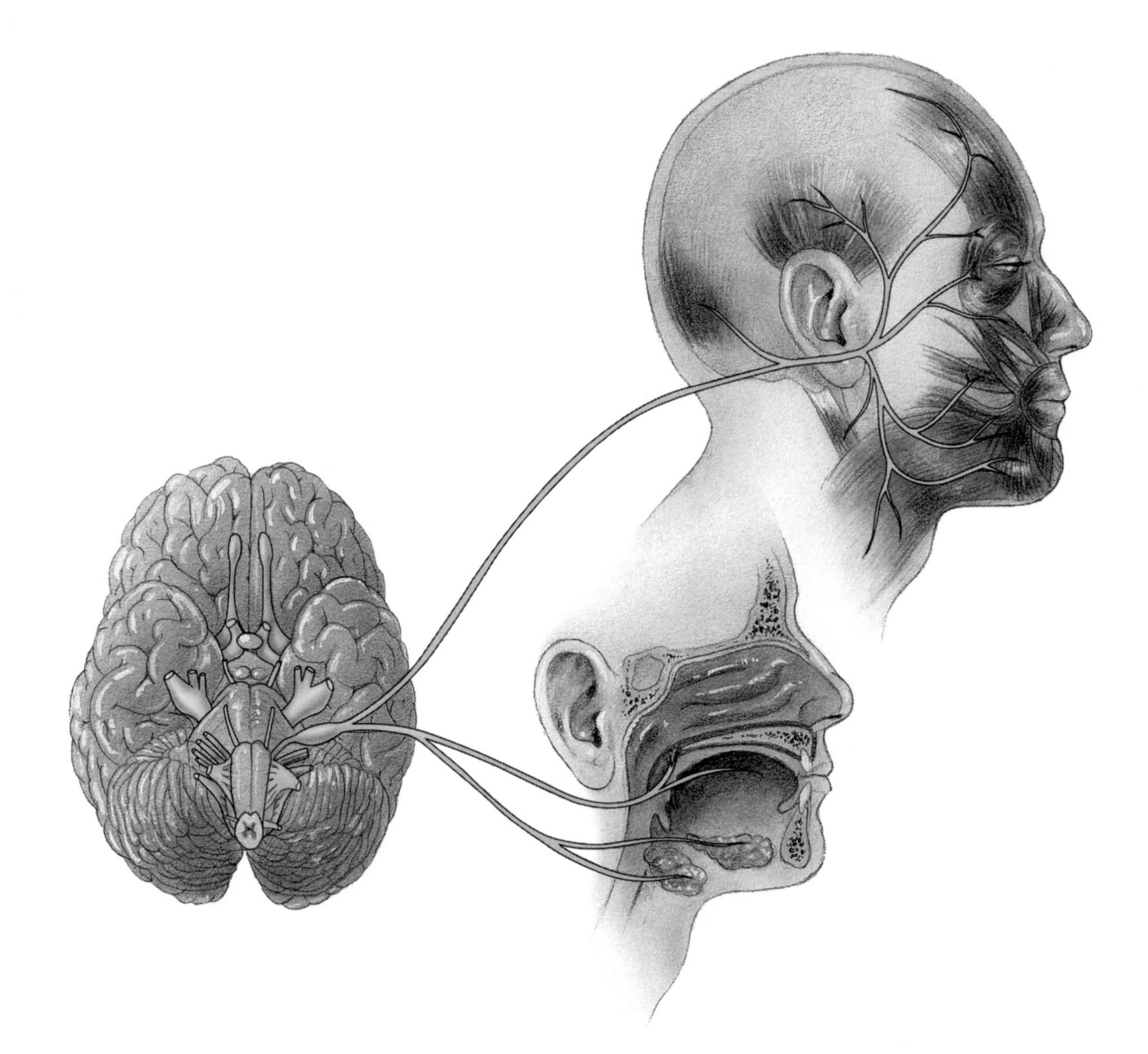

NOTES

18

**Figure 18.20** Glossopharyngeal (IX) nerve (page 576).

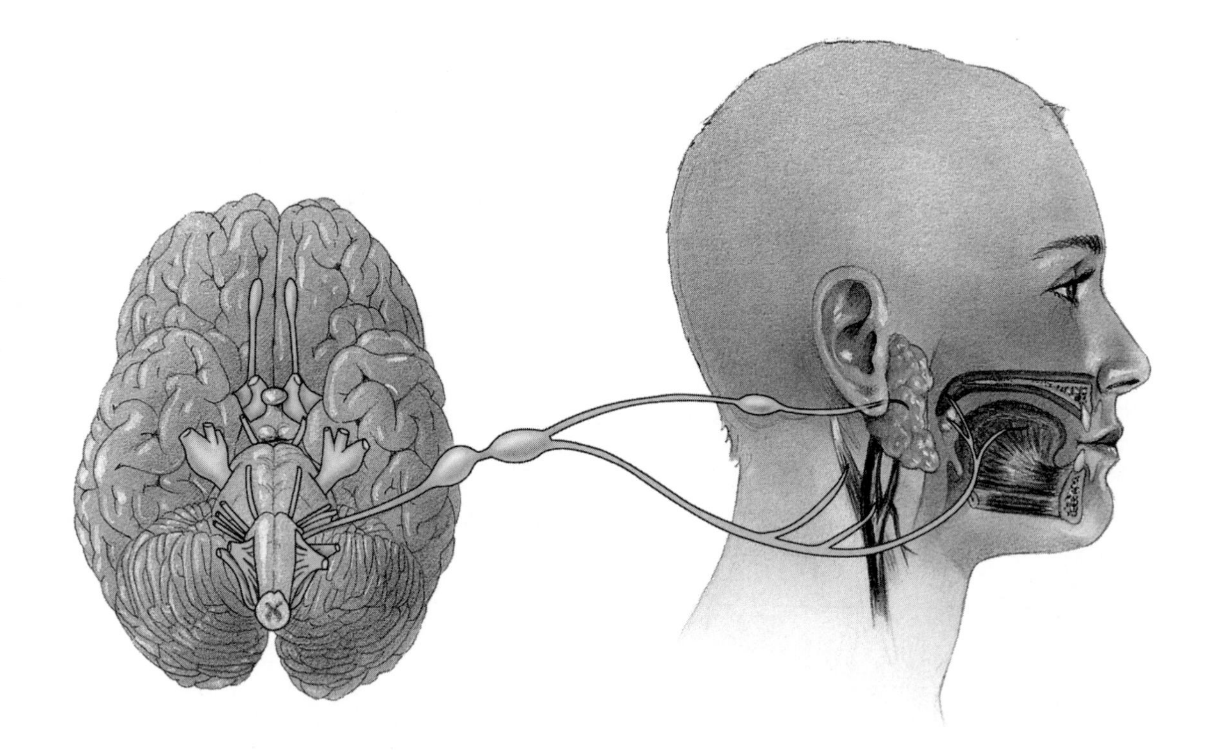

NOTES

18

**Figure 18.21** Vagus (X) nerve (page 577).

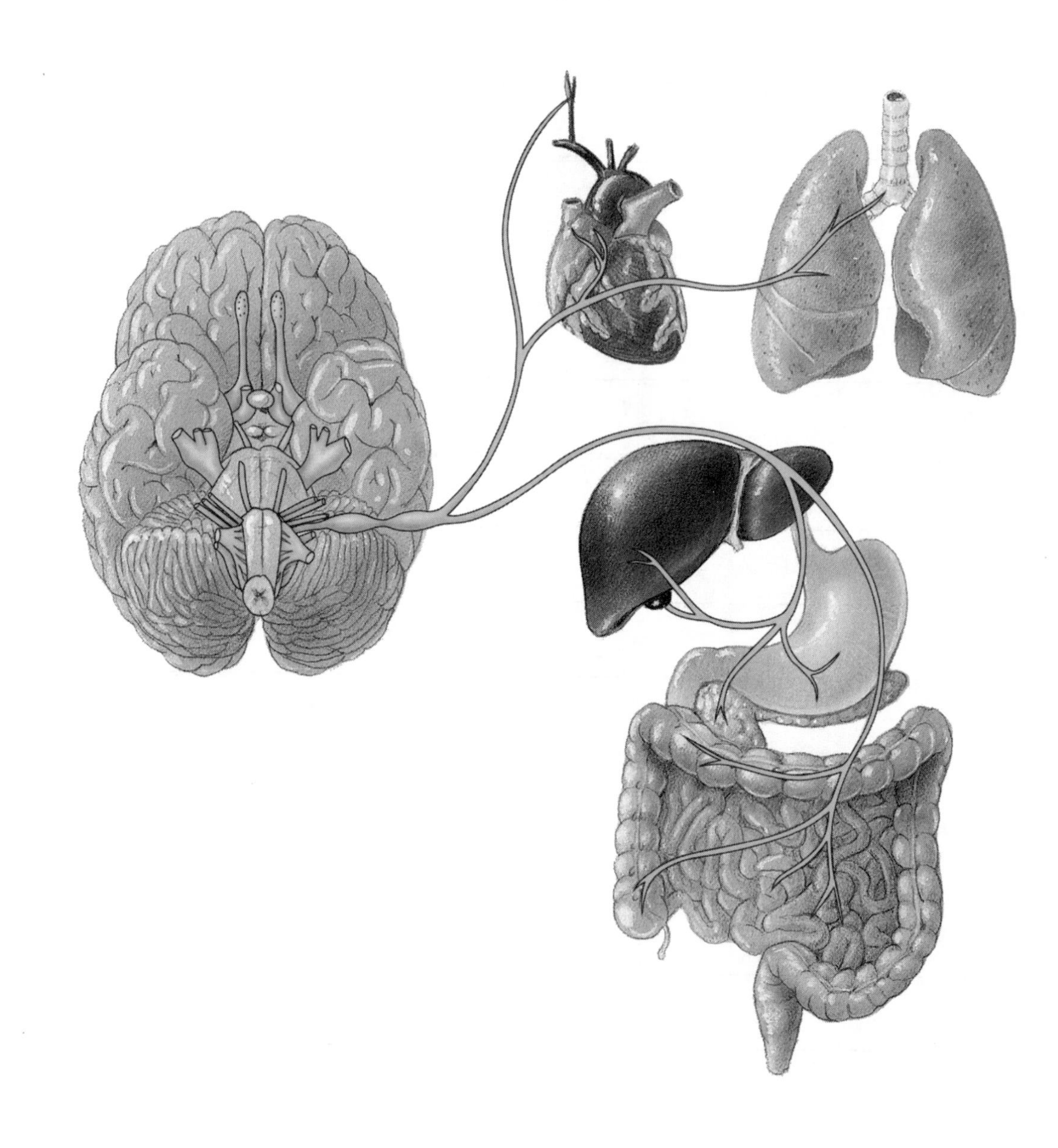

NOTES

18

**Figure 18.22** Accessory (XI) nerve (page 578).

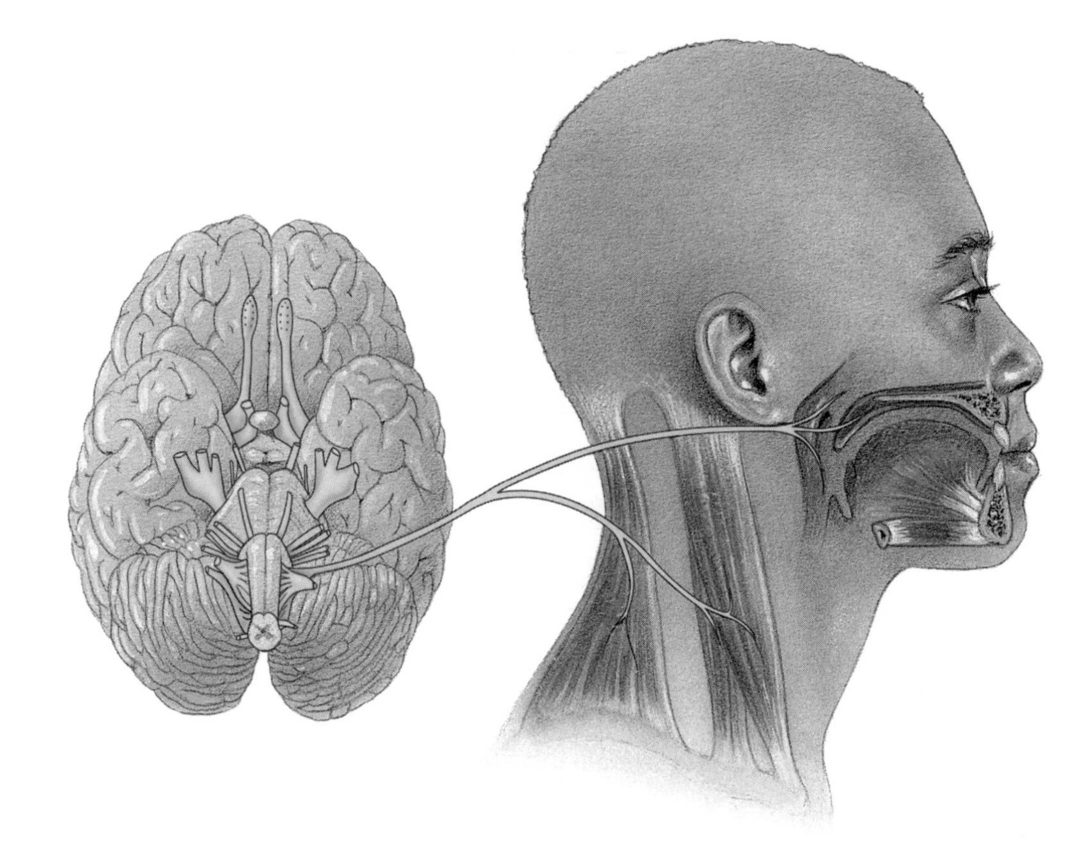

**Figure 18.23** Hypoglossal (XII) nerve (page 578).

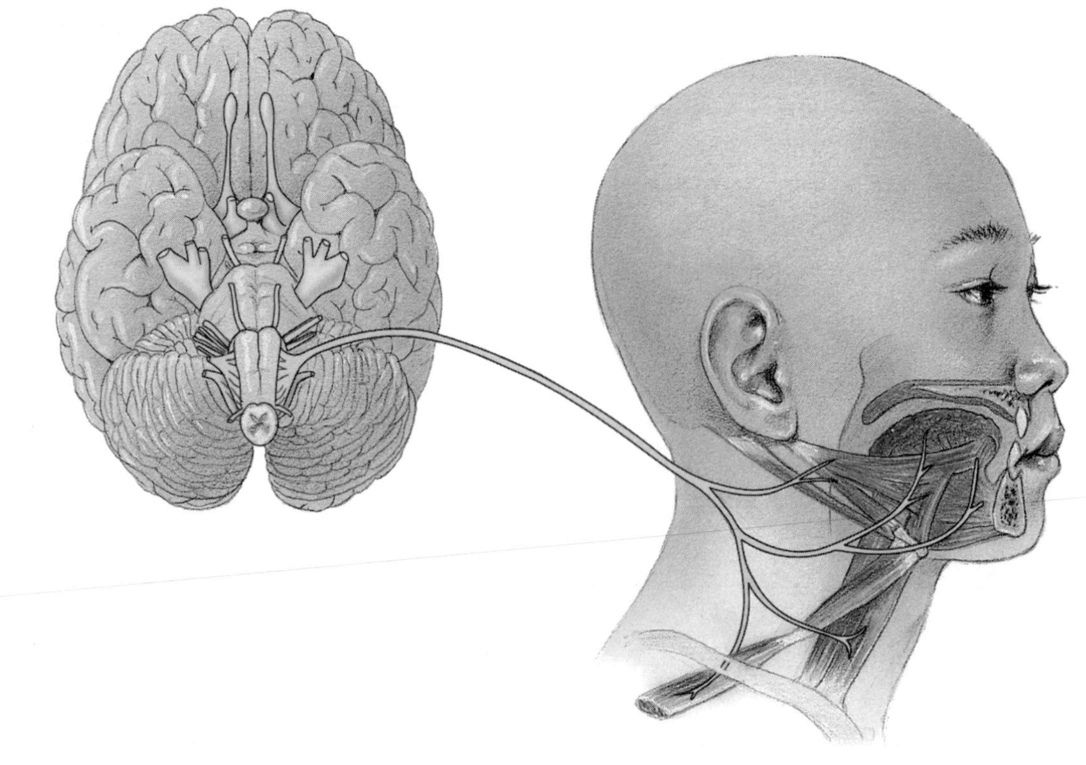

NOTES

18

**Figure 18.24** Origin of the nervous system (page 583).

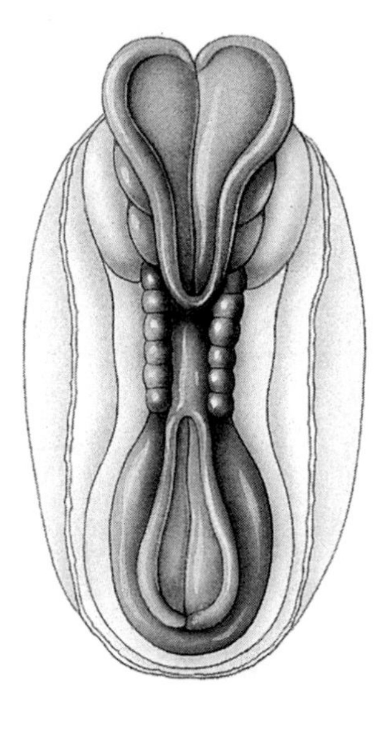

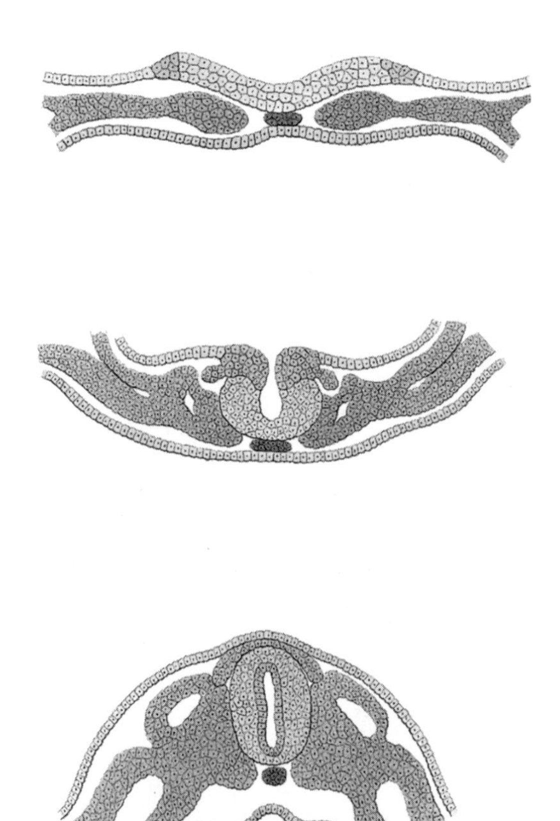

NOTES

18

**Figure 18.25** Development of the brain and spinal cord (page 584).

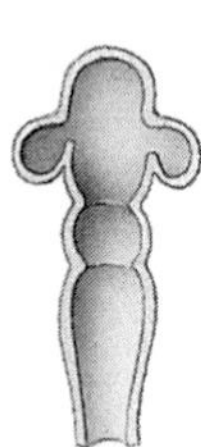

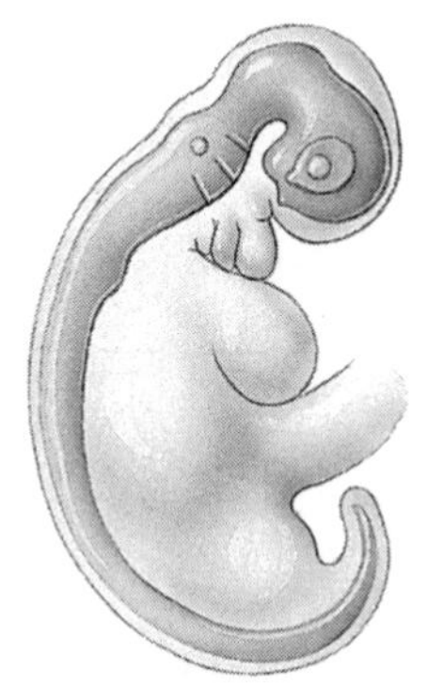

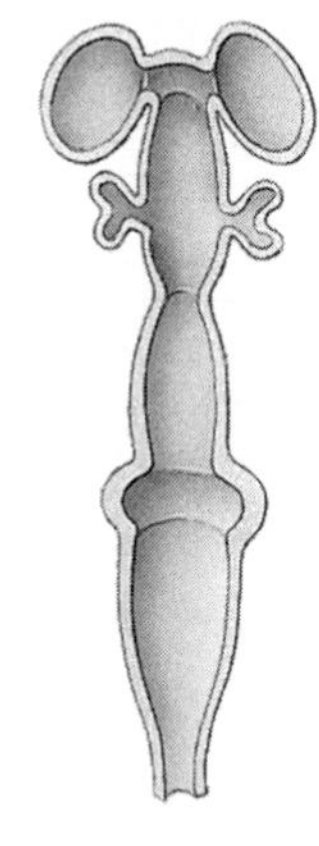

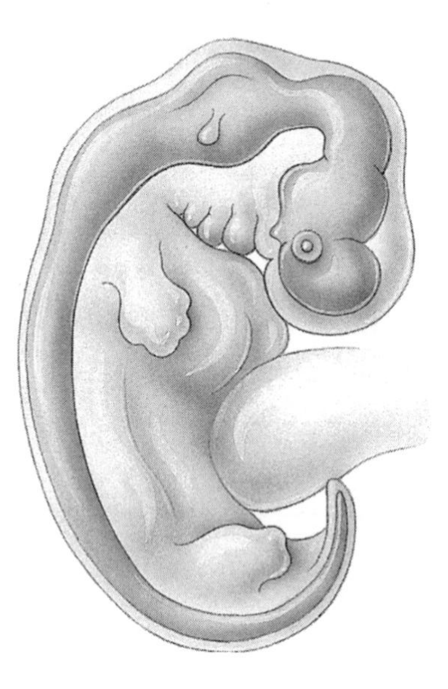

NOTES

18

**Figure 19.1** Structure and location of sensory receptors in the skin (page 593).

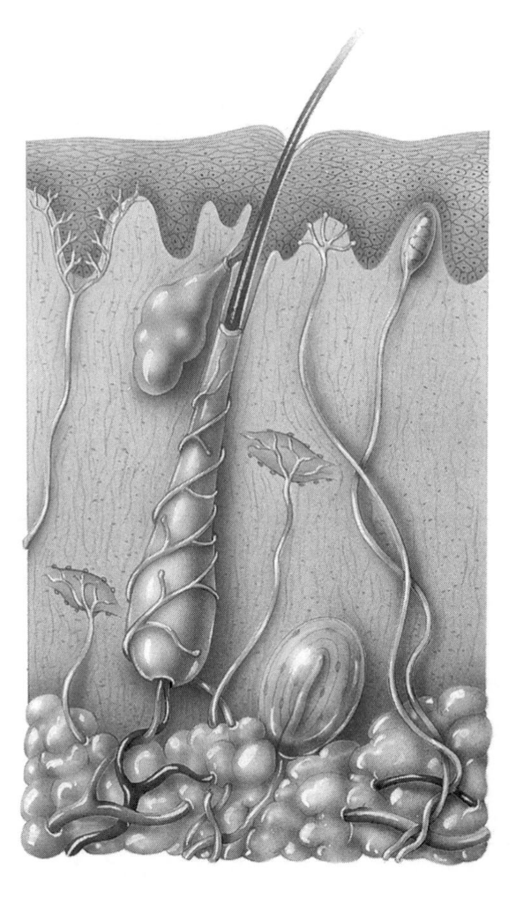

**Figure 19.2** Distribution of referred pain (page 594).

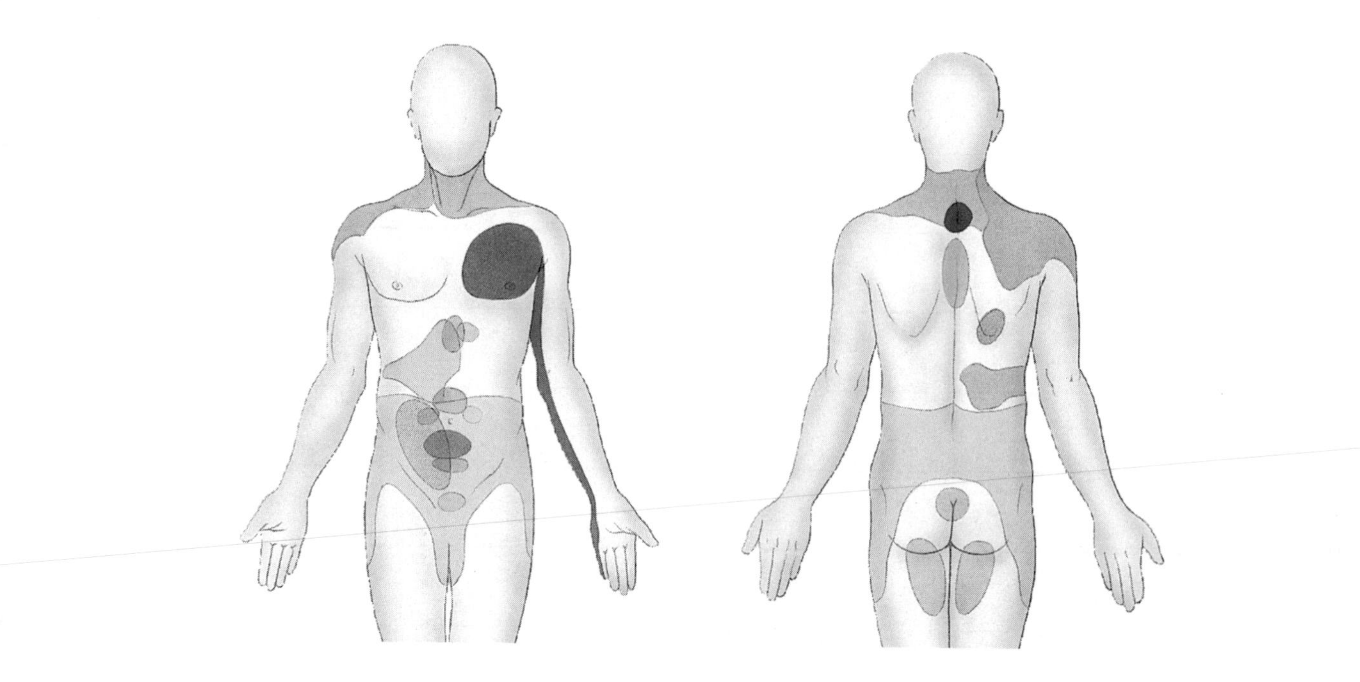

NOTES

19

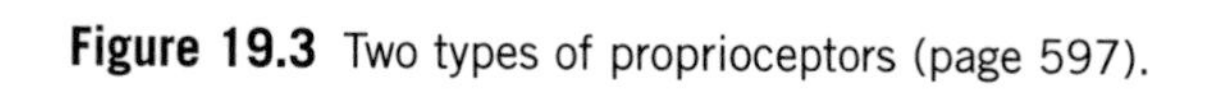
**Figure 19.3** Two types of proprioceptors (page 597).

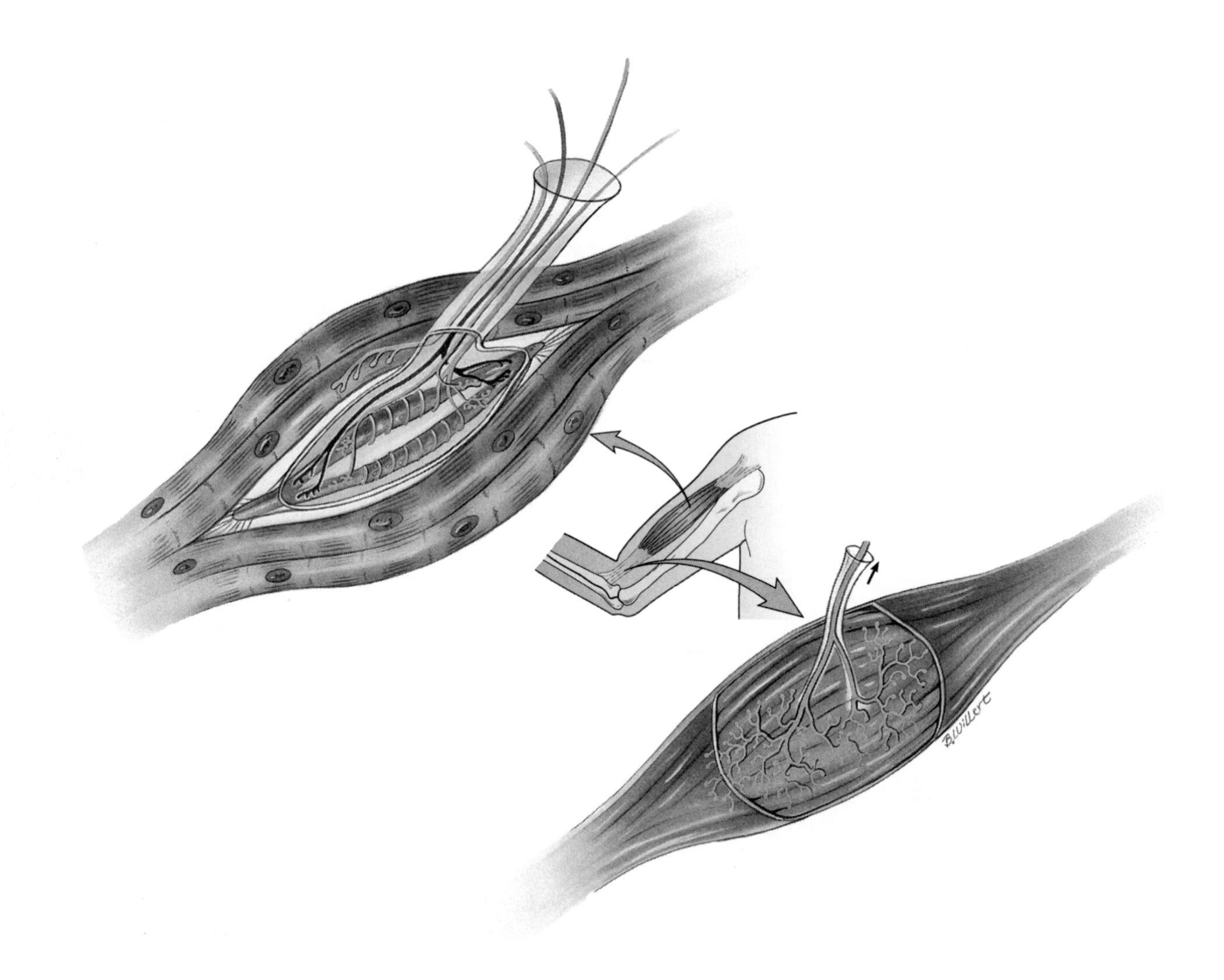

NOTES

19

**Figure 19.4** Somatic sensory pathways (page 599).

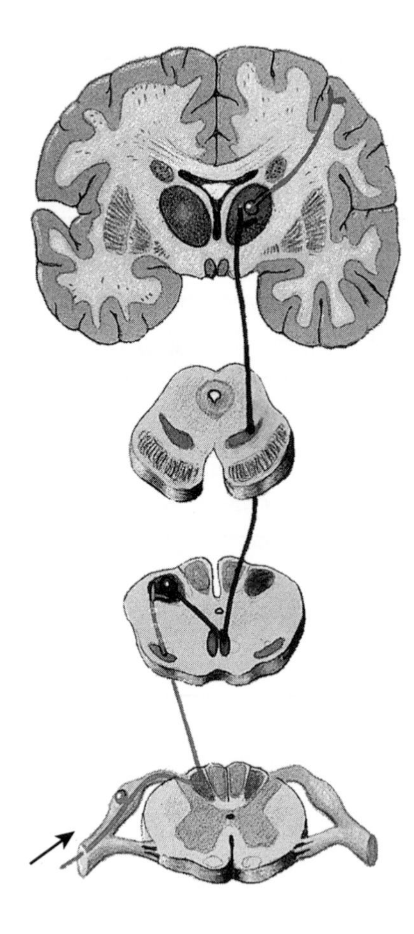

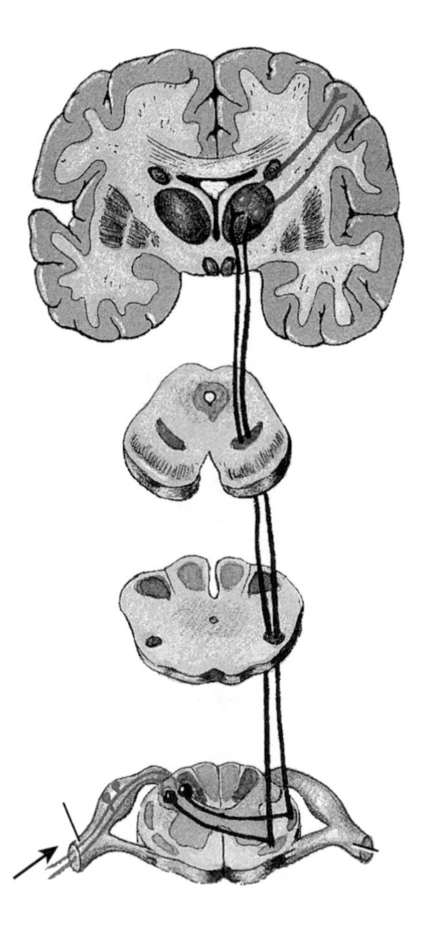

NOTES

19

**Figure 19.5** Primary somatosensory area and primary motor area (page 601).

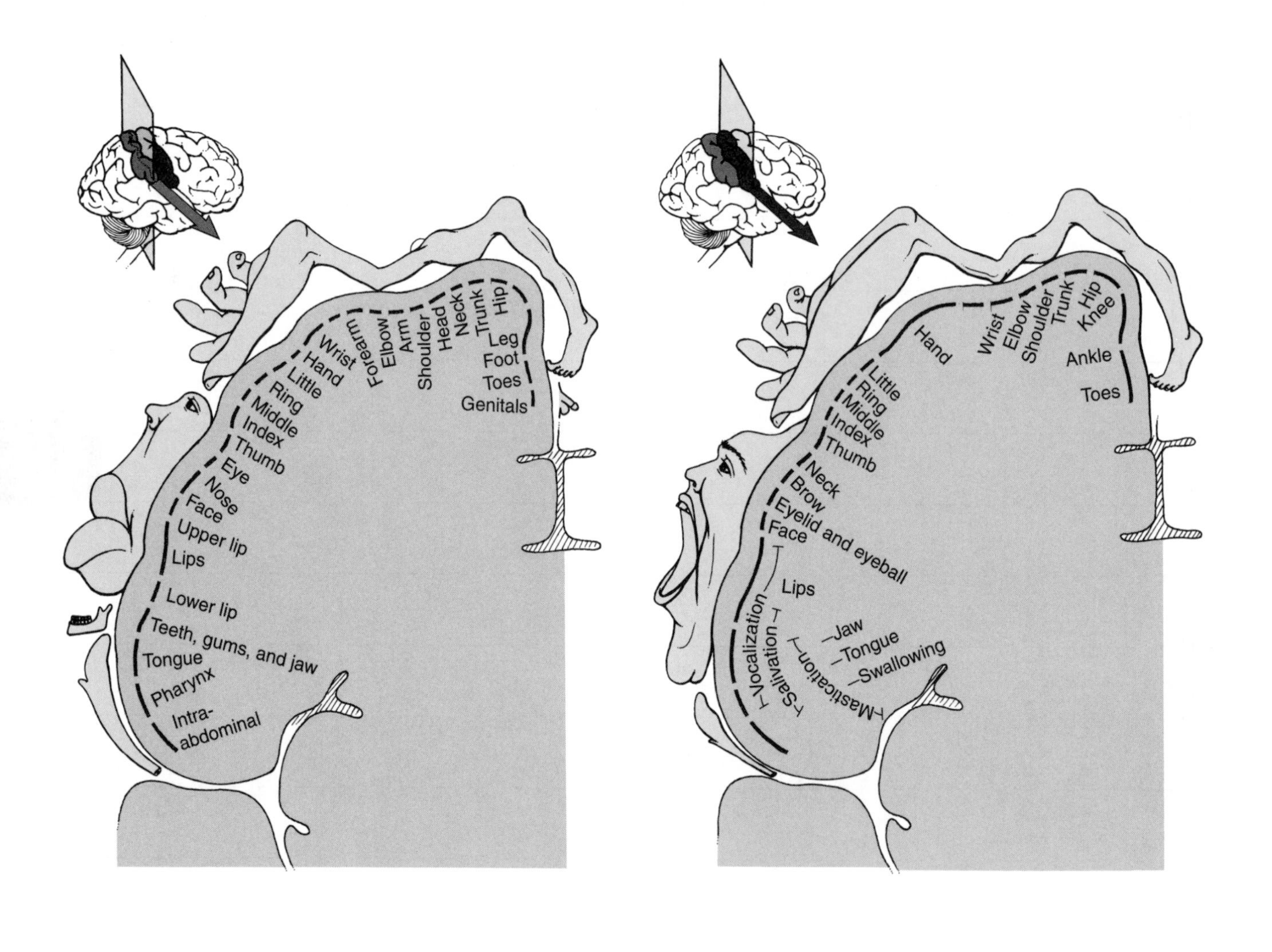

NOTES

19

**Figure 19.6** Direct motor pathways (page 603).

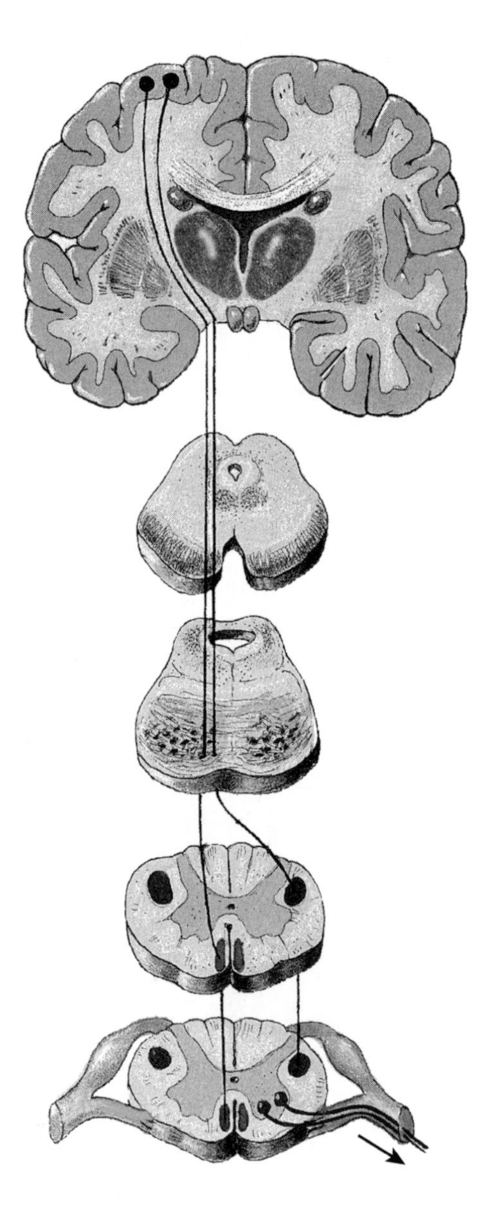

NOTES

19

**Figure 20.1a,b** Olfactory epithelium and olfactory receptors (page 612).

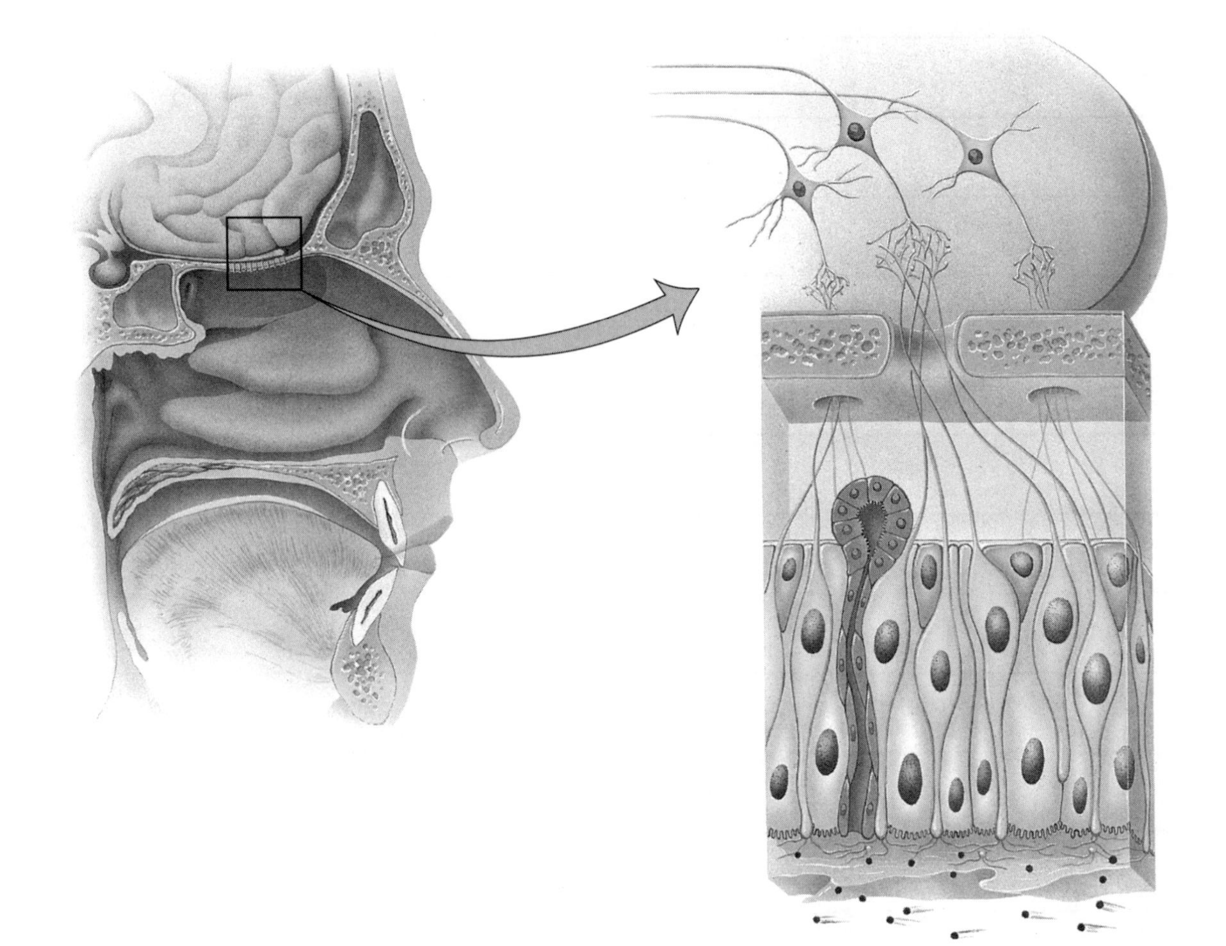

NOTES

20

**Figure 20.2a–c** The relationship of gustatory receptor cells in taste buds to tongue papillae (page 614).

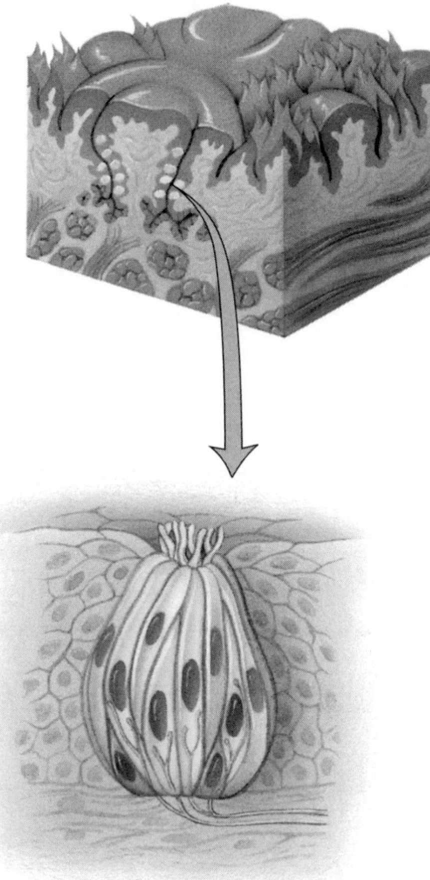

NOTES

20

**Figure 20.3** Accessory structures of the eye (page 616).

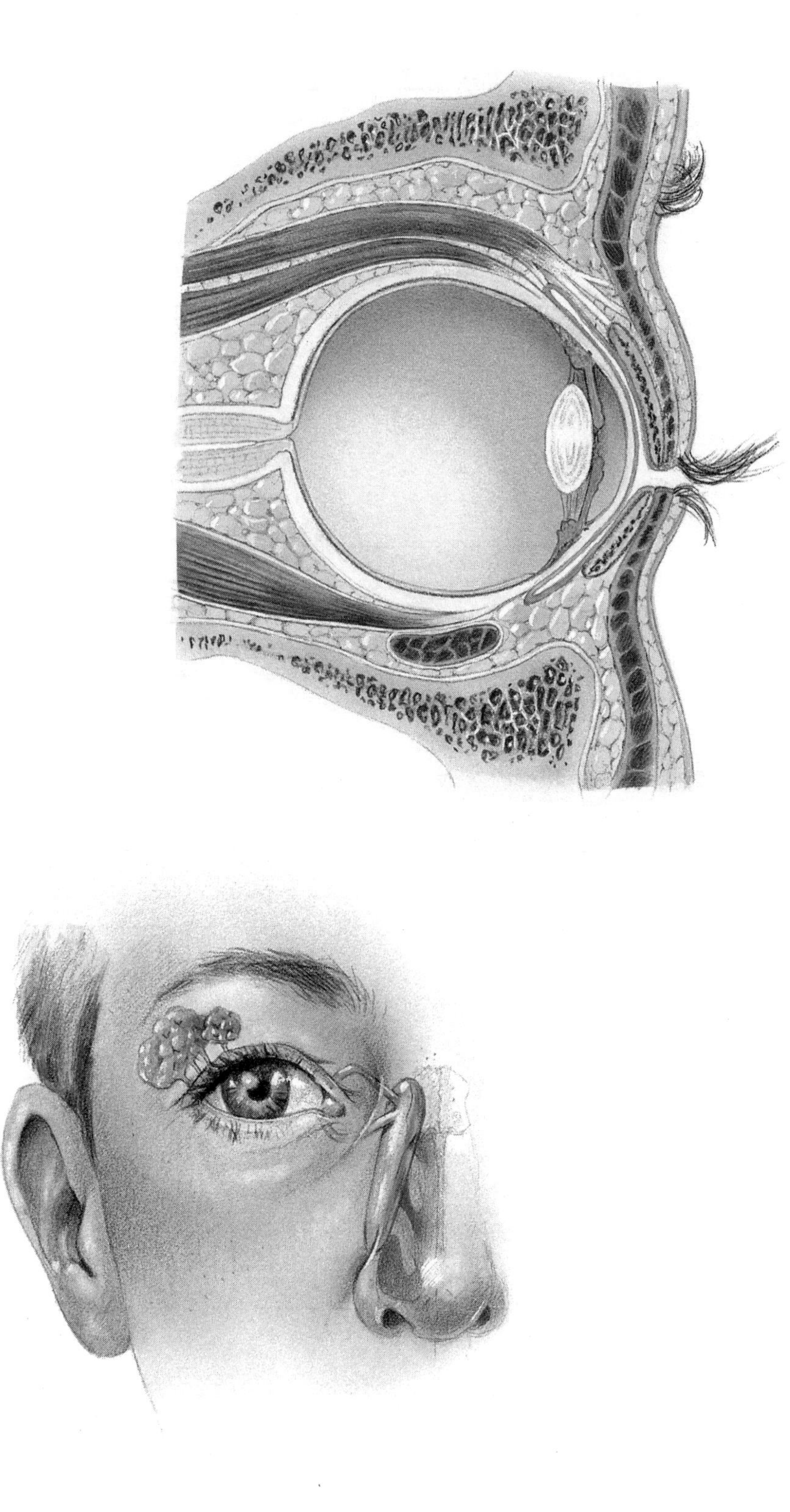

## NOTES

20

**Figure 20.4** Gross structure of the eyeball (page 617).

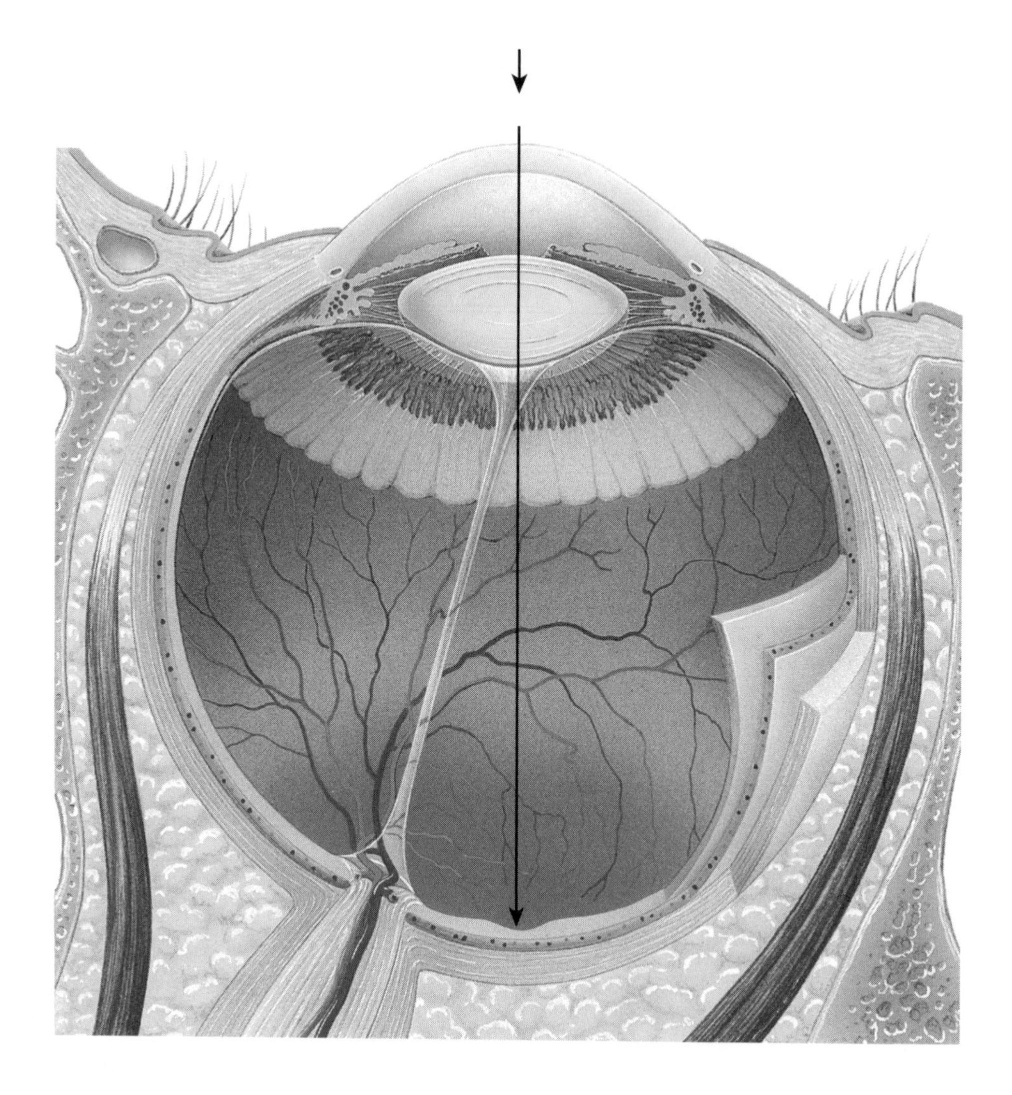

## NOTES

20

**Figure 20.5** Responses of the pupil to light of varying brightness (page 618).

**Figure 20.7a** Microscopic structure of the retina (page 619).

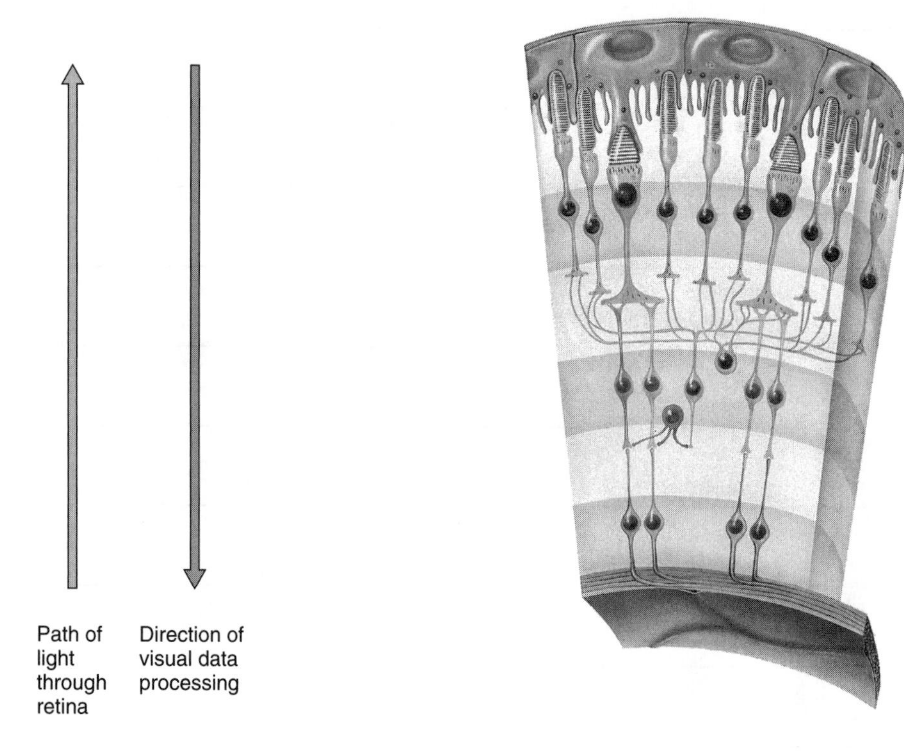

NOTES

20

**Figure 20.8** The anterior and posterior chambers of the eye (page 621).

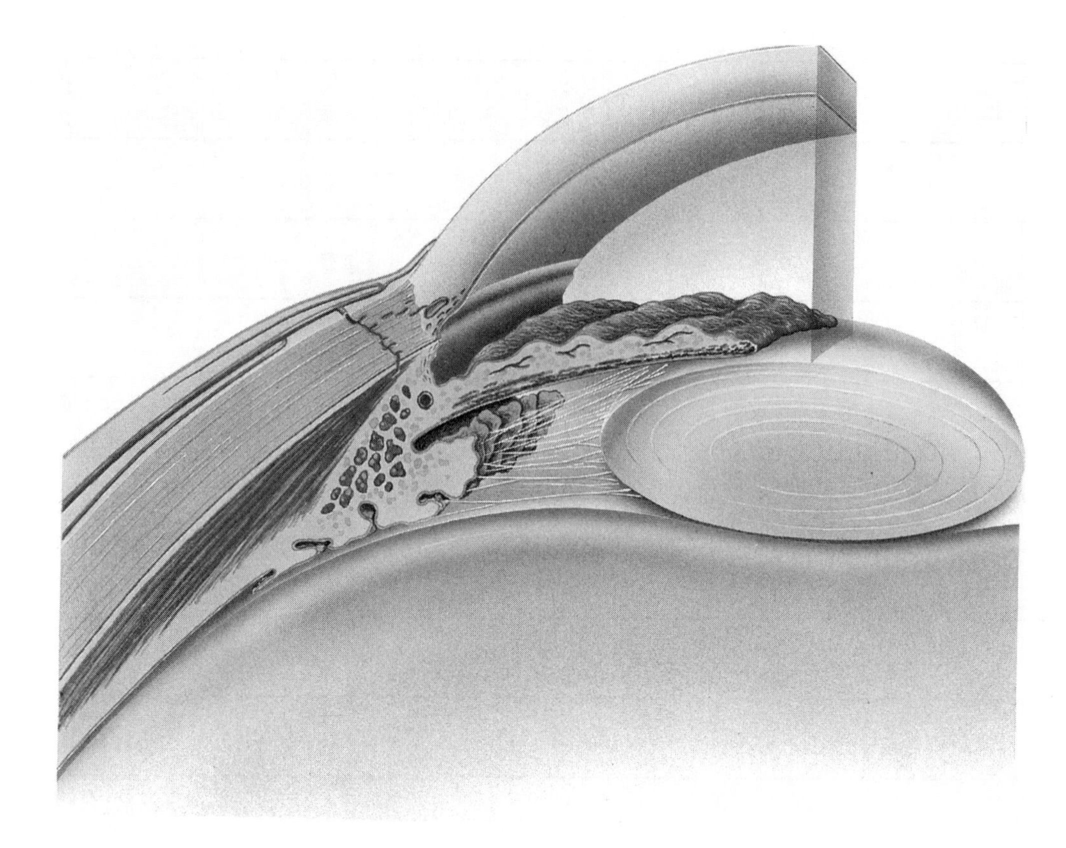

**Figure 20.9b** The visual pathway (page 622).

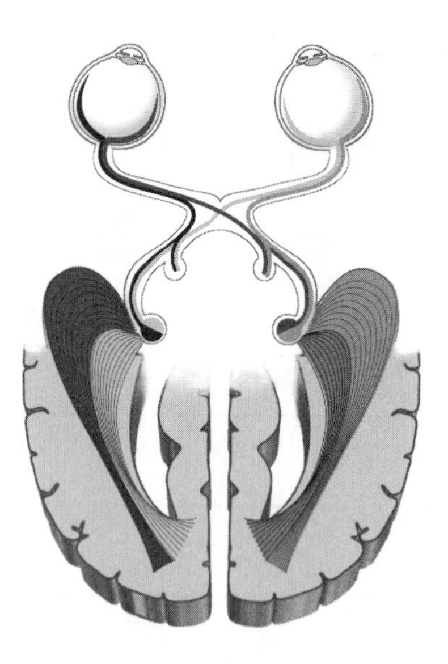

NOTES

20

**Figure 20.10** Structure of the ear (page 623).

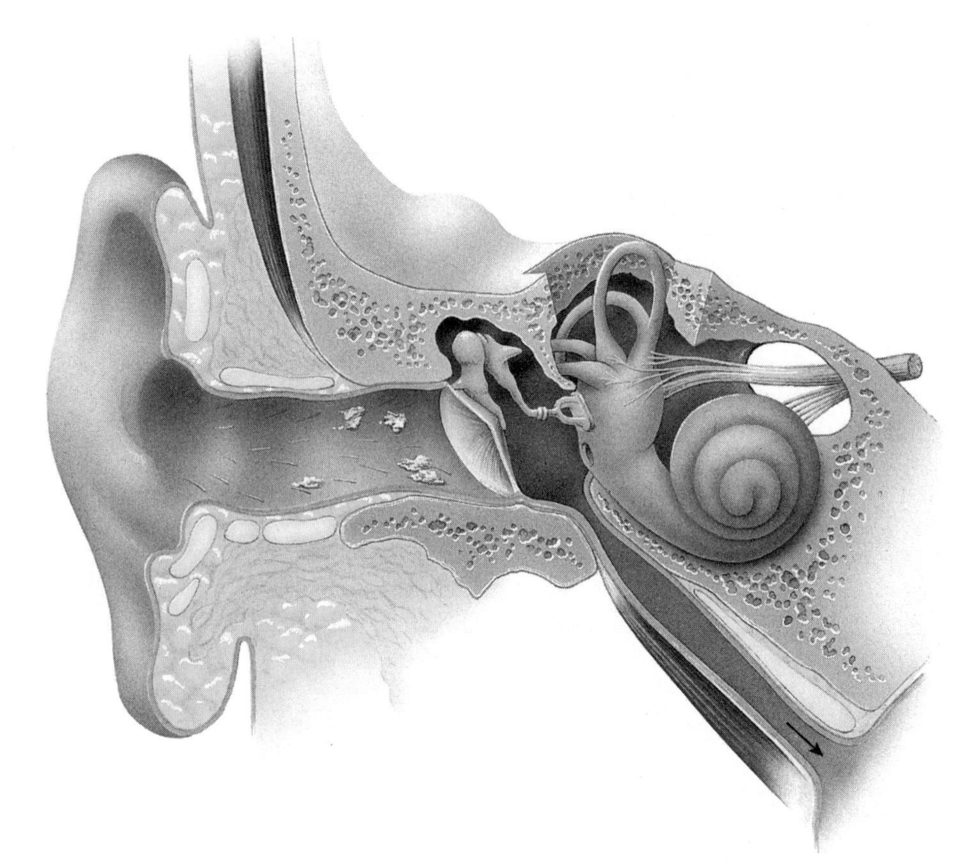

**Figure 20.11** The right middle ear containing the auditory ossicles (page 625).

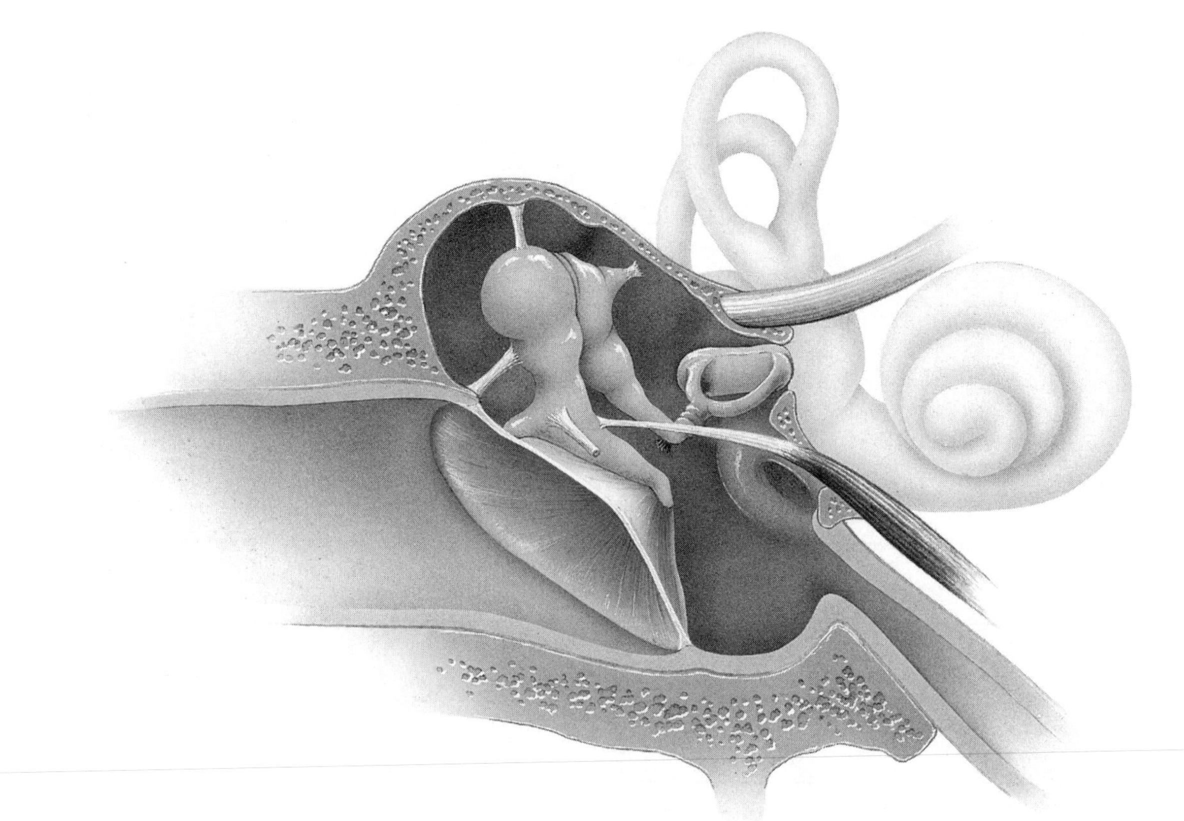

## NOTES

20

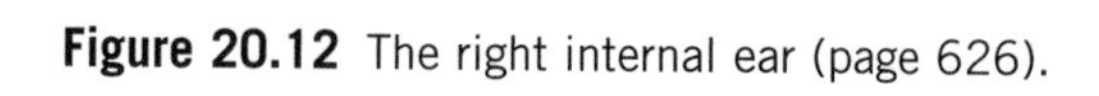

**Figure 20.12** The right internal ear (page 626).

**Figure 20.13a** Semicircular canals, vestibule, and cochlea of the right ear (page 627).

## NOTES

20

**Figure 20.13b,c** Semicircular canals, vestibule, and cochlea of the right ear (page 628).

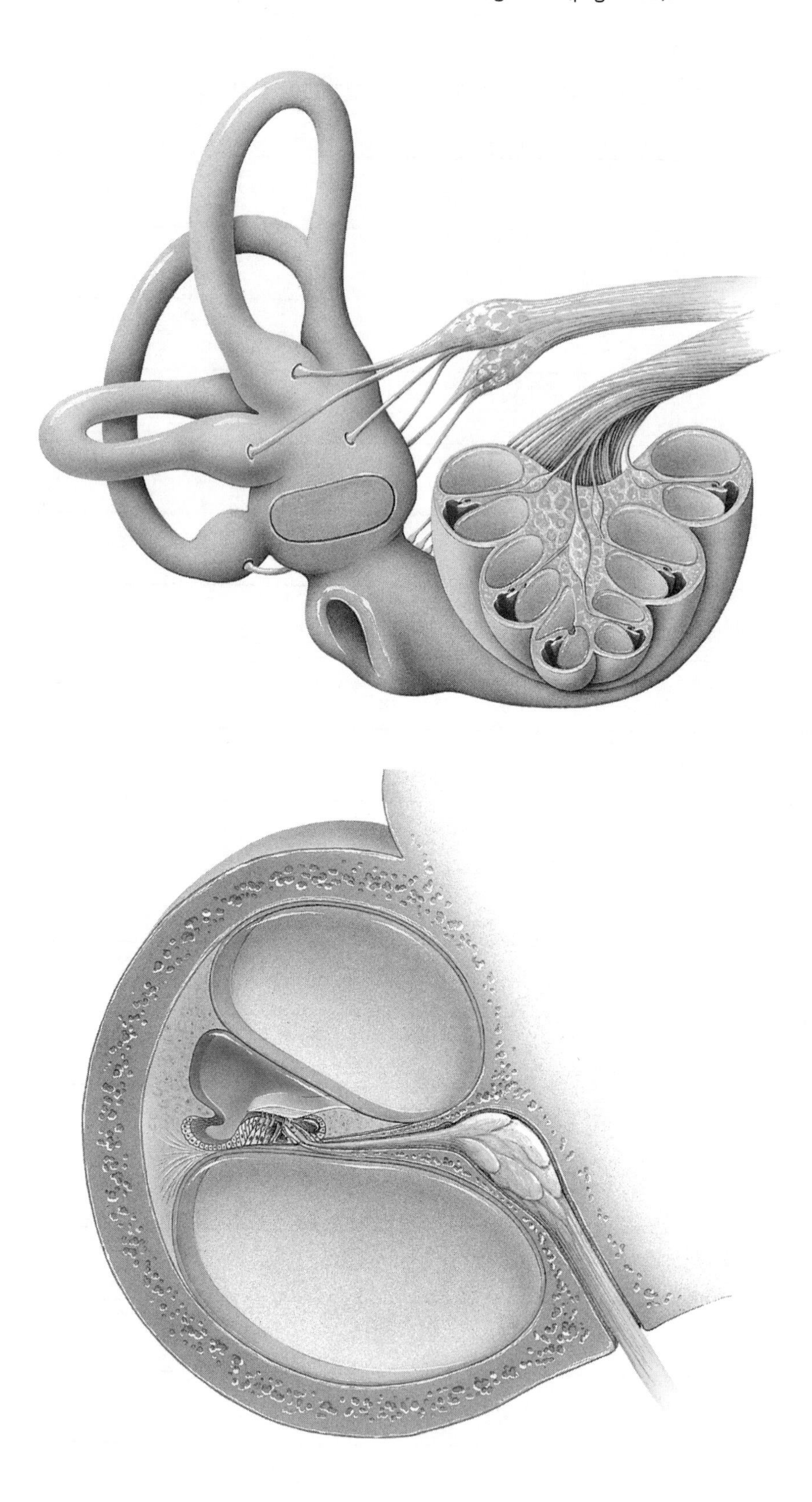

NOTES

20

**Figure 20.13d** Semicircular canals, vestibule, and cochlea of the right ear (page 629).

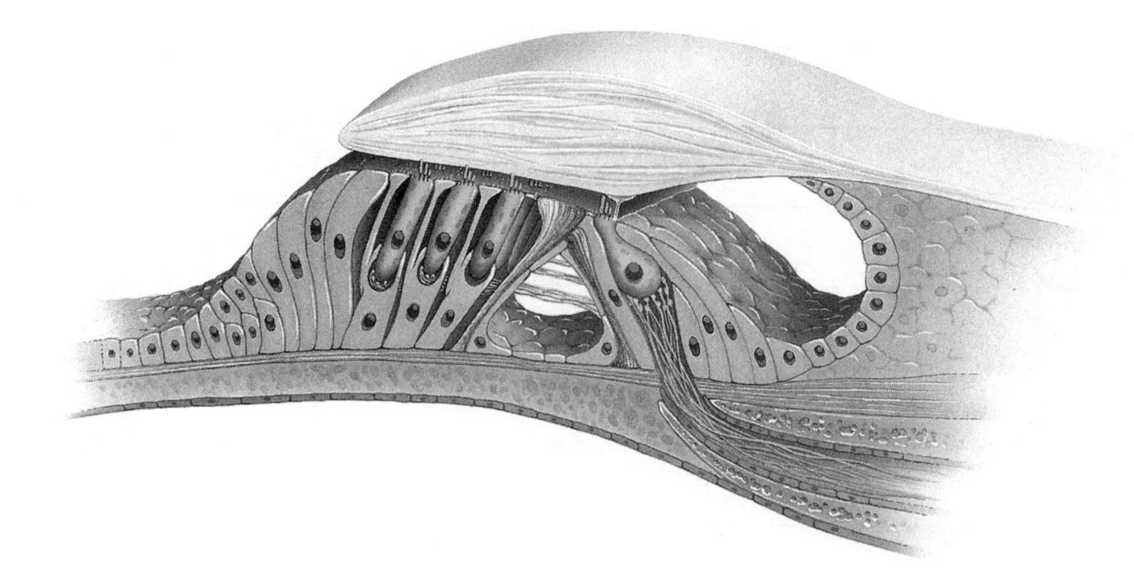

**Figure 20.14** Events in the stimulation of auditory receptors in the right ear (page 630).

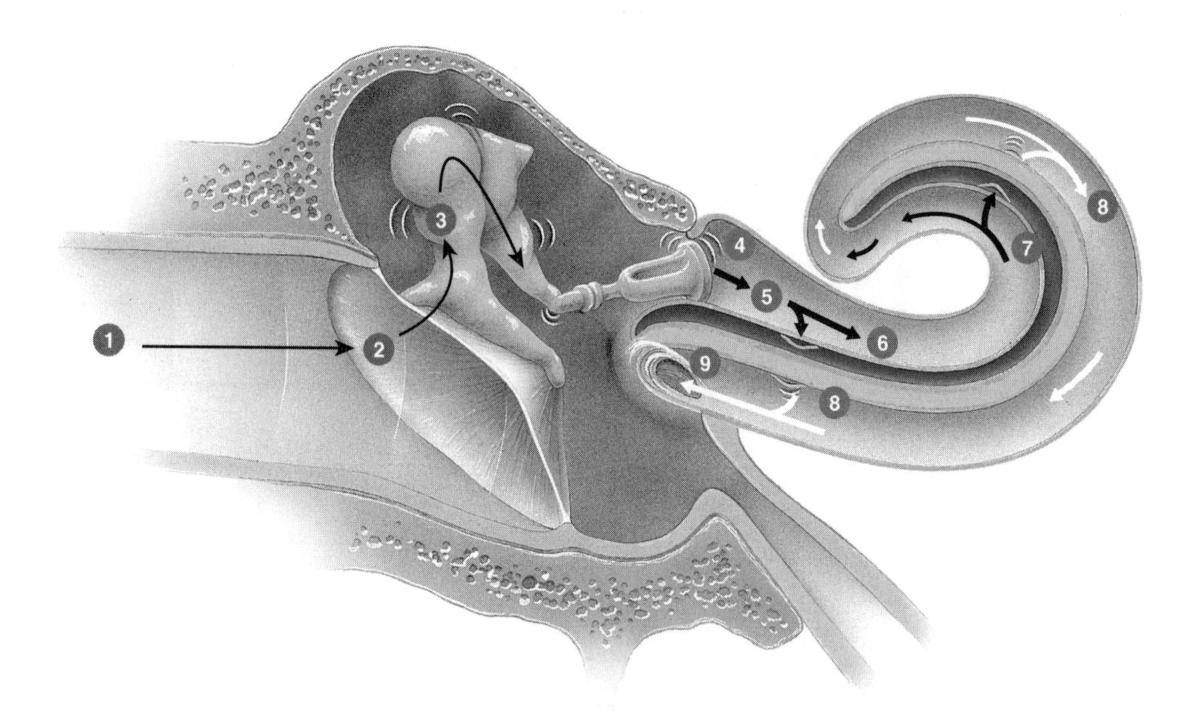

# NOTES

20

**Figure 20.15** Location and structure of receptors in the maculae of the right ear (page 632).

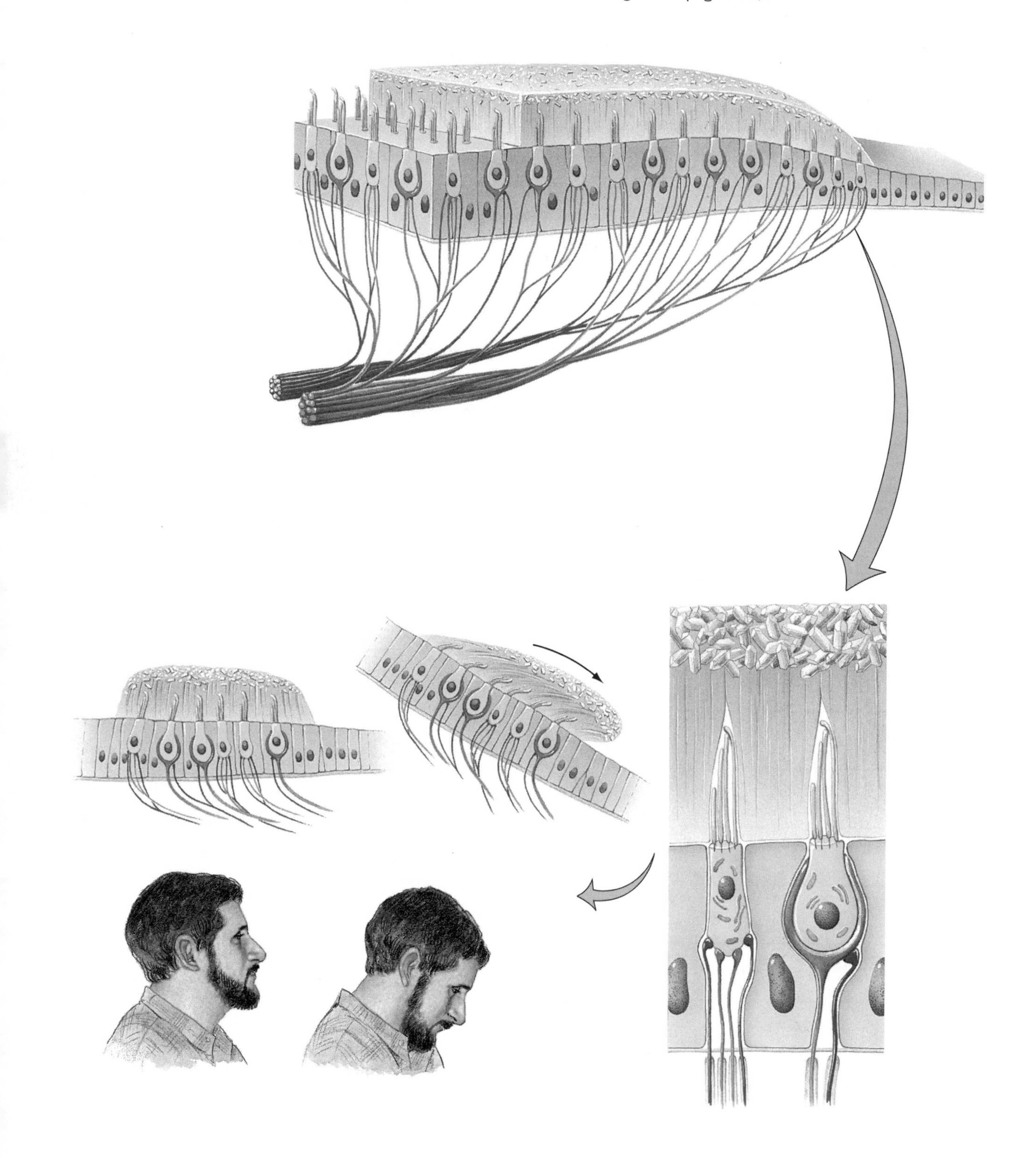

## NOTES

20

**Figure 20.16** Location and structure of the membranous semicircular ducts of the right ear (page 633).

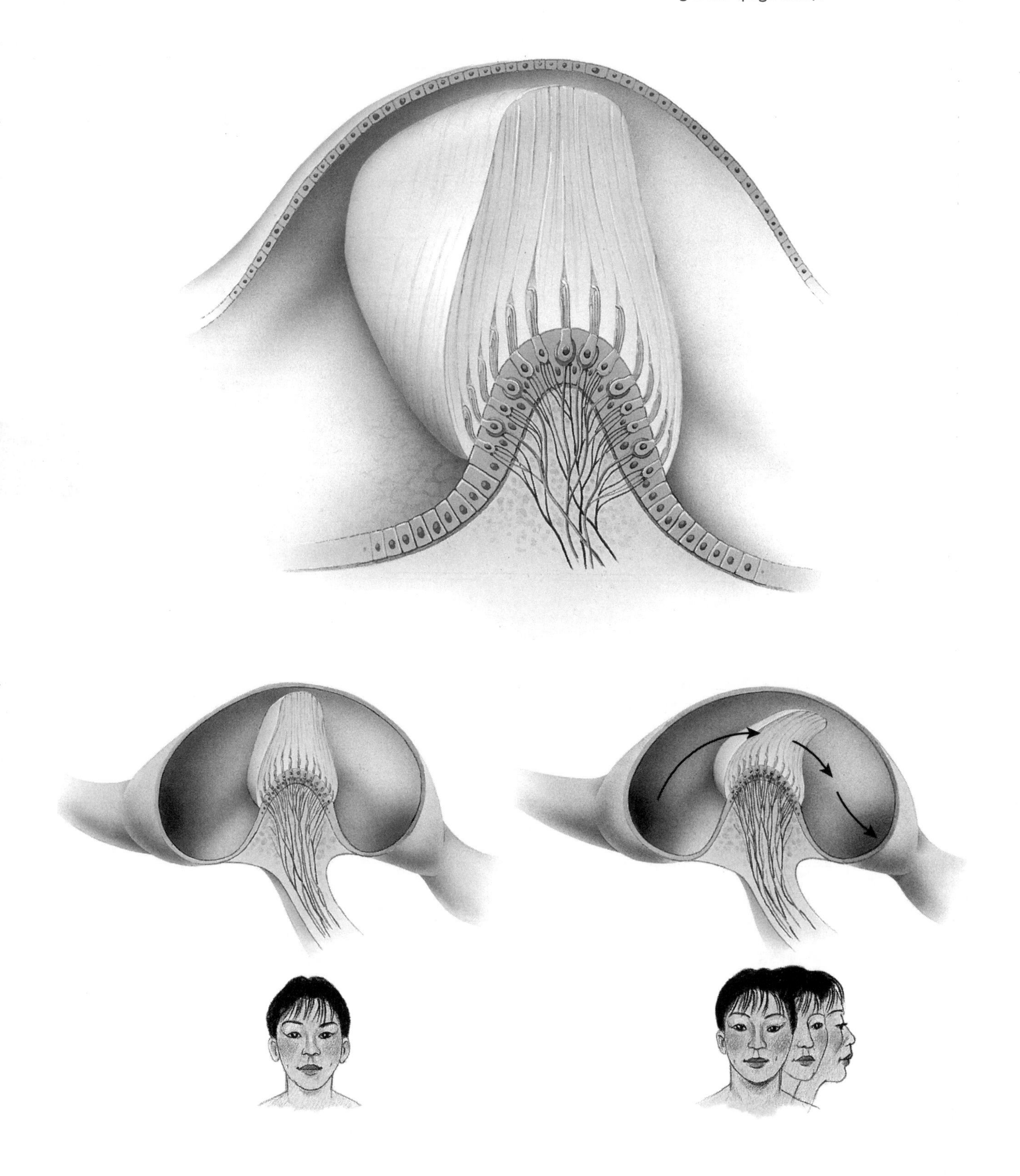

NOTES

20

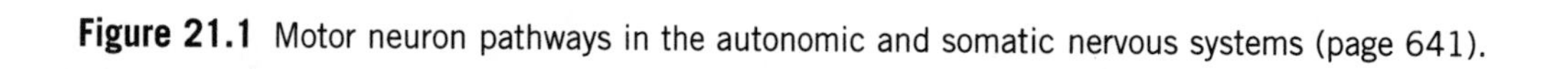

**Figure 21.1** Motor neuron pathways in the autonomic and somatic nervous systems (page 641).

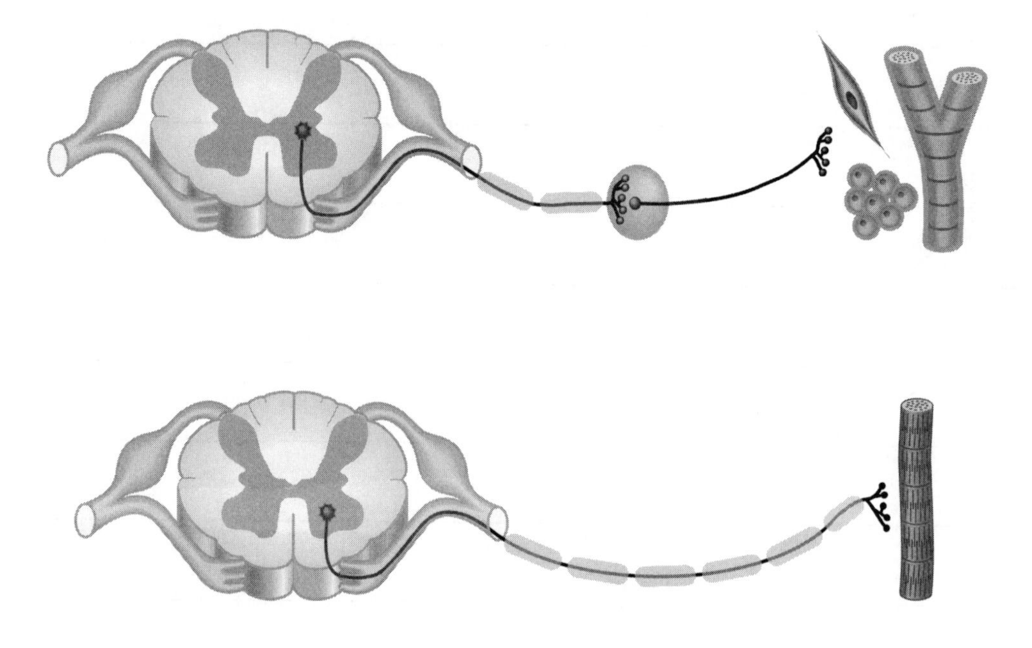

NOTES

21

**Figure 21.2** Structure of the sympathetic and parasympathetic divisions of the autonomic nervous system (page 643).

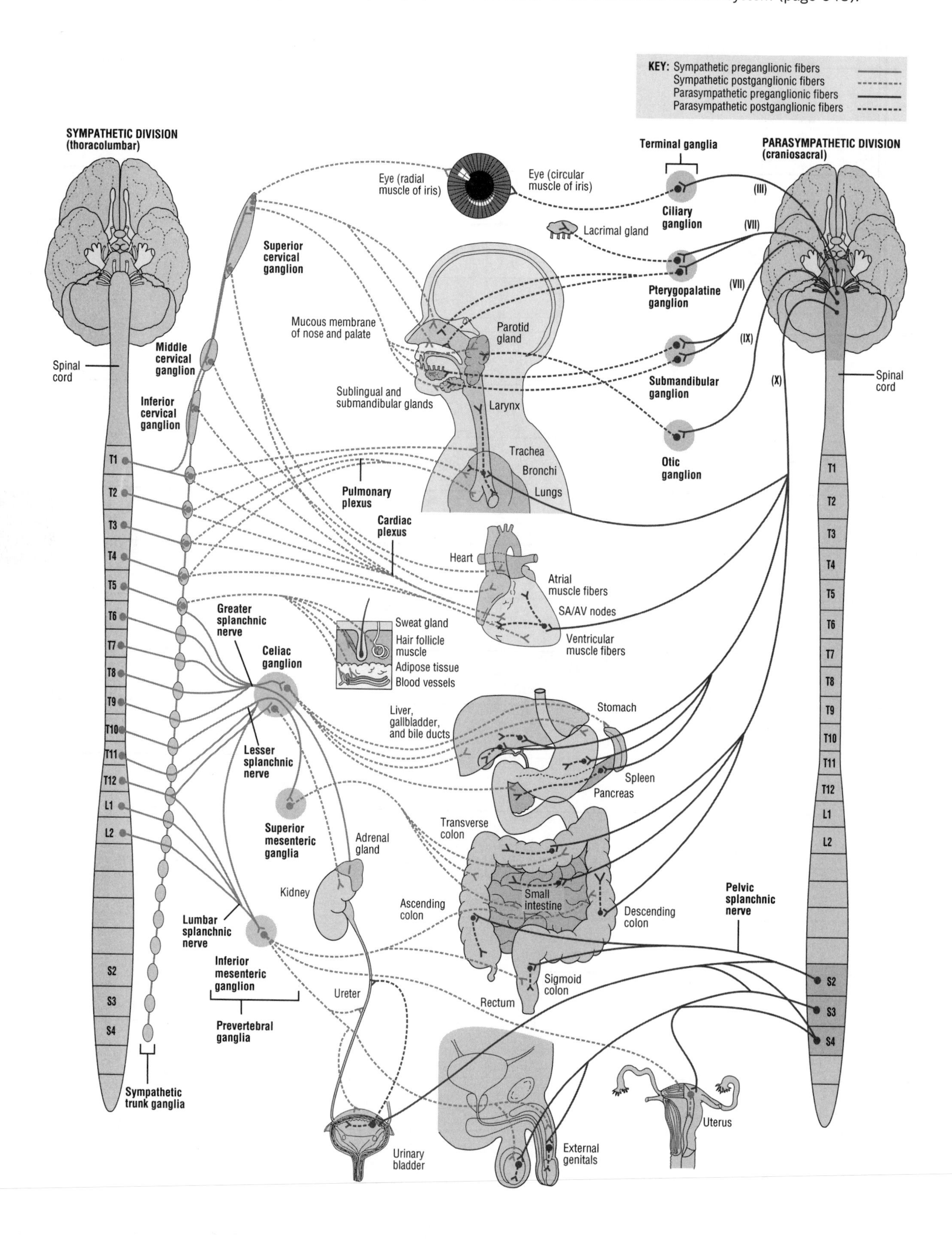

NOTES

21

**Figure 21.3** Autonomic plexuses in the thorax, abdomen and pelvis (page 644).

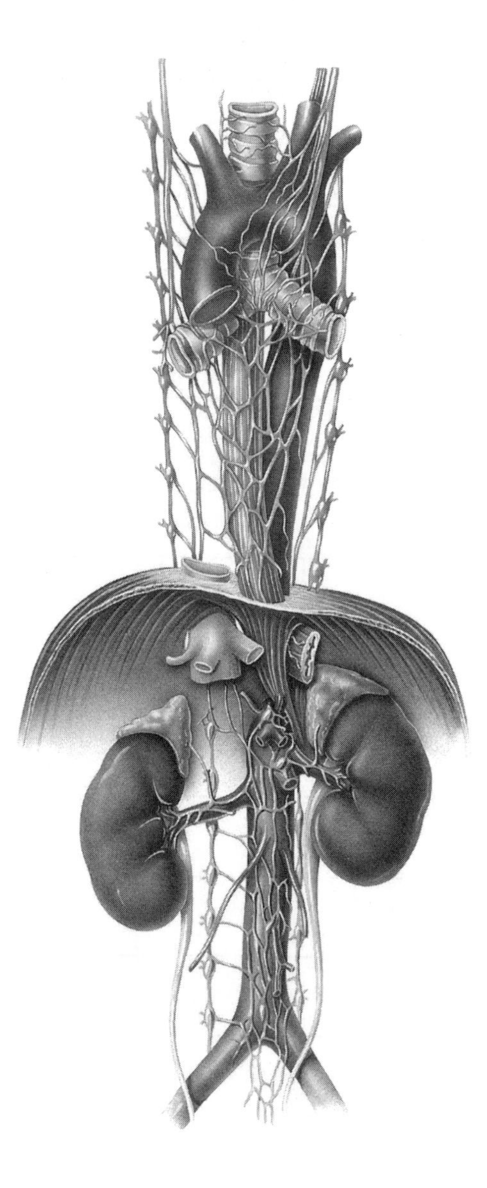

## NOTES

21

**Figure 21.4** Types of connections between ganglia and postganglionic neurons in the sympathetic division of the autonomic nervous system (page 645).

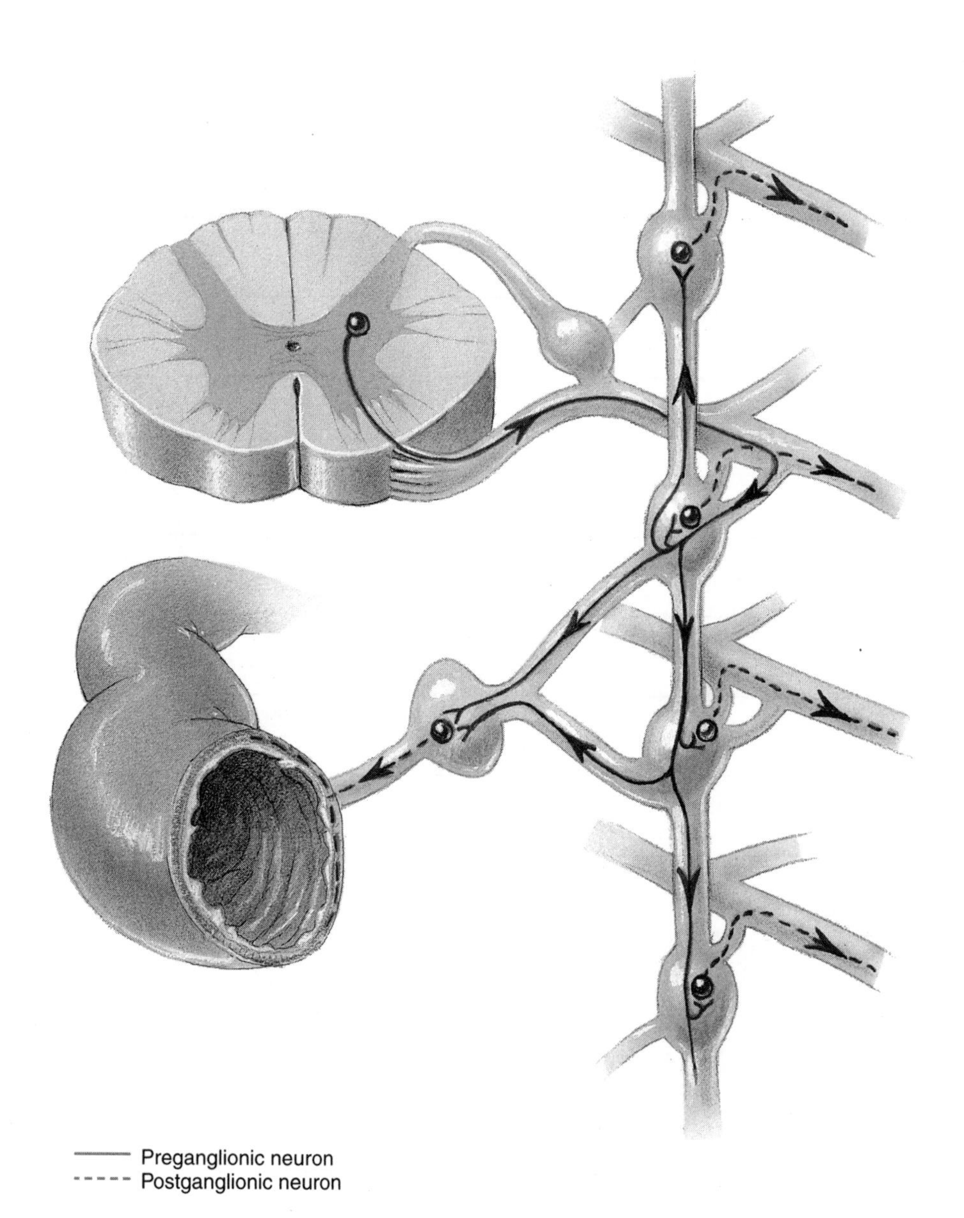

NOTES

21

**Figure 21.5** Cholinergic neurons and adrenergic neurons in the sympathetic and parasympathetic divisions (page 648).

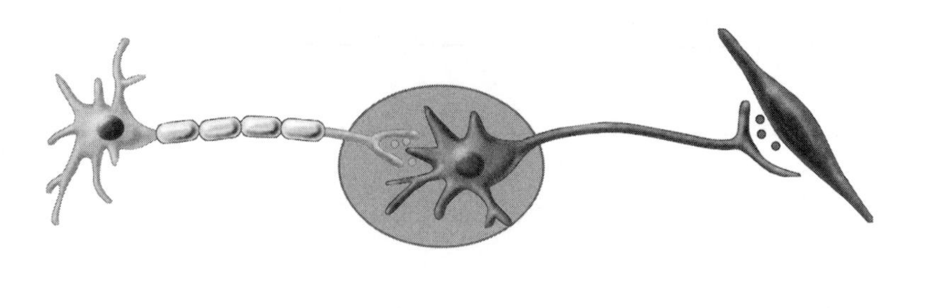

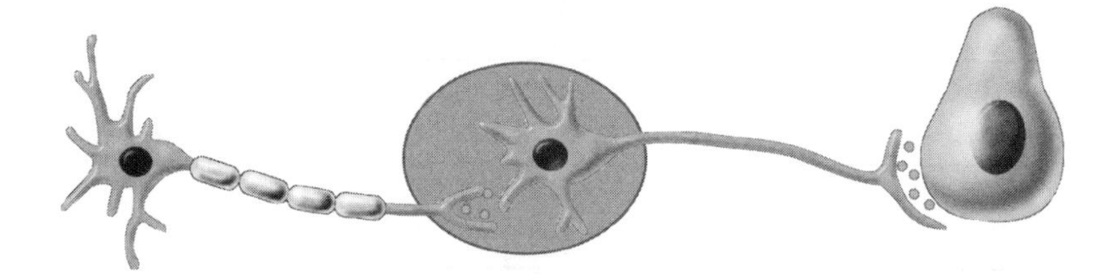

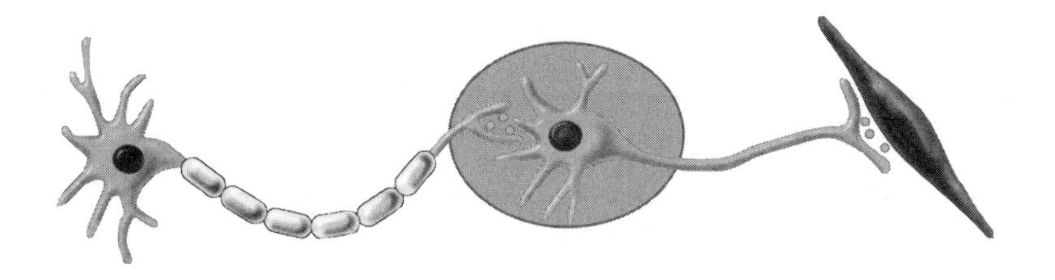

NOTES

21

...tion of many endocrine glands (page 660).

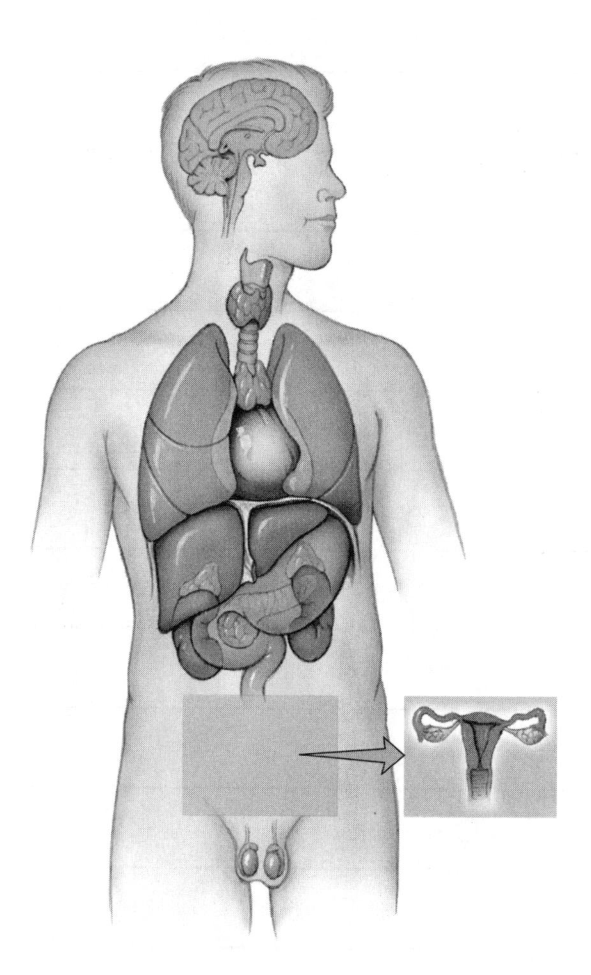

NOTES

22

and pituitary gland, and their blood supply (page 663).

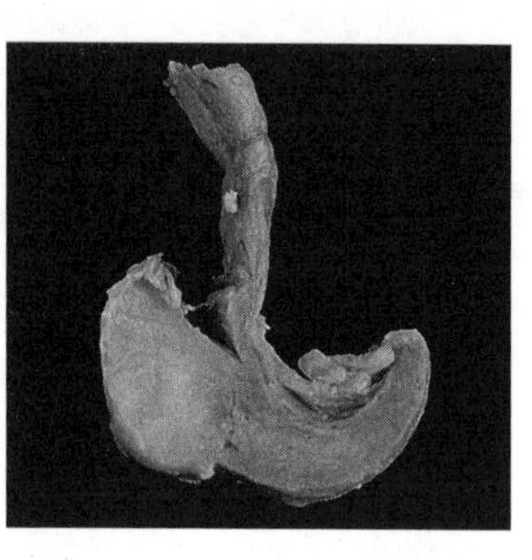

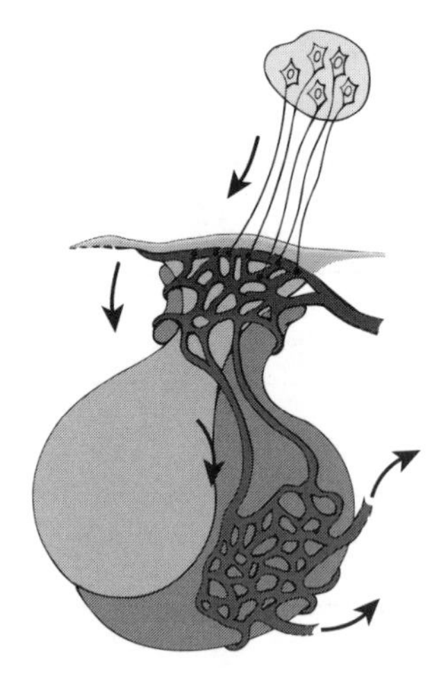

NOTES

22

**Figure 22.3** Axons of hypothalamic neurosecretory cells form the hypothalamohypophlyseal tract (page 665).

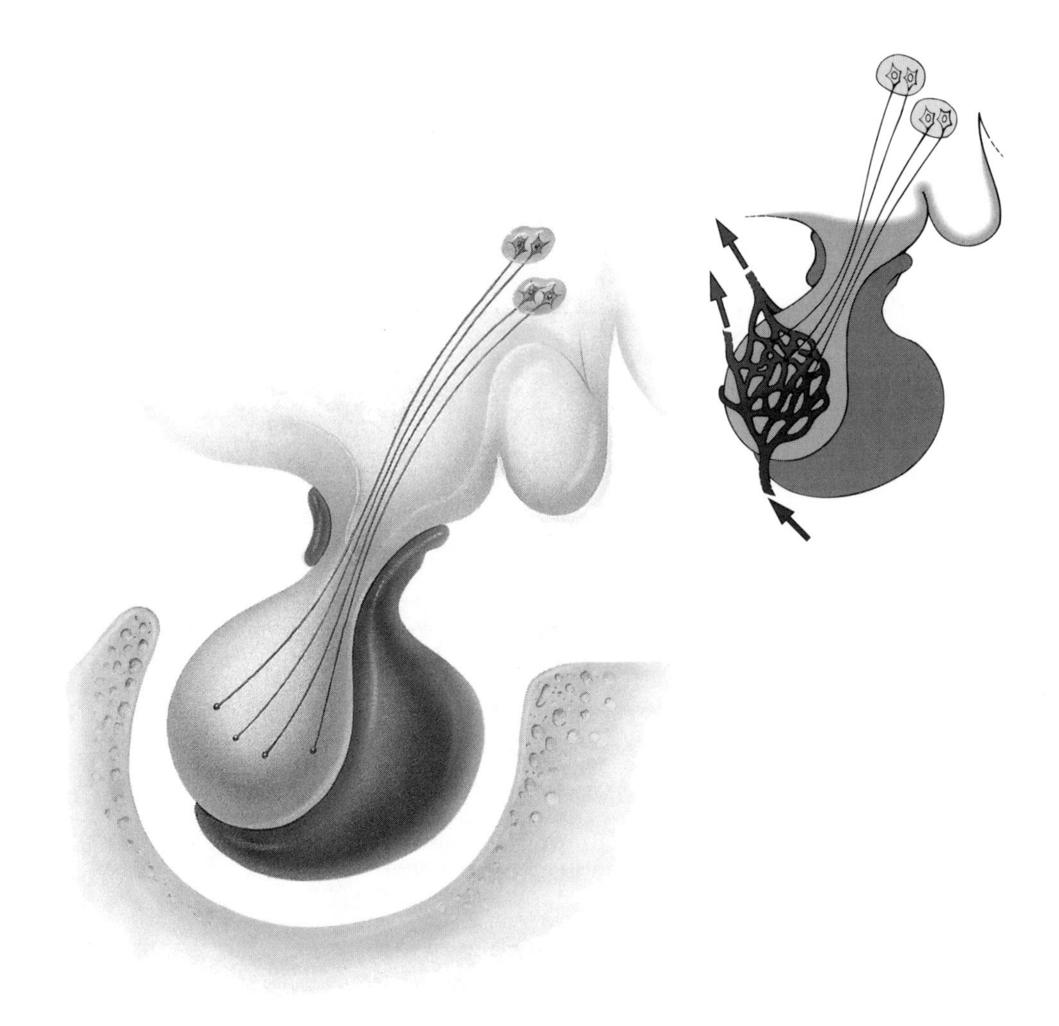

**Figure 22.4a** Location and blood supply of the thyroid gland (page 666).

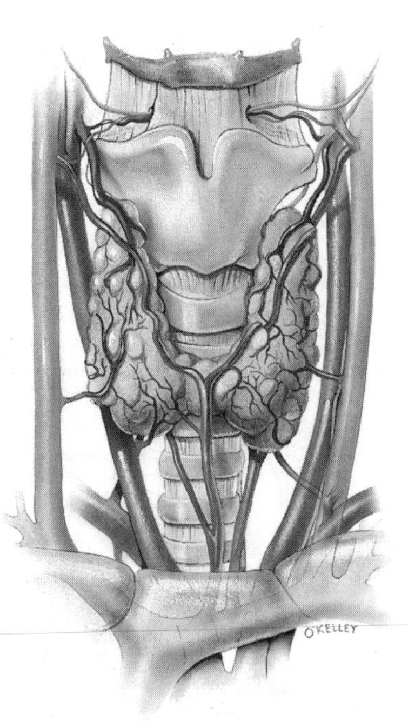

NOTES

22

**Figure 22.5a** Location and blood supply of the parathyroid glands (page 668).

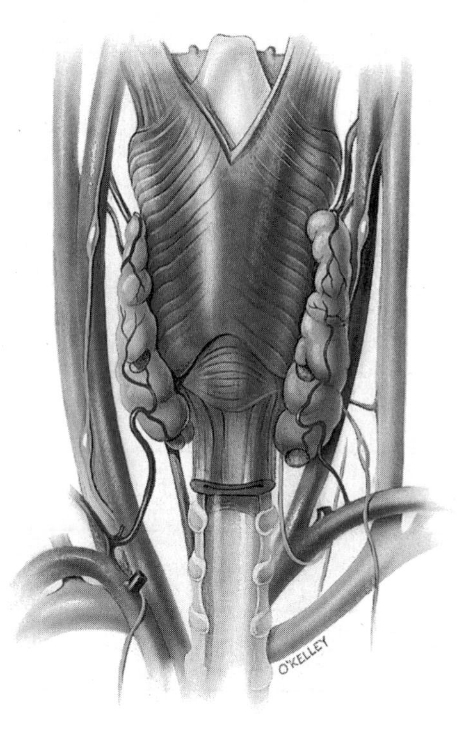

**Figure 22.6a** Location and blood supply of the adrenal (suprarenal) glands (page 670).

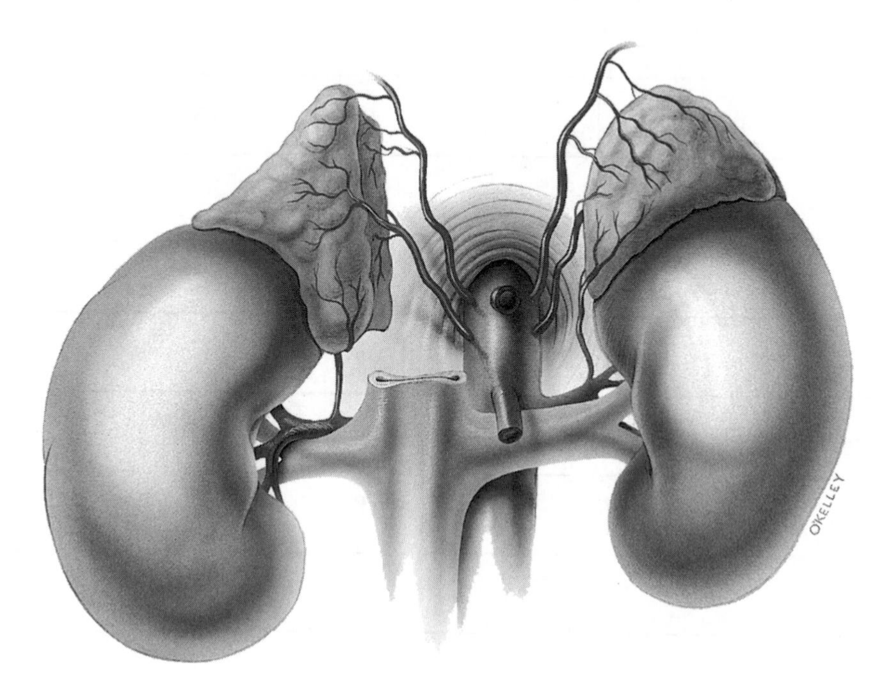

NOTES

**Figure 22.7a,b** Location and blood supply of the pancreas (page 672).

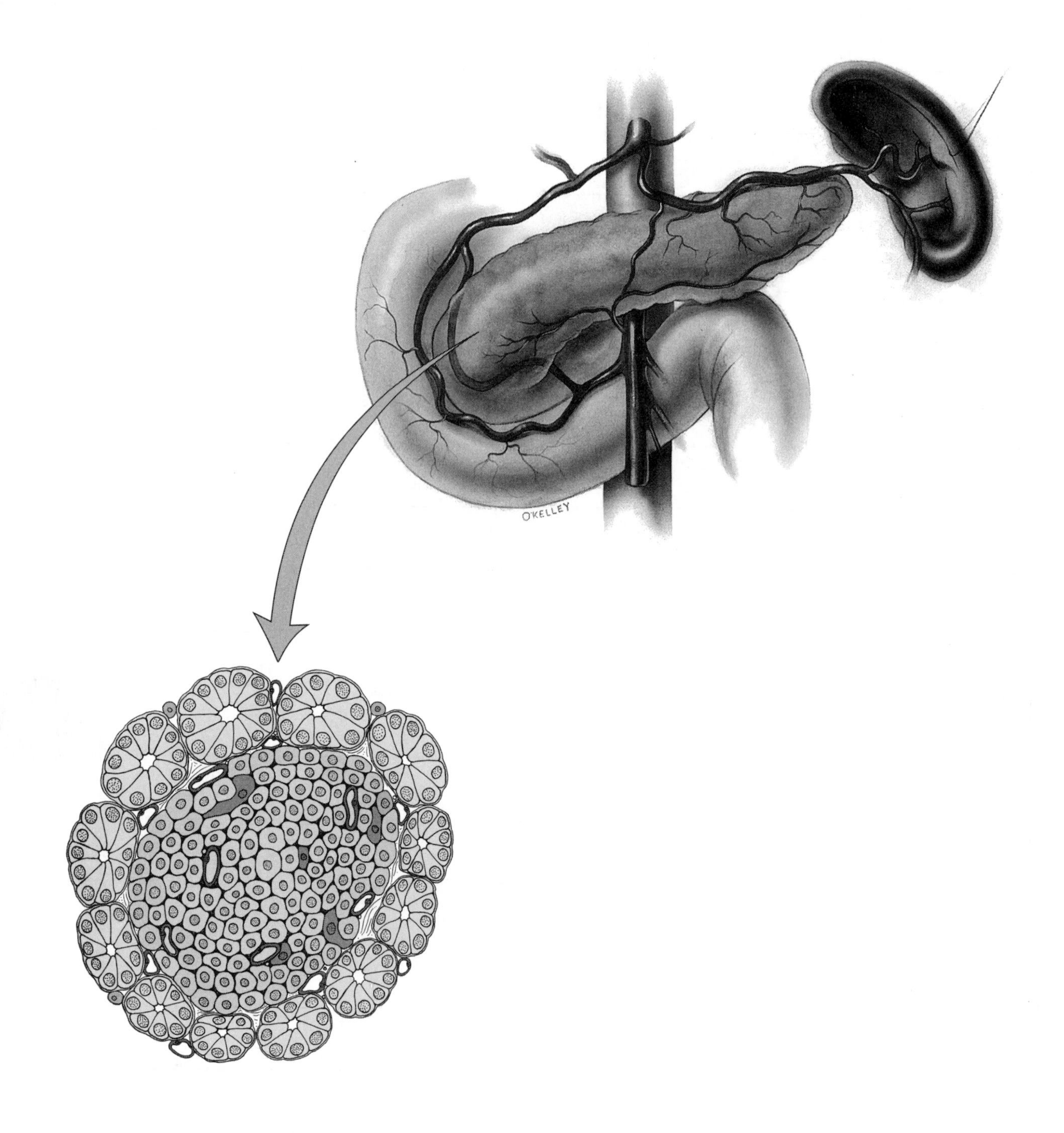

NOTES

22

**Figure 22.8** Development of the endocrine system (page 676).

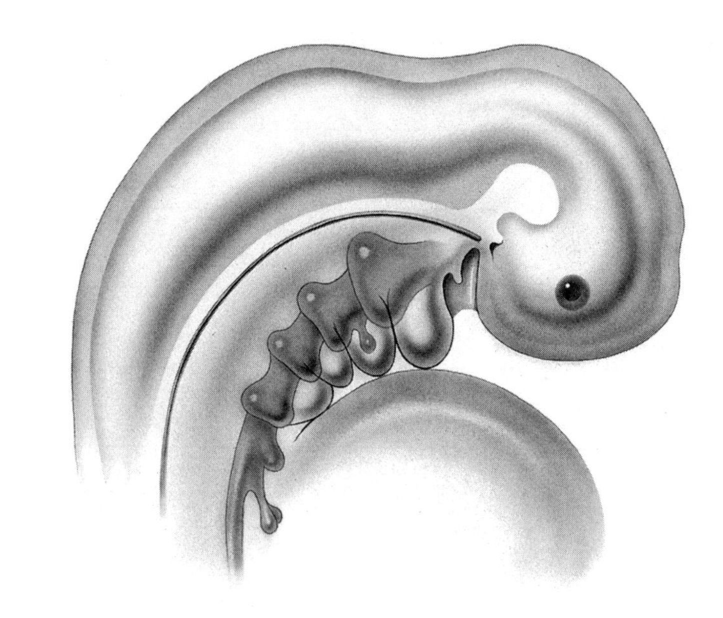

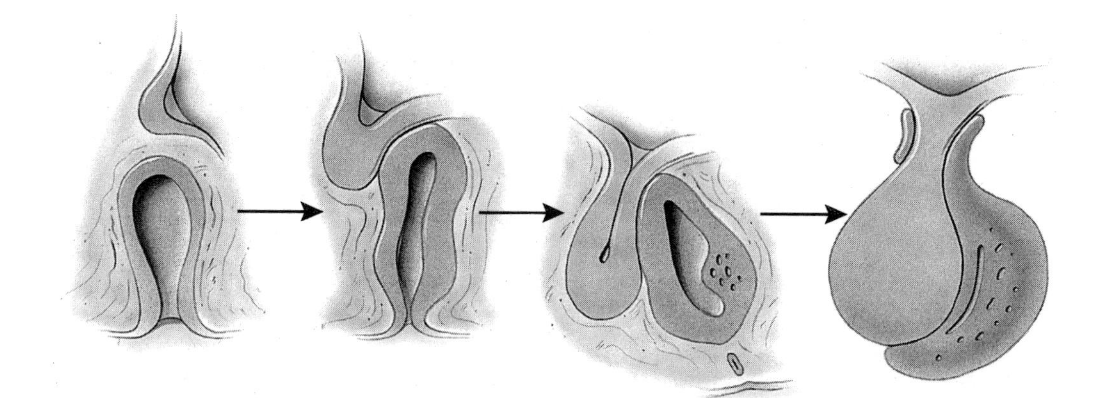

NOTES

22

**Figure 23.1a** Structures of the respiratory system (page 686).

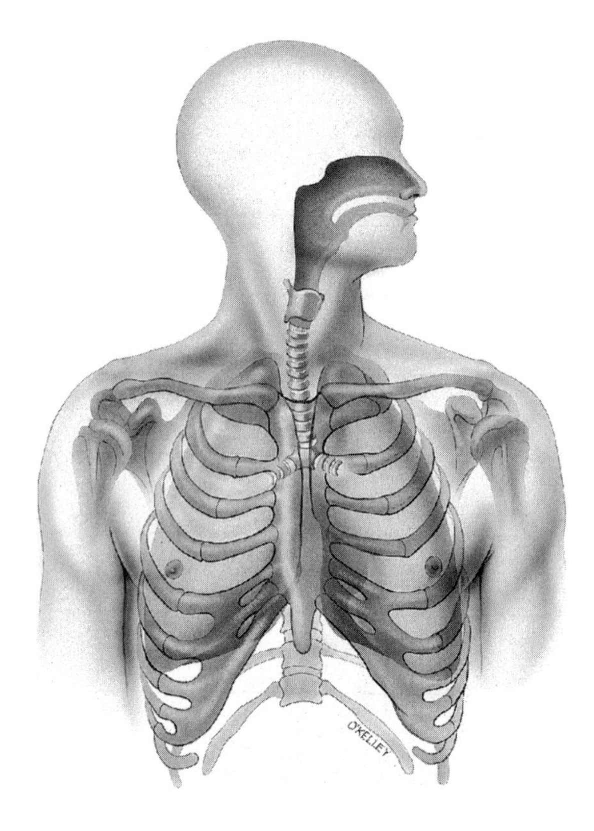

NOTES

23

**Figure 23.2a,b** Respiratory structures in the head and neck (page 687).

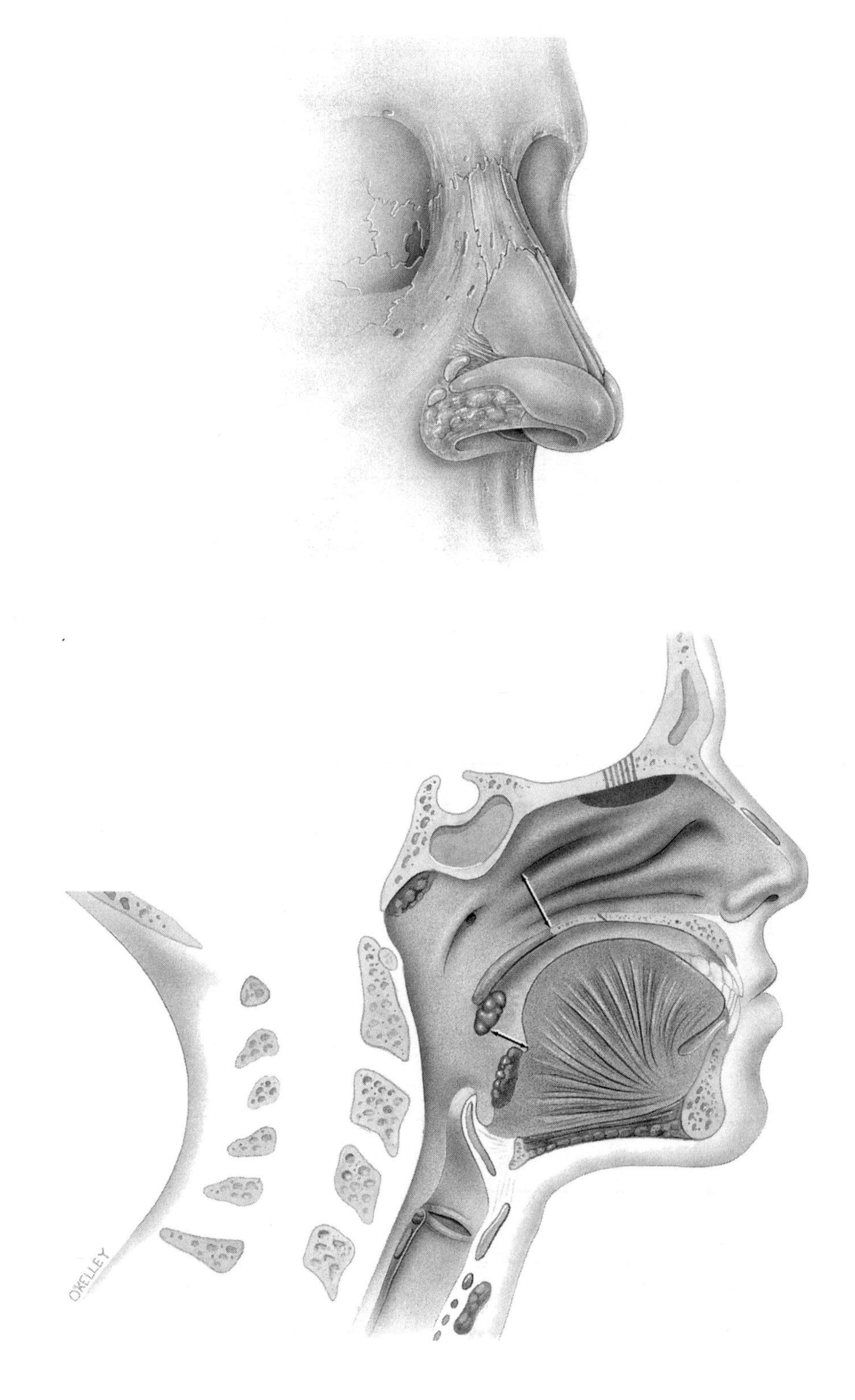

NOTES

23

**Figure 23.3** Pharynx (page 689).

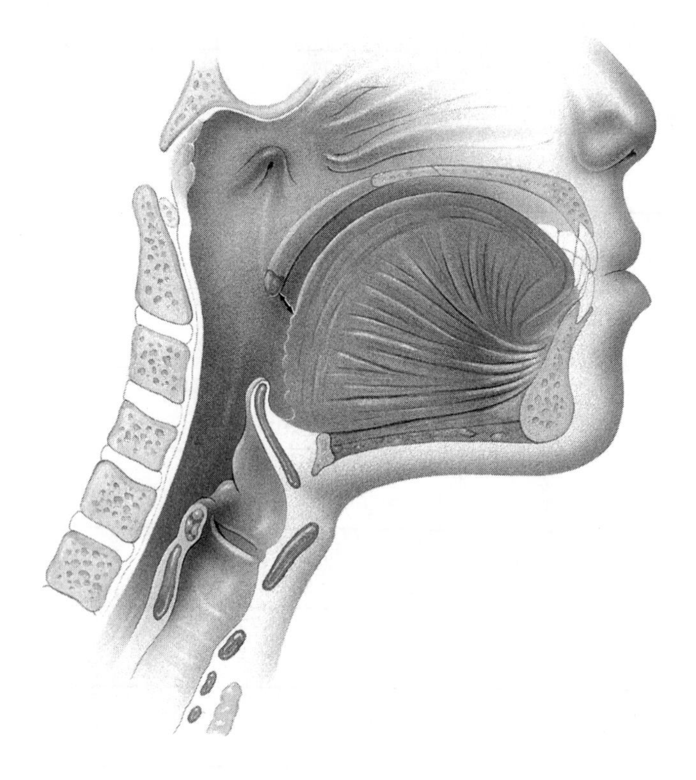

NOTES

23

**Figure 23.4** Larynx (page 690).

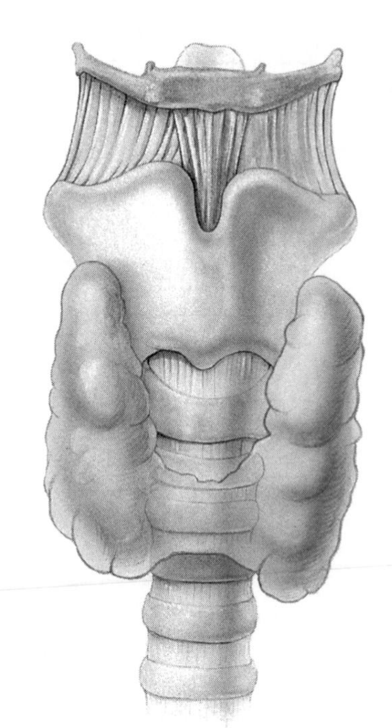

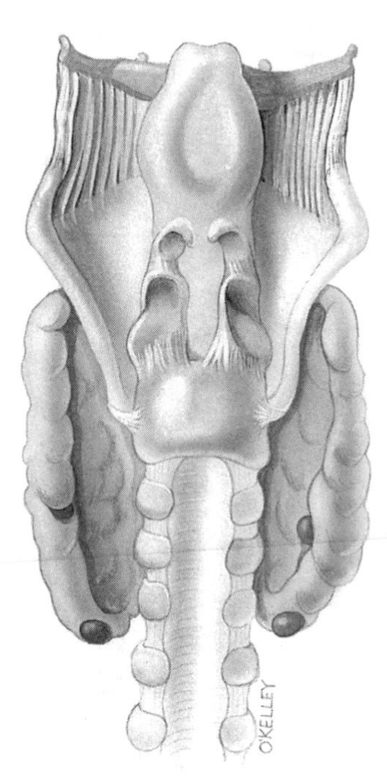


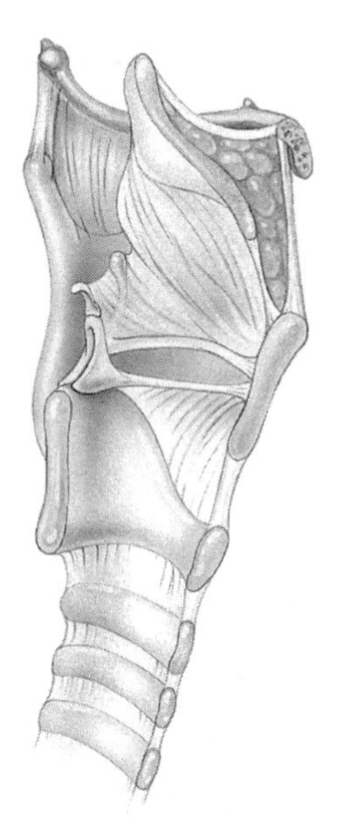

NOTES

23

**Figure 23.5a,b** Vocal folds (page 693).

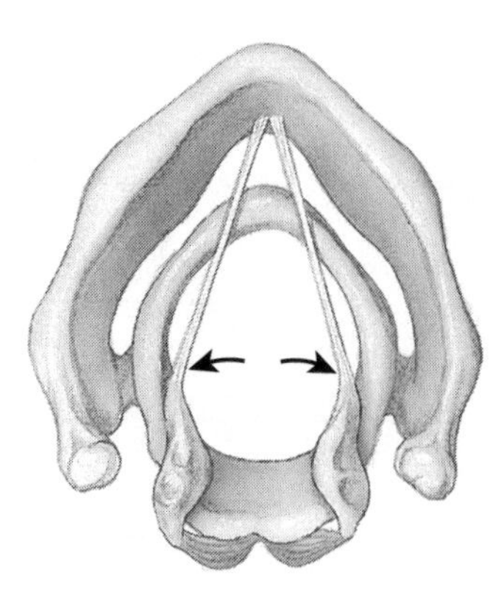

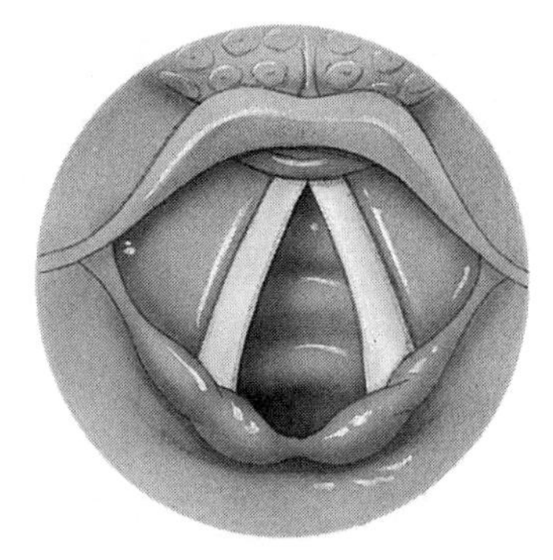

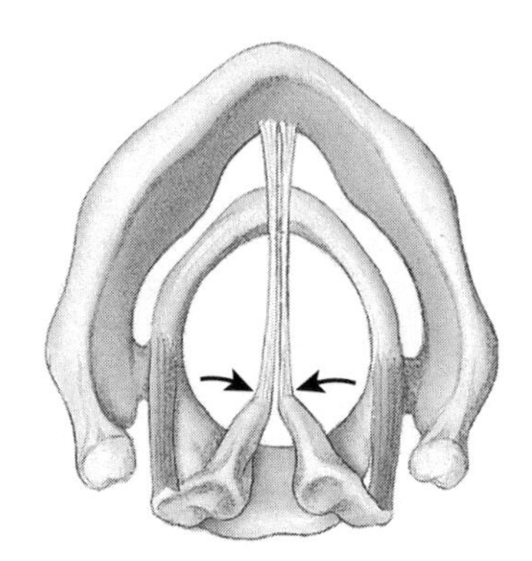

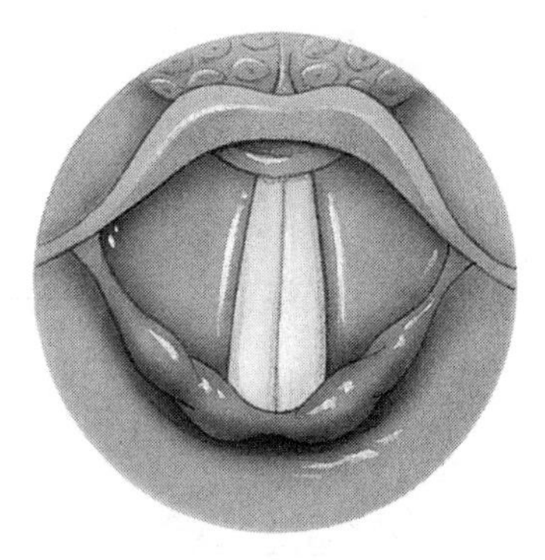

NOTES

23

**Figure 23.7a** Branching of airways from the trachea: the bronchial tree (page 695).

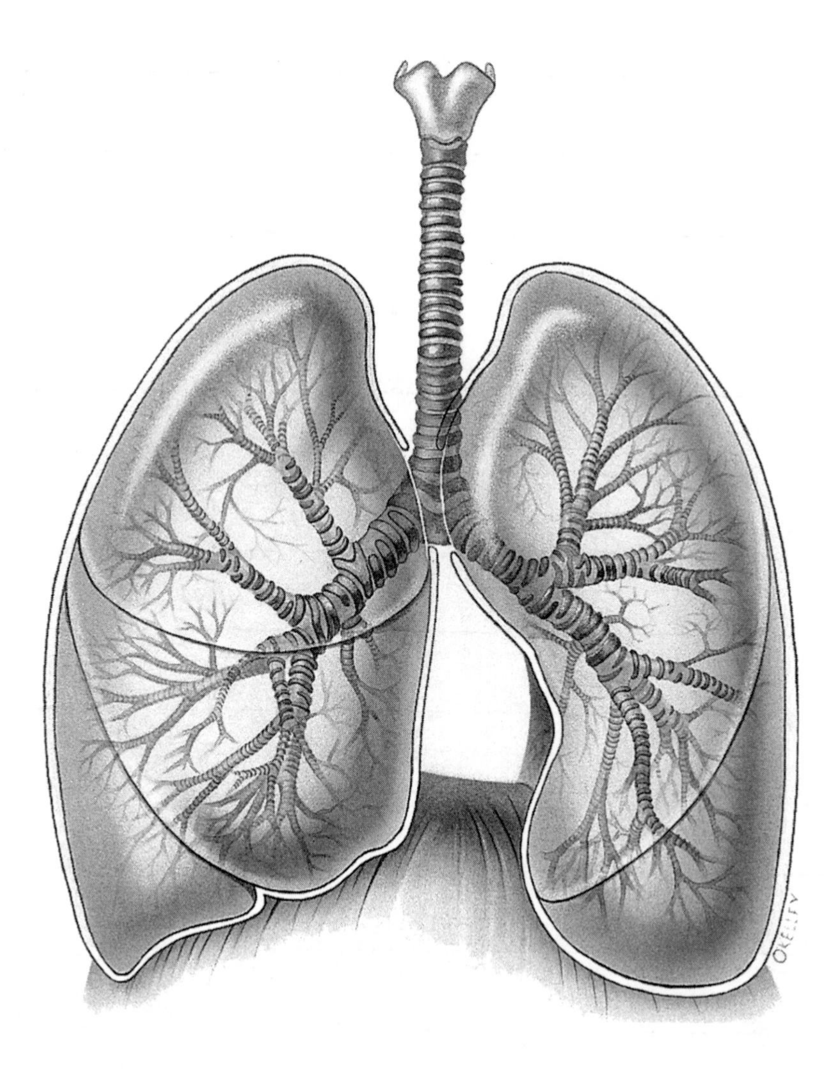

NOTES

23

**Figure 23.9a–e** Surface anatomy of the lungs (page 698).

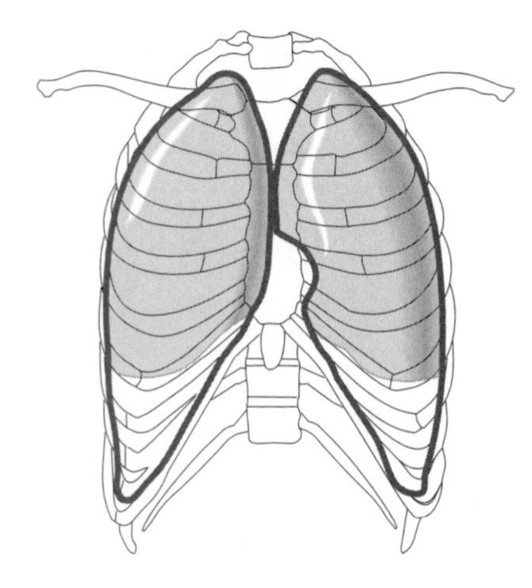

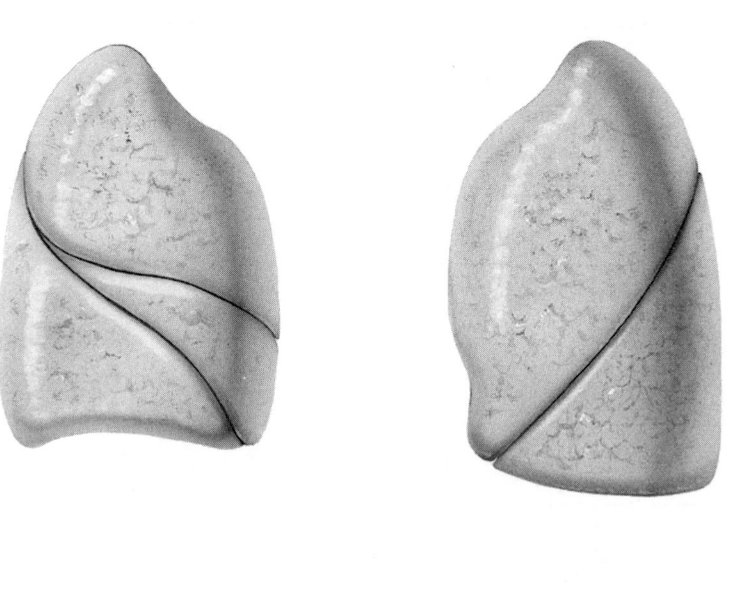

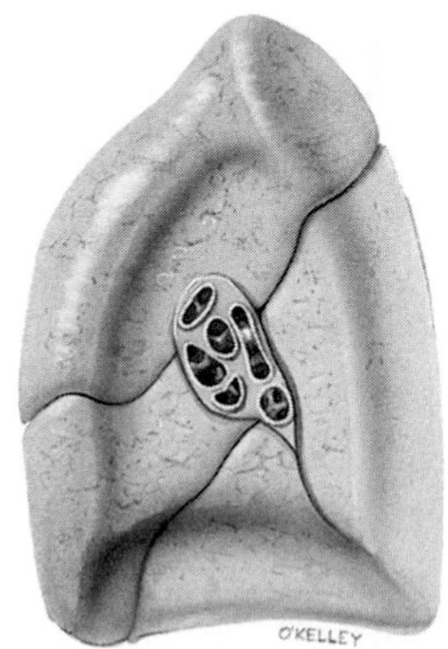


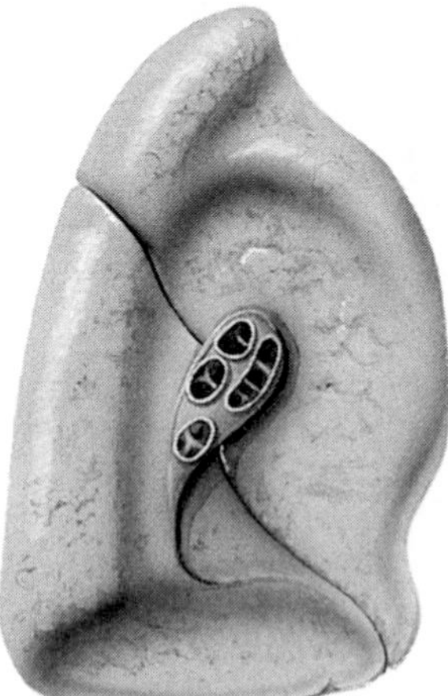

## NOTES

23

**Figure 23.10** Bronchopulmonary segments of the lungs (page 700).

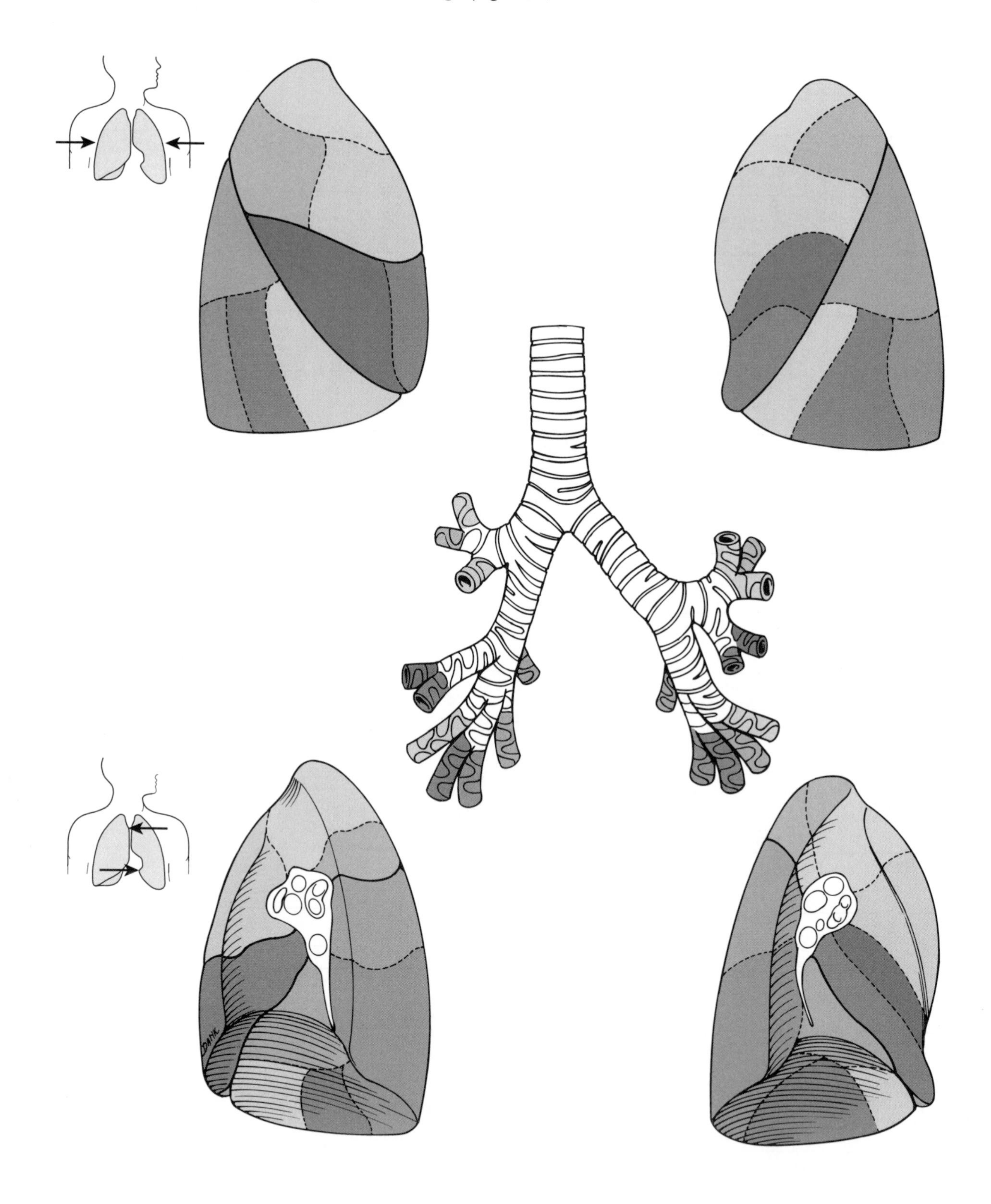

NOTES

23

**Figure 23.11a** Microscopic anatomy of a lobule of the lungs (page 701).

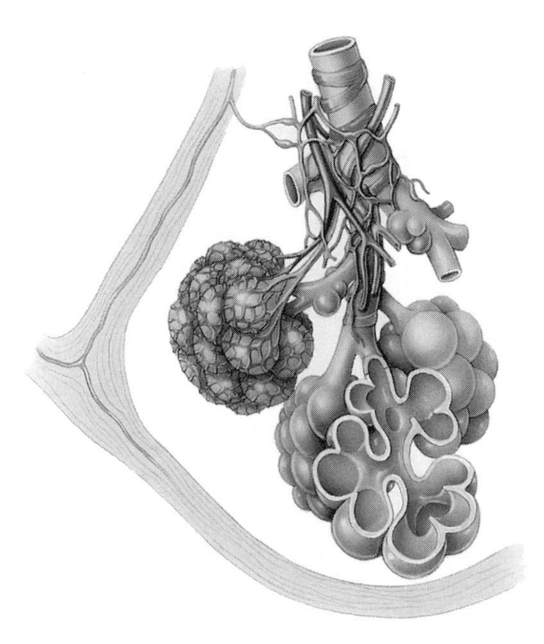

**Figure 23.12** Structural components and function of an alveolus (page 702).

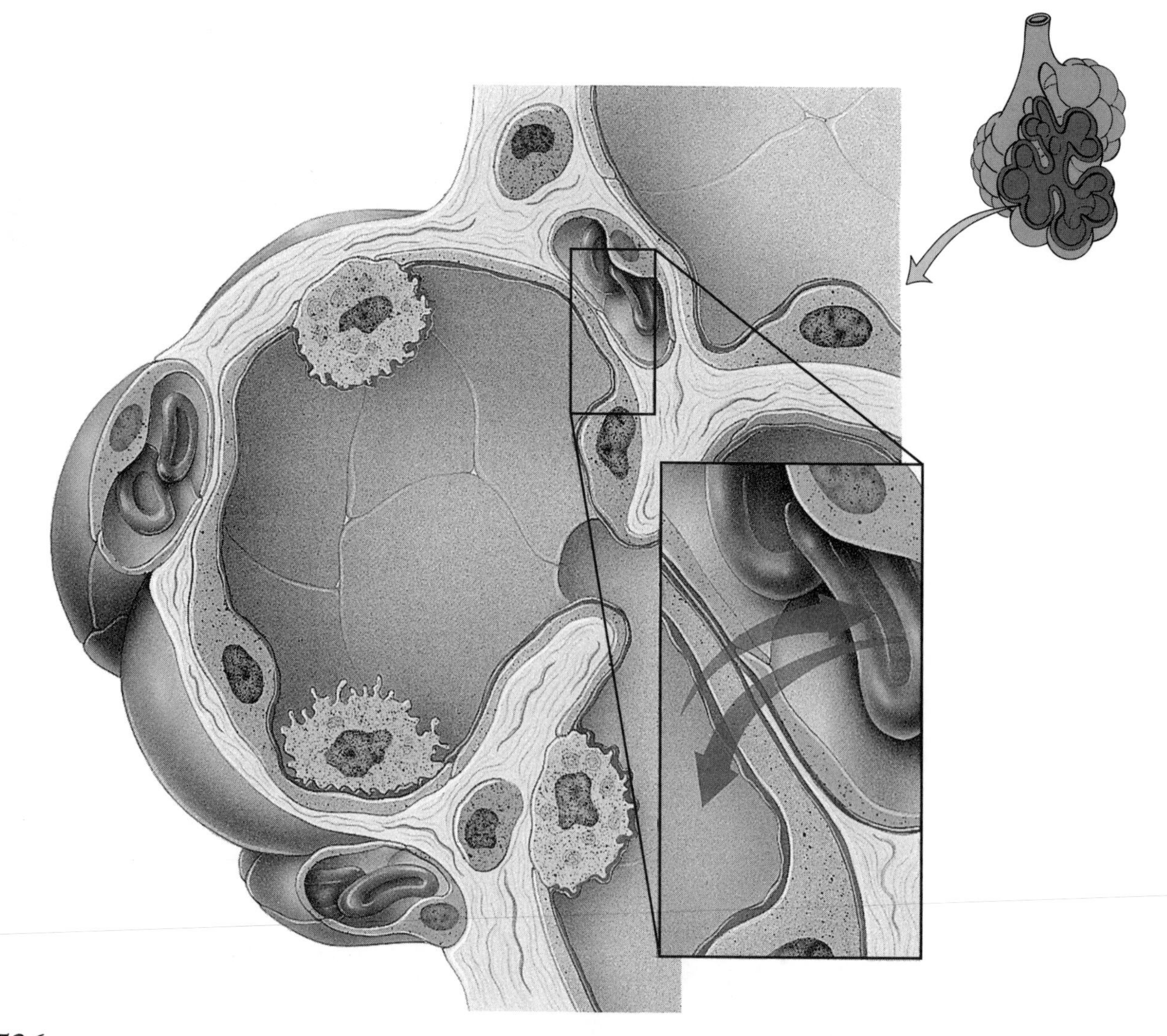

NOTES

23

**Figure 23.13** Muscles of inspiration and expiration and their roles in pulmonary ventilation (page 704).

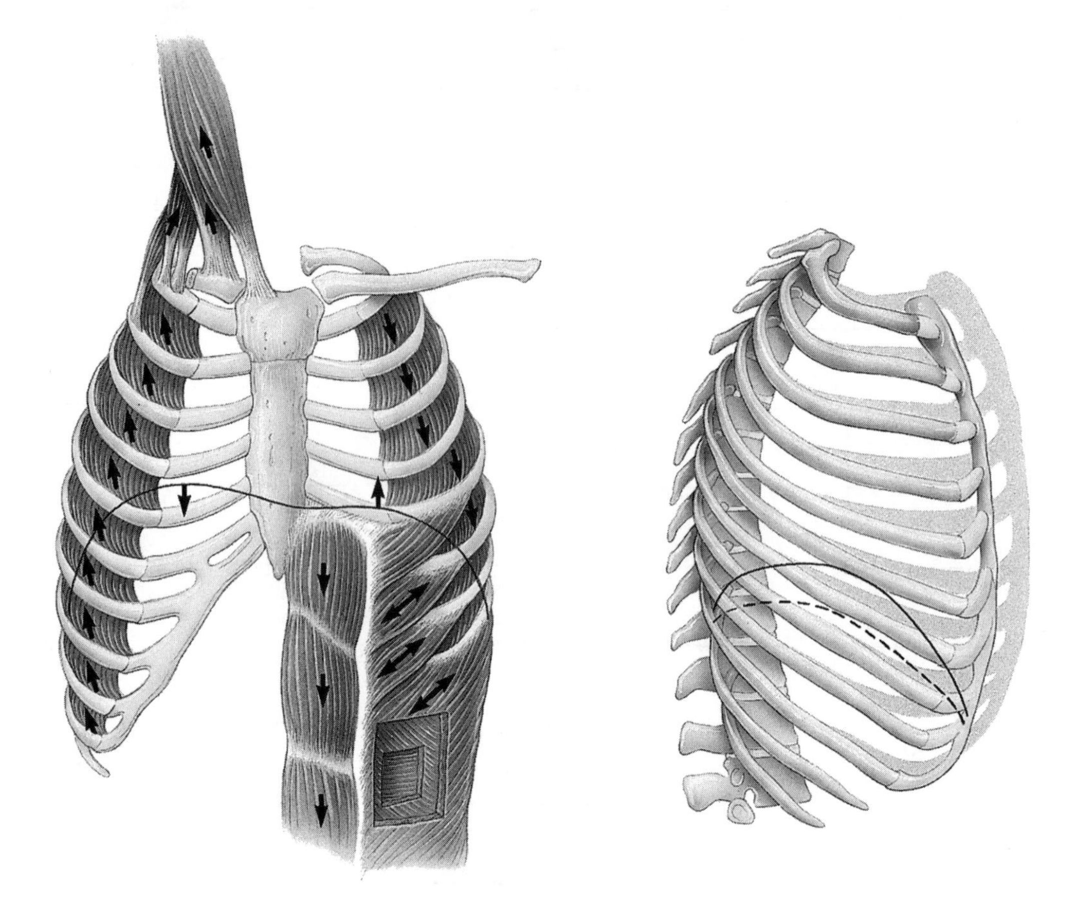

**Figure 23.14** Locations of areas of the respiratory center (page 705).

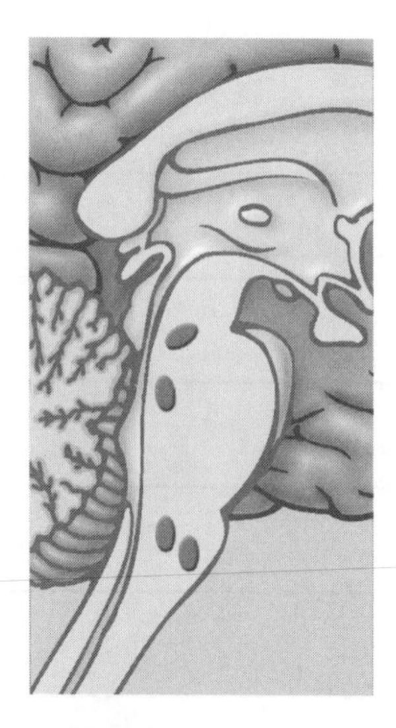

NOTES

23

**Figure 23.15** Development of the bronchial tubes and lungs (page 707).

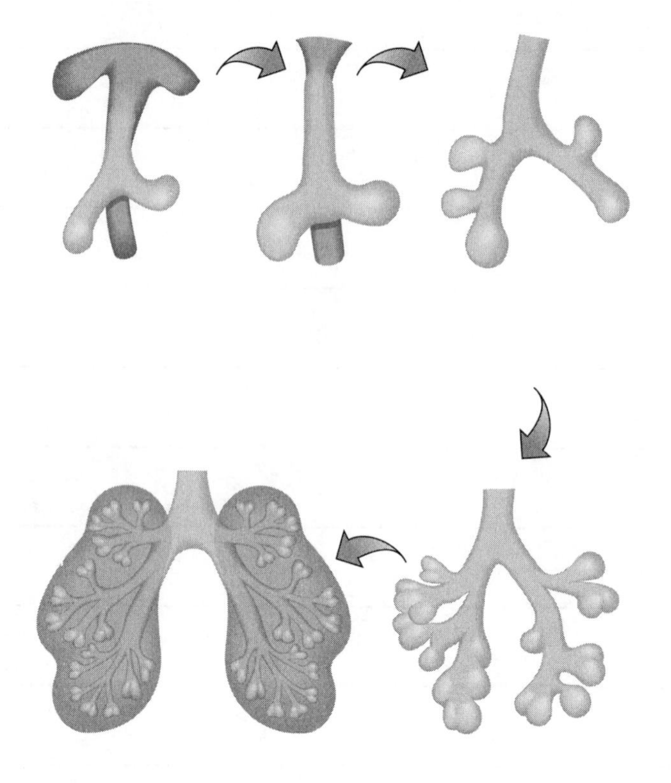

NOTES

23

**Figure 24.1a** Organs of the digestive system (page 716).

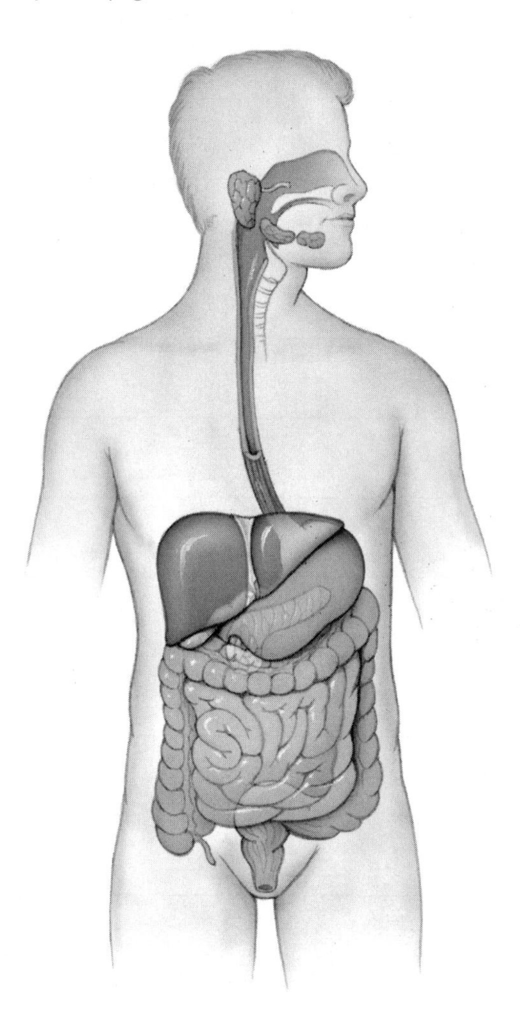

**Figure 24.2** Three-dimensional depiction of the various layers of the gastrointestinal tract (page 717).

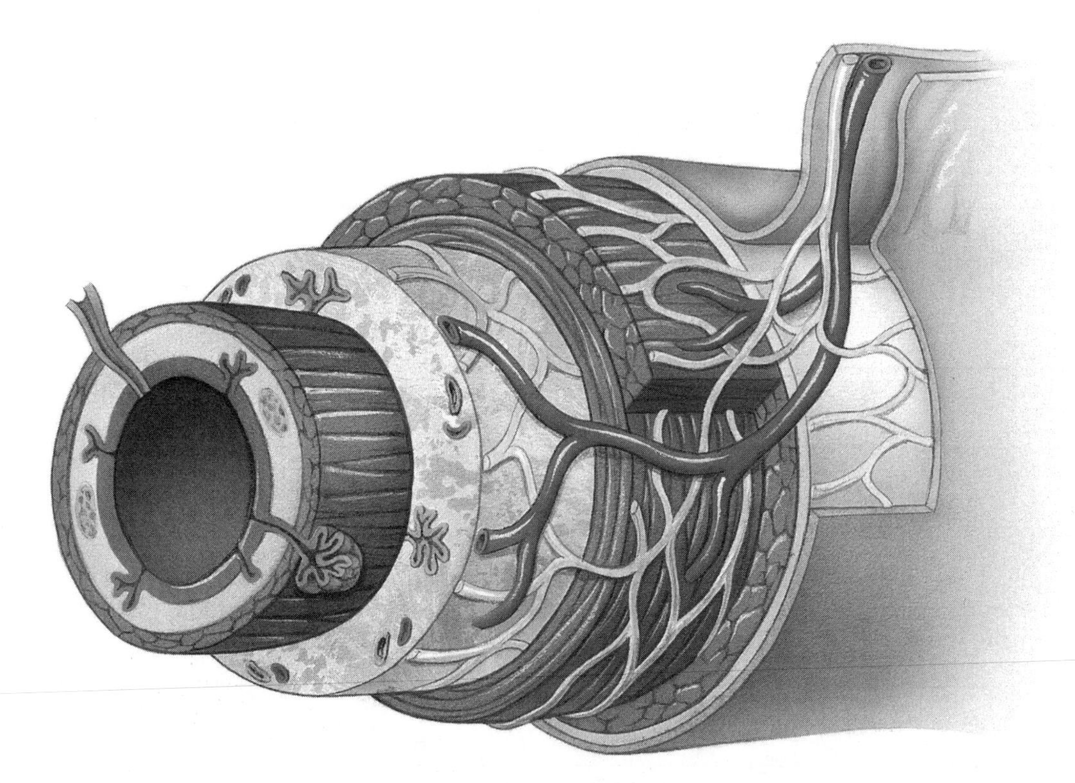

NOTES

24

**Figure 24.3a** Relationship of the peritoneal folds to each other and to organs of the digestive system (page 719).

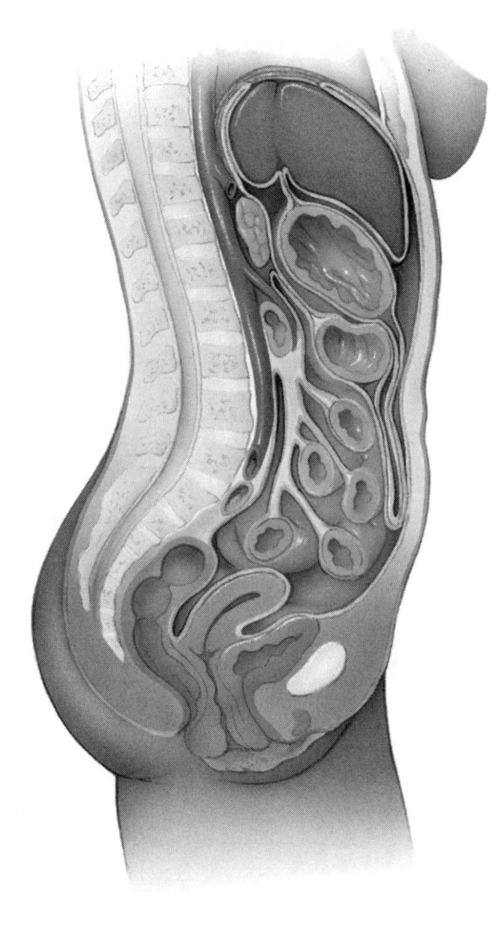

NOTES

24

**Figure 24.3b–d** Relationship of the peritoneal folds to each other and to organs of the digestive system (page 720).

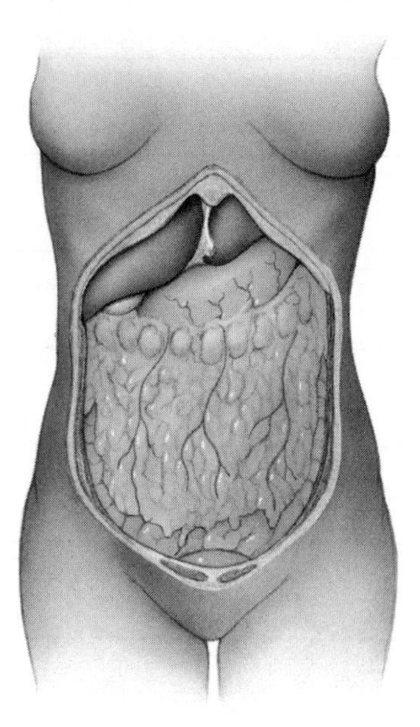

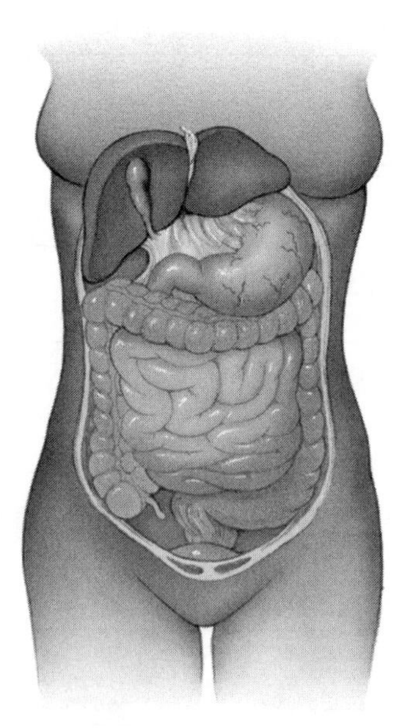

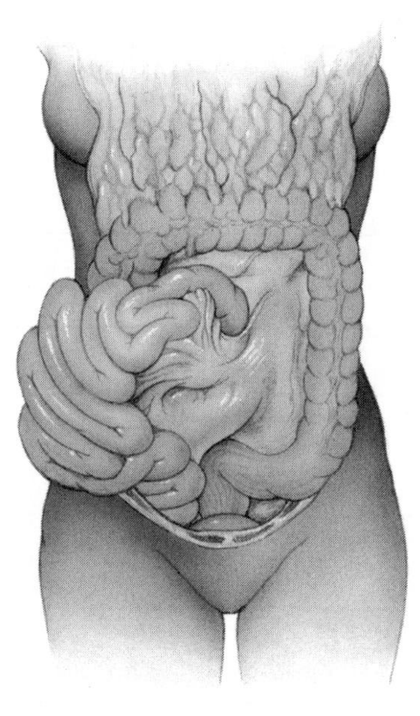

NOTES

24

**Figure 24.4** Structures of the mouth (oral cavity) (page 721).

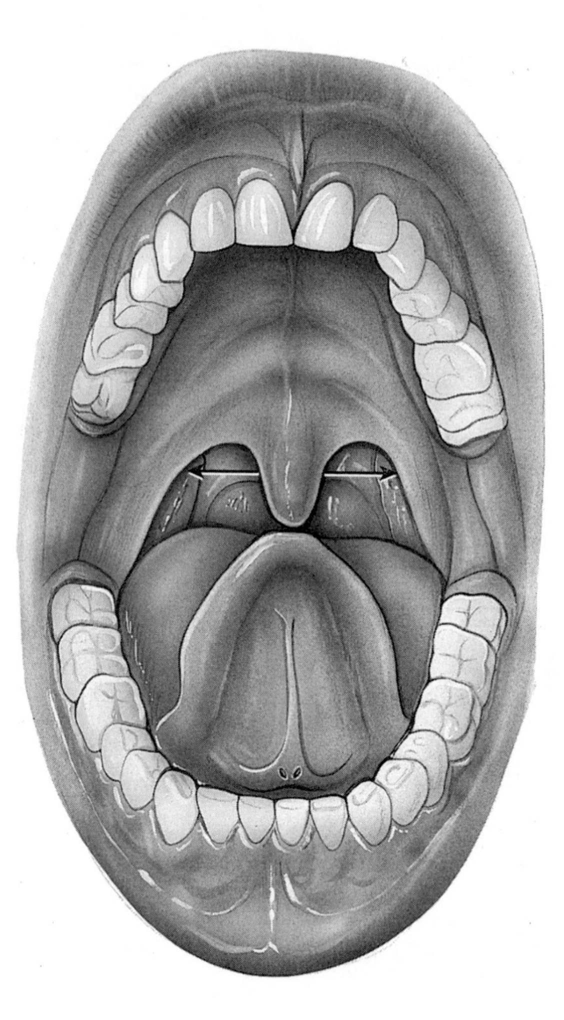

NOTES

24

**Figure 24.5** Tongue (page 722).

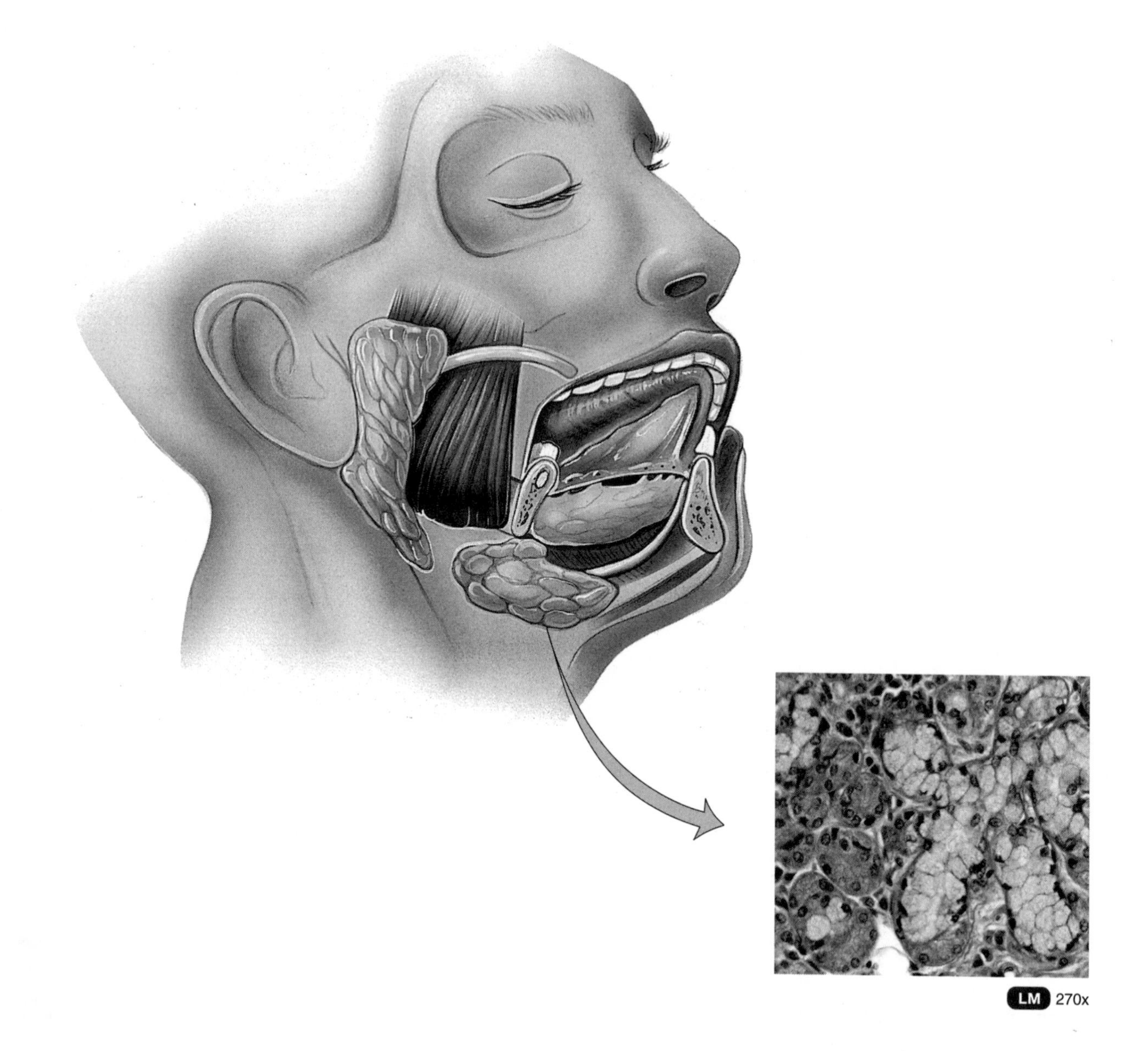

NOTES

24

**Figure 24.6** A typical tooth and surrounding structures (page 724).

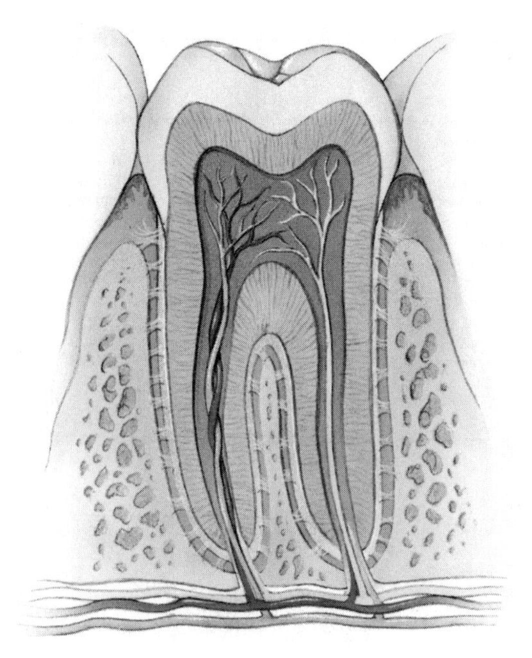

**Figure 24.7a,b** Dentitions and times of eruptions (page 725).

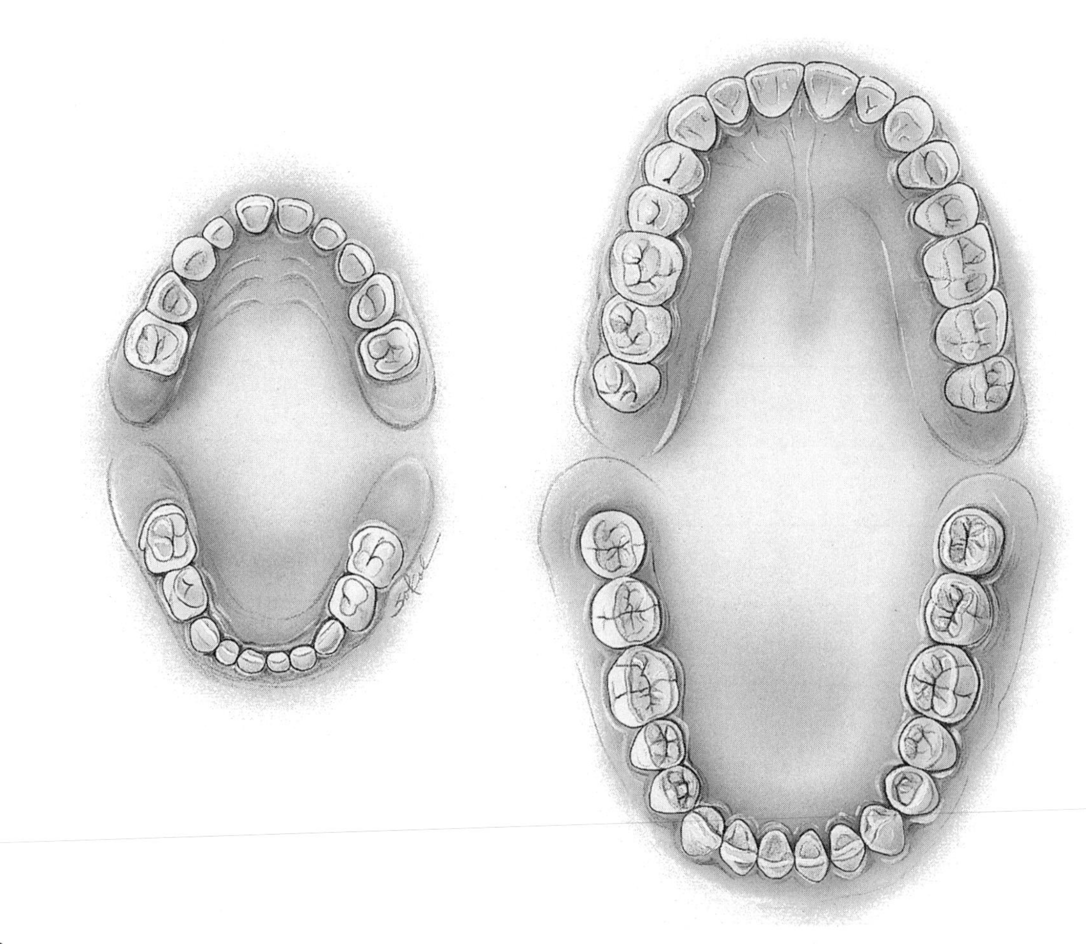

NOTES

24

**Figure 24.9** Peristalsis during deglutition (swallowing) (page 727).

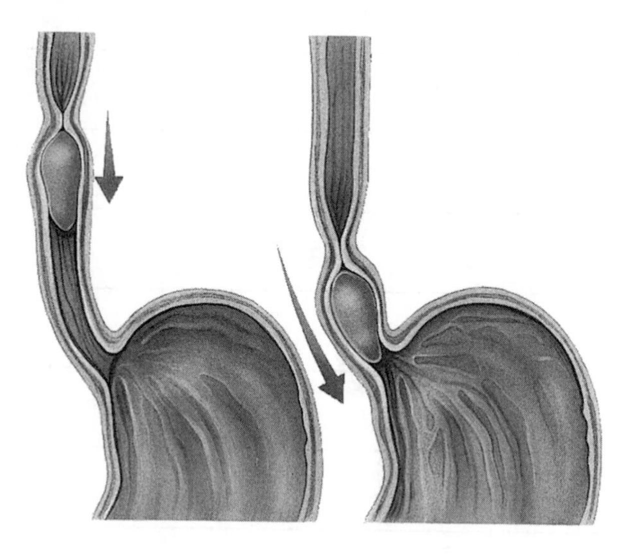

**Figure 24.10a** External and internal anatomy of the stomach (page 729).

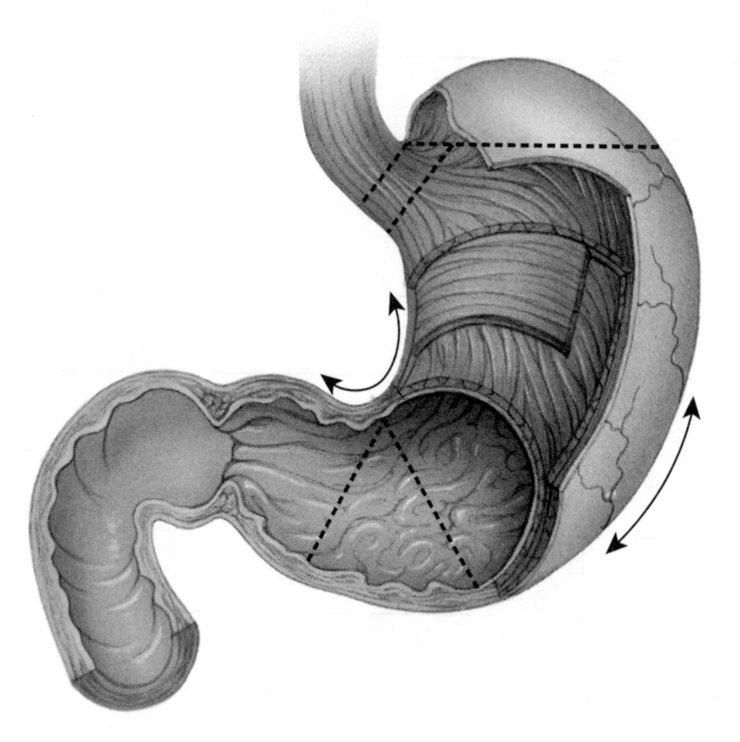

NOTES

24

**Figure 24.11a,b** Histology of the stomach (page 730).

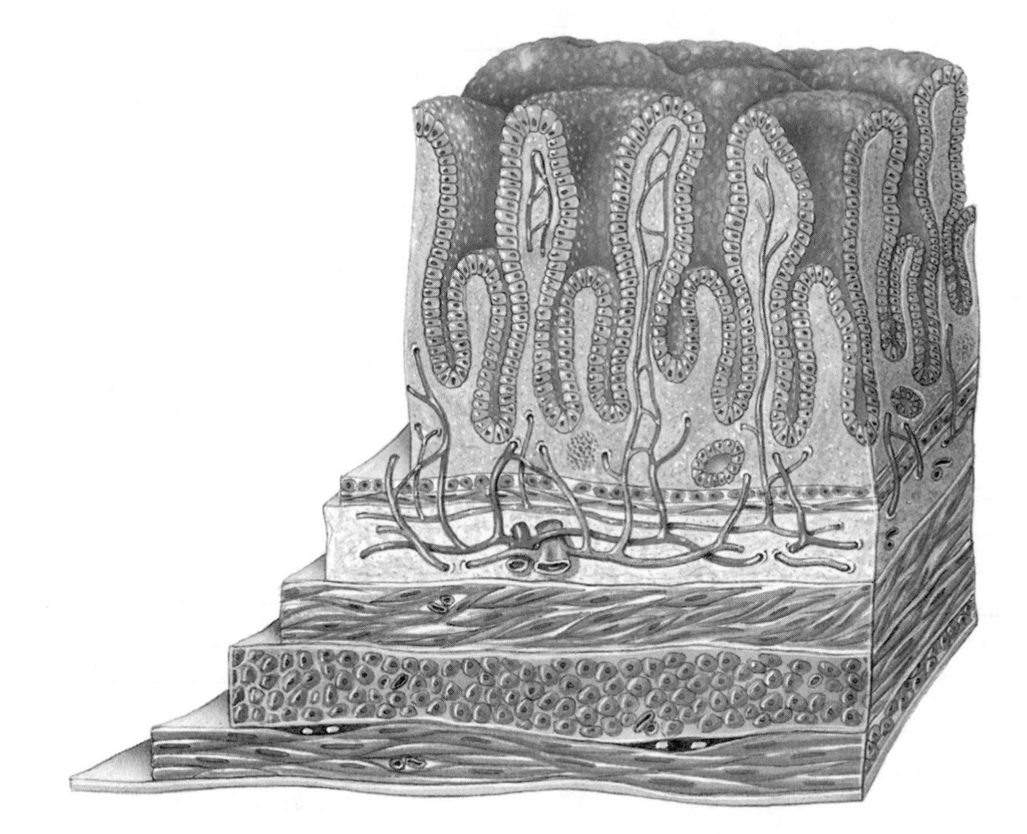

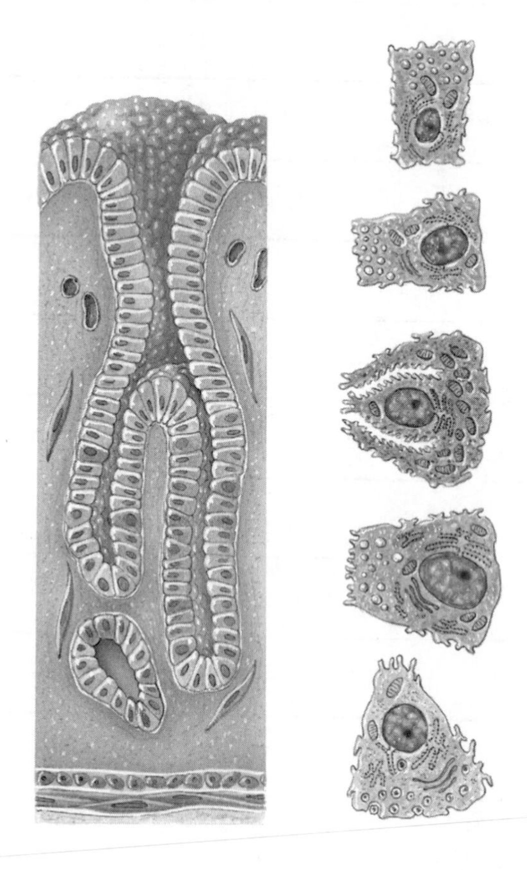

## NOTES

**Figure 24.12a** Relation of the pancreas to the liver, gallbladder and duodenum (page 733).

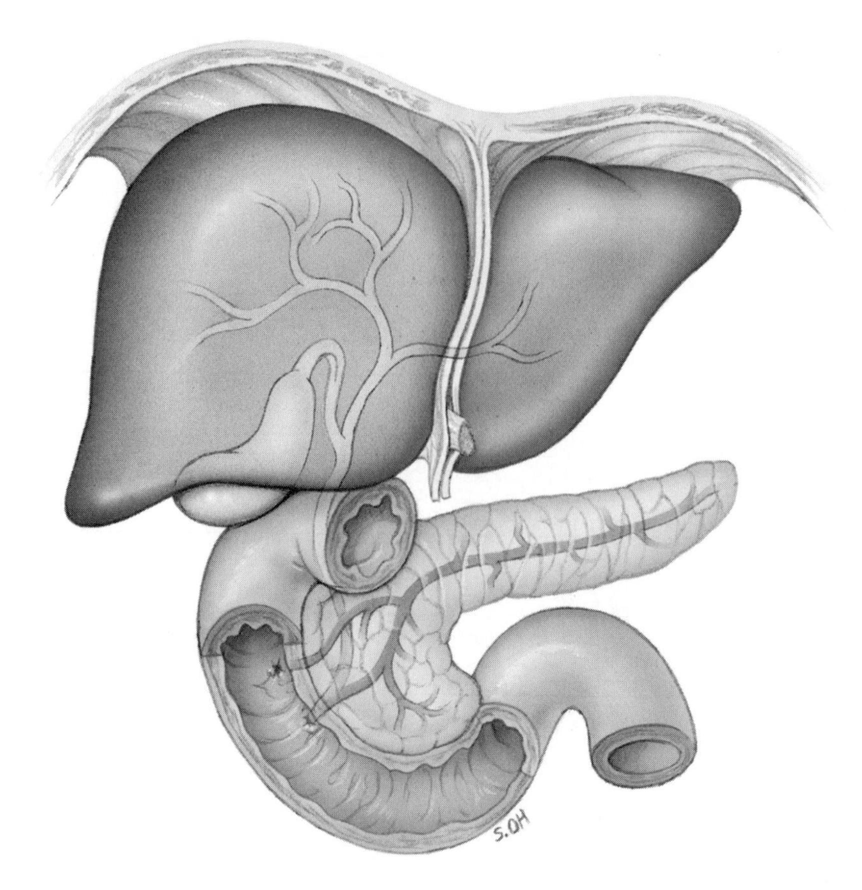


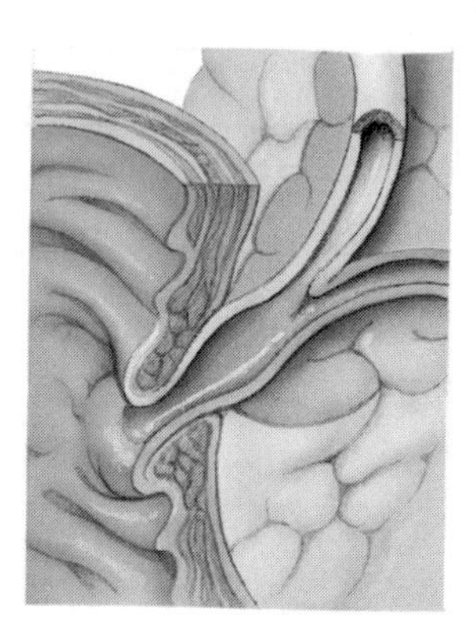

**Figure 24.13a** External anatomy of the liver (page 736).

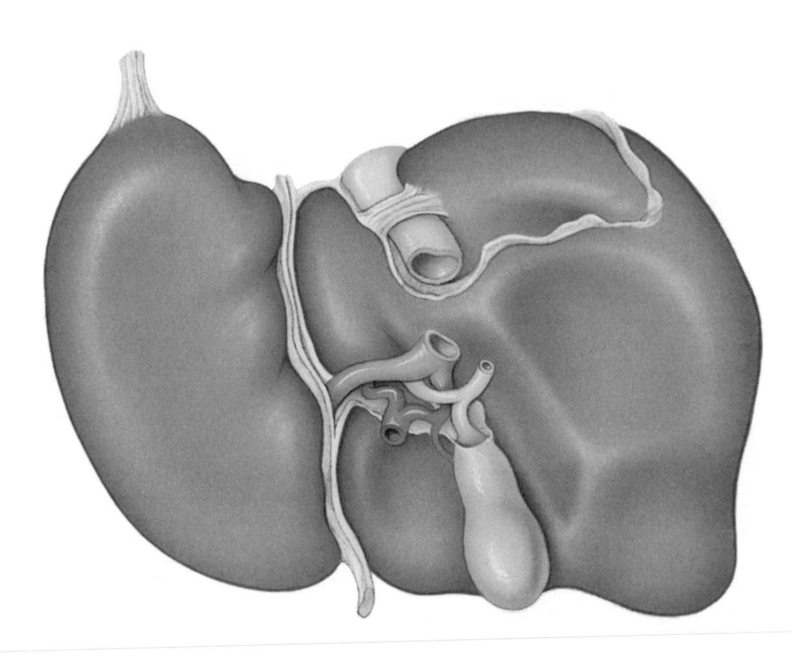

NOTES

24

**Figure 24.14a,b** Histology of a lobule, the functional unit of the liver (page 737).

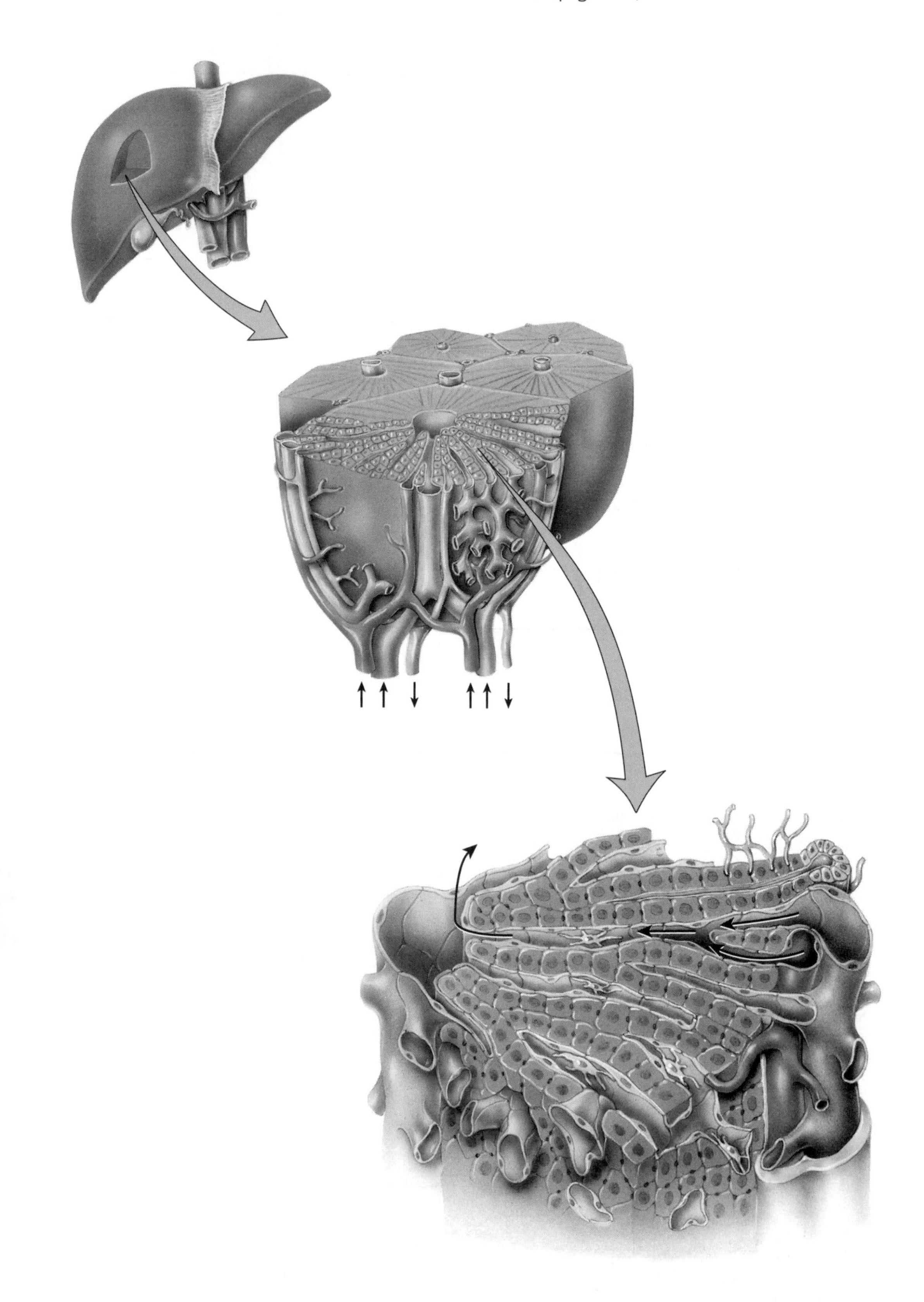

NOTES

24

**Figure 24.15** Hepatic blood flow: sources, path through the liver, and return to the heart (page 738).

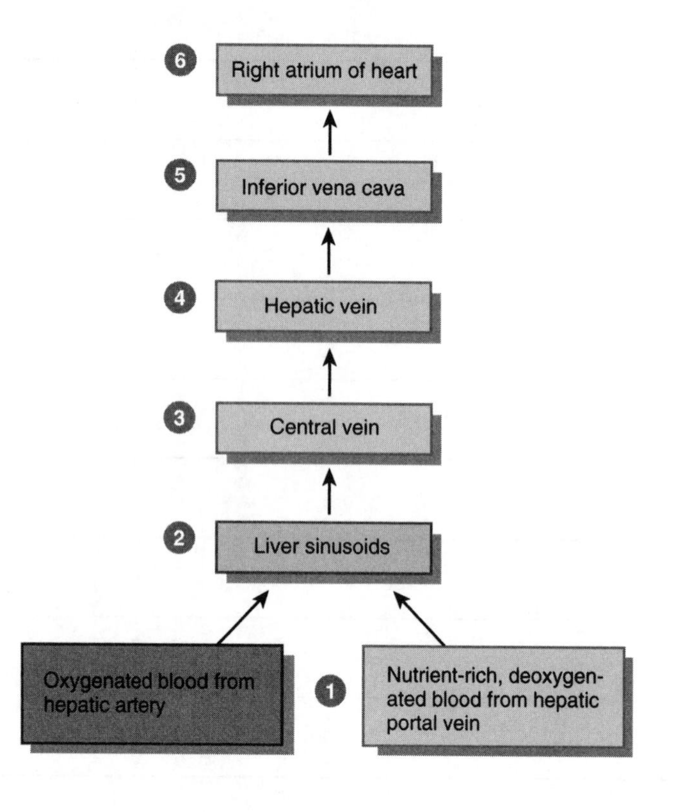


**Figure 24.16** Regions of the small intestine (page 739).

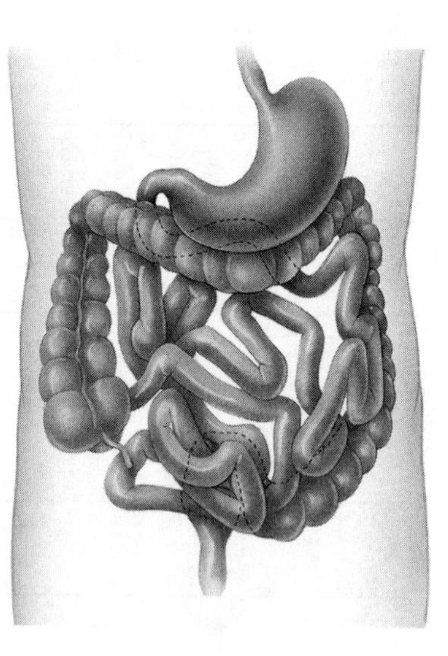

NOTES

24

**Figure 24.17a** Anatomy of the small intestine (page 740).

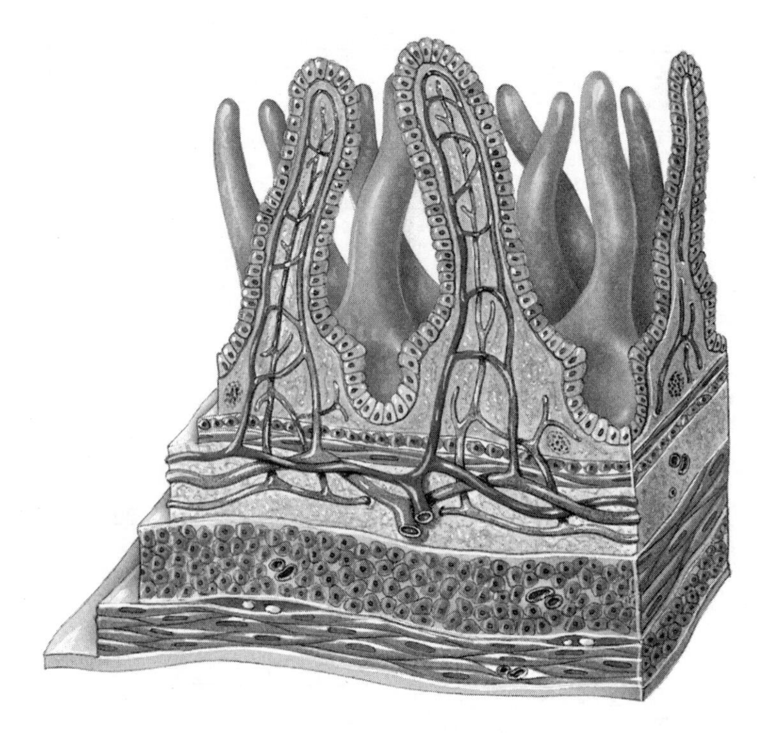

NOTES

24

**Figure 24.17b,c** Anatomy of the small intestine (page 741).

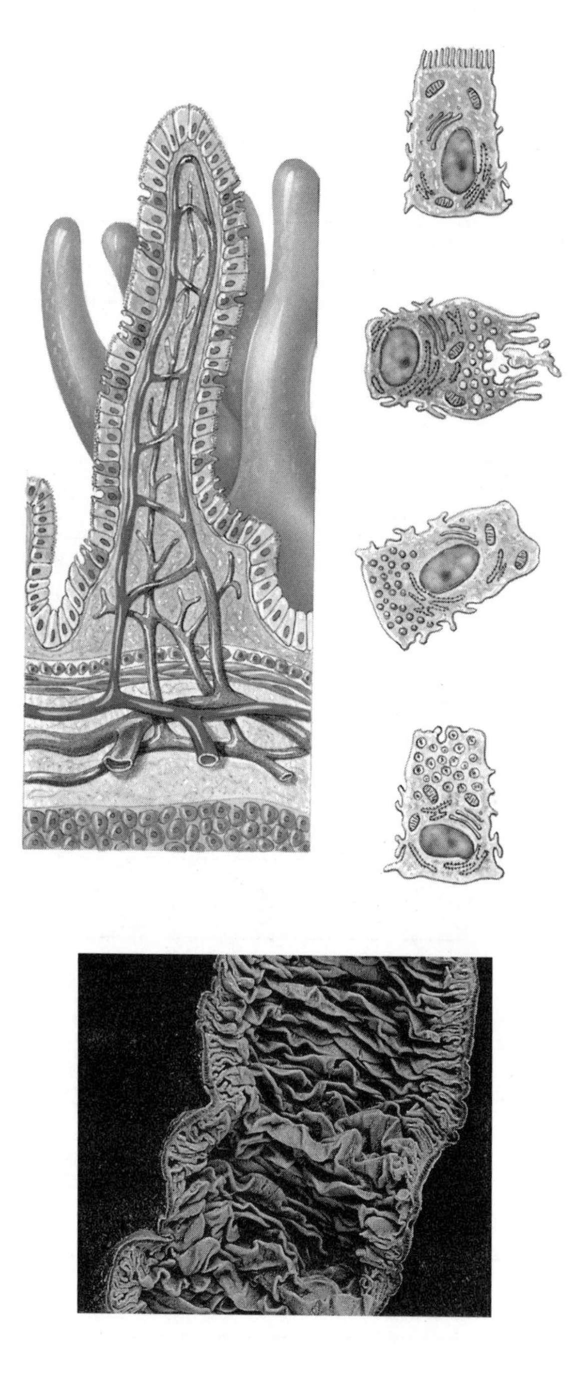

NOTES

24

**Figure 24.19a,b** Anatomy of the large intestine (page 745).

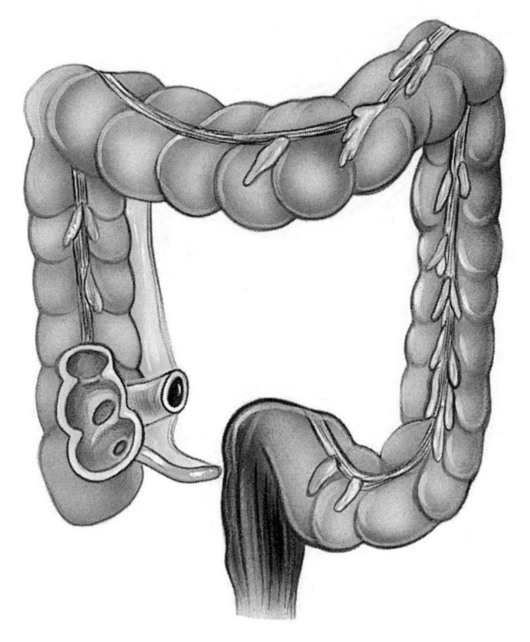

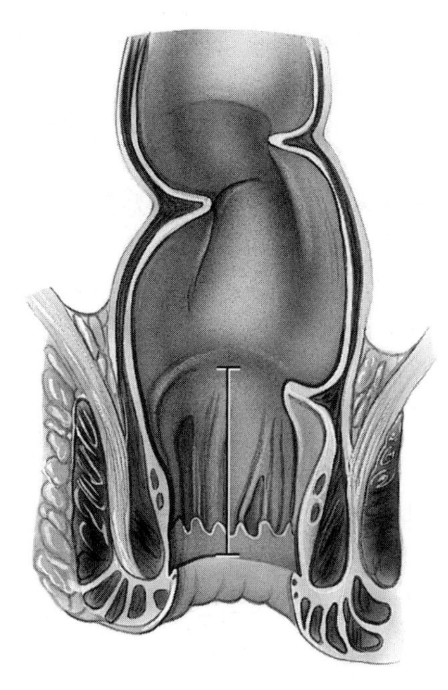

NOTES

24

**Figure 24.20a,b** Histology of the large intestine (page 746).

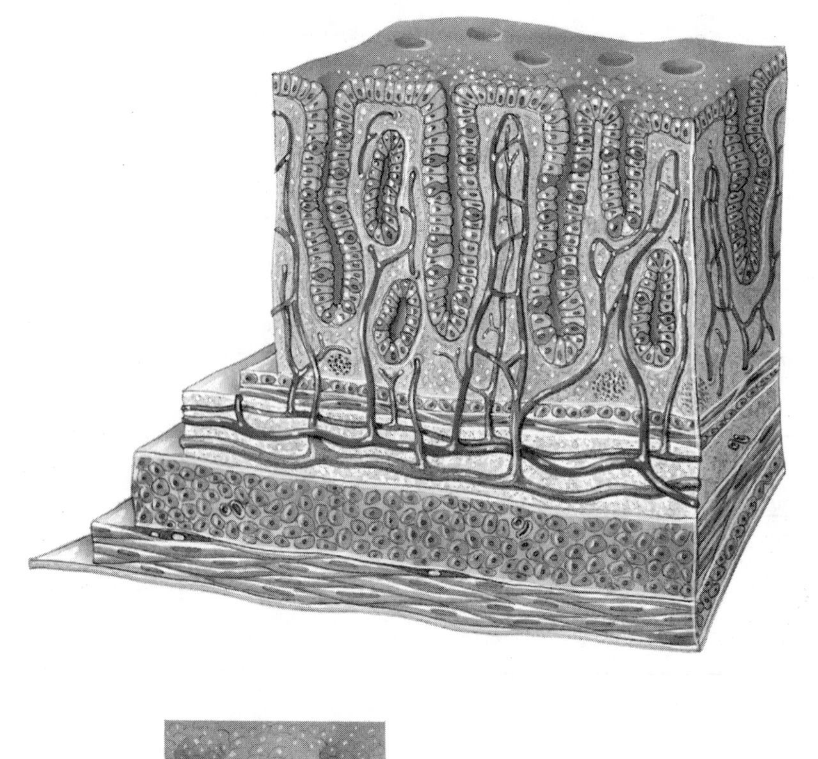

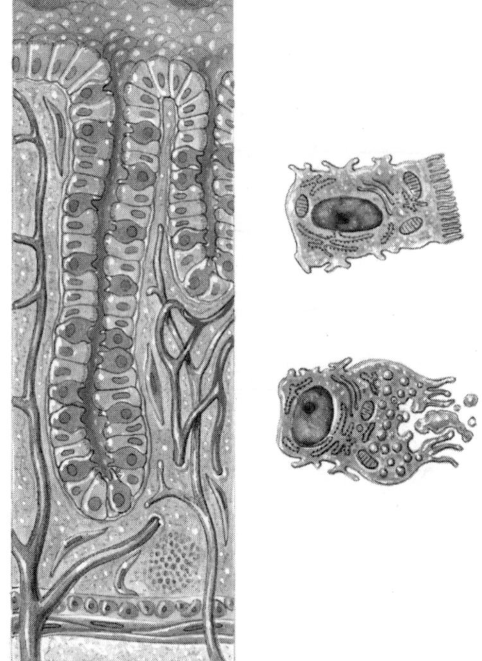

NOTES

24

**Figure 24.21a,b** Development of the digestive system (page 749).

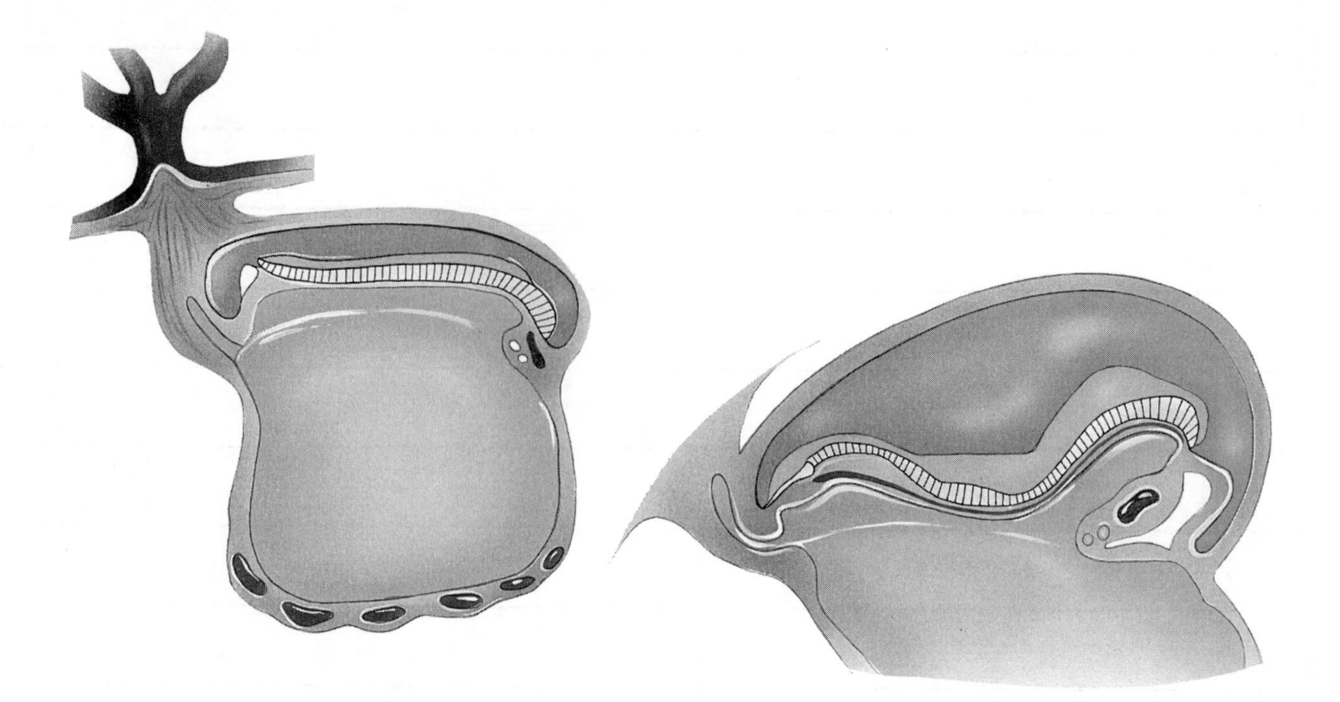

NOTES

24

**Figure 24.21c,d** Development of the digestive system (page 750).

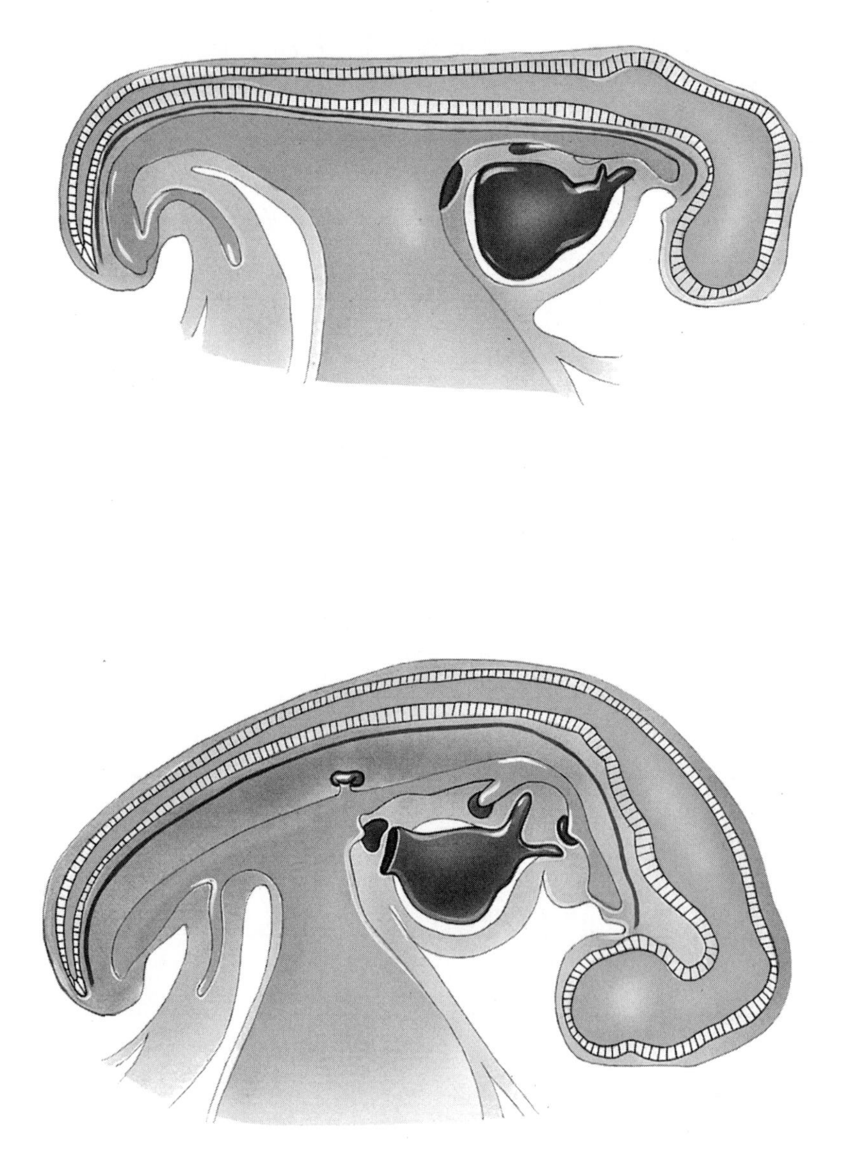

NOTES

24

**Figure 25.1a** Organs of the female urinary system (page 758).

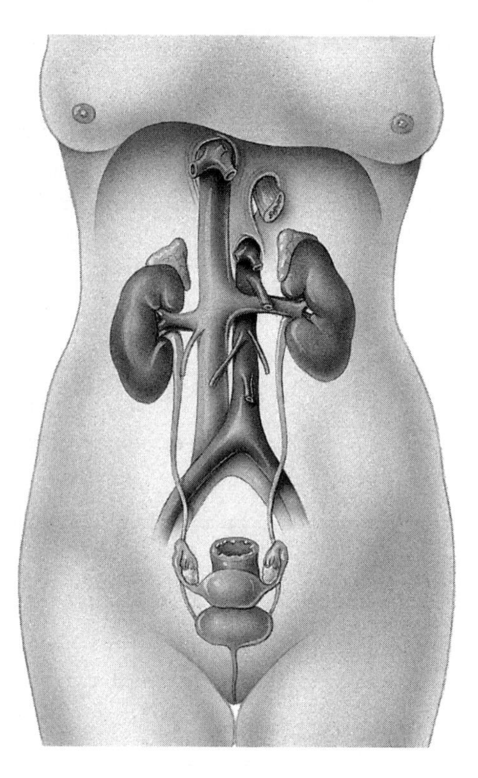

NOTES

25

**Figure 25.2** Position and coverings of the kidneys (page 760).

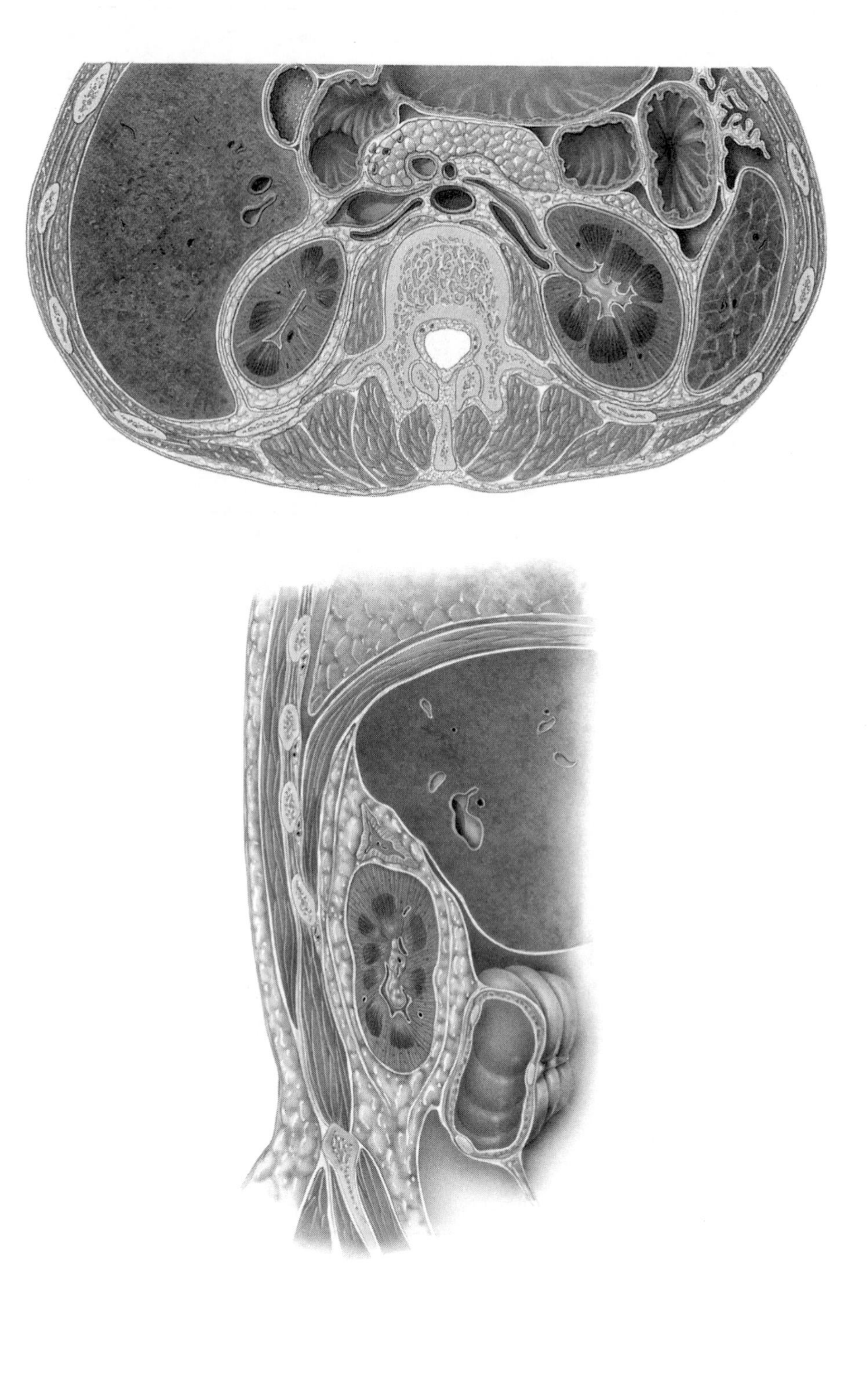

NOTES

25

**Figure 25.3a** Internal anatomy of the kidneys (page 761).

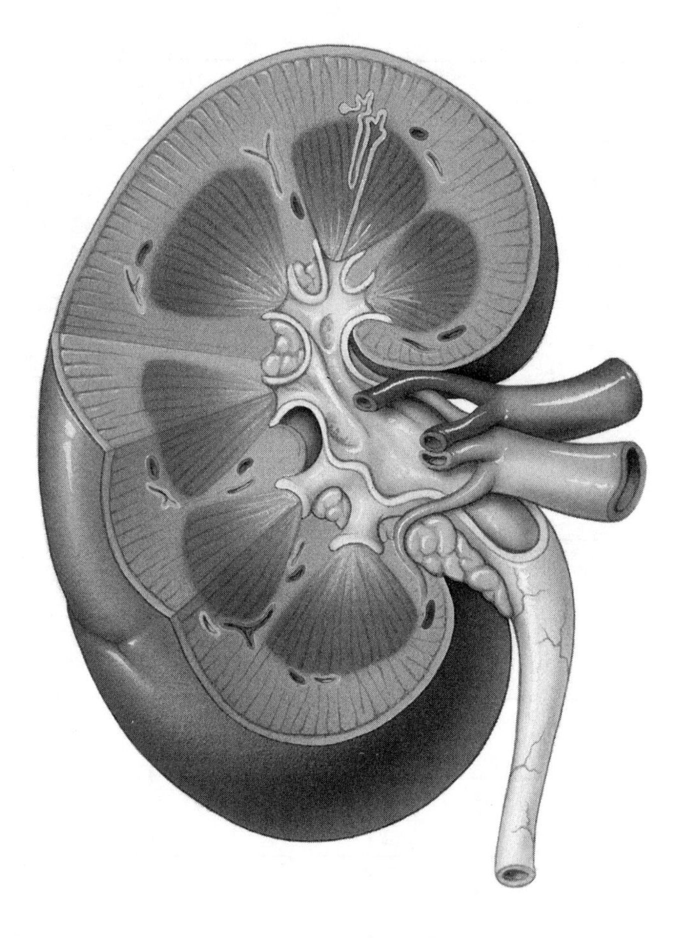

## NOTES

25

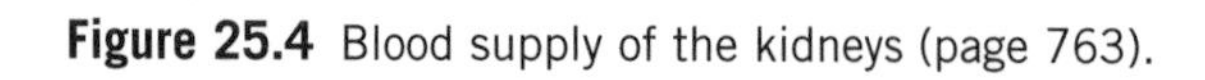

**Figure 25.4** Blood supply of the kidneys (page 763).

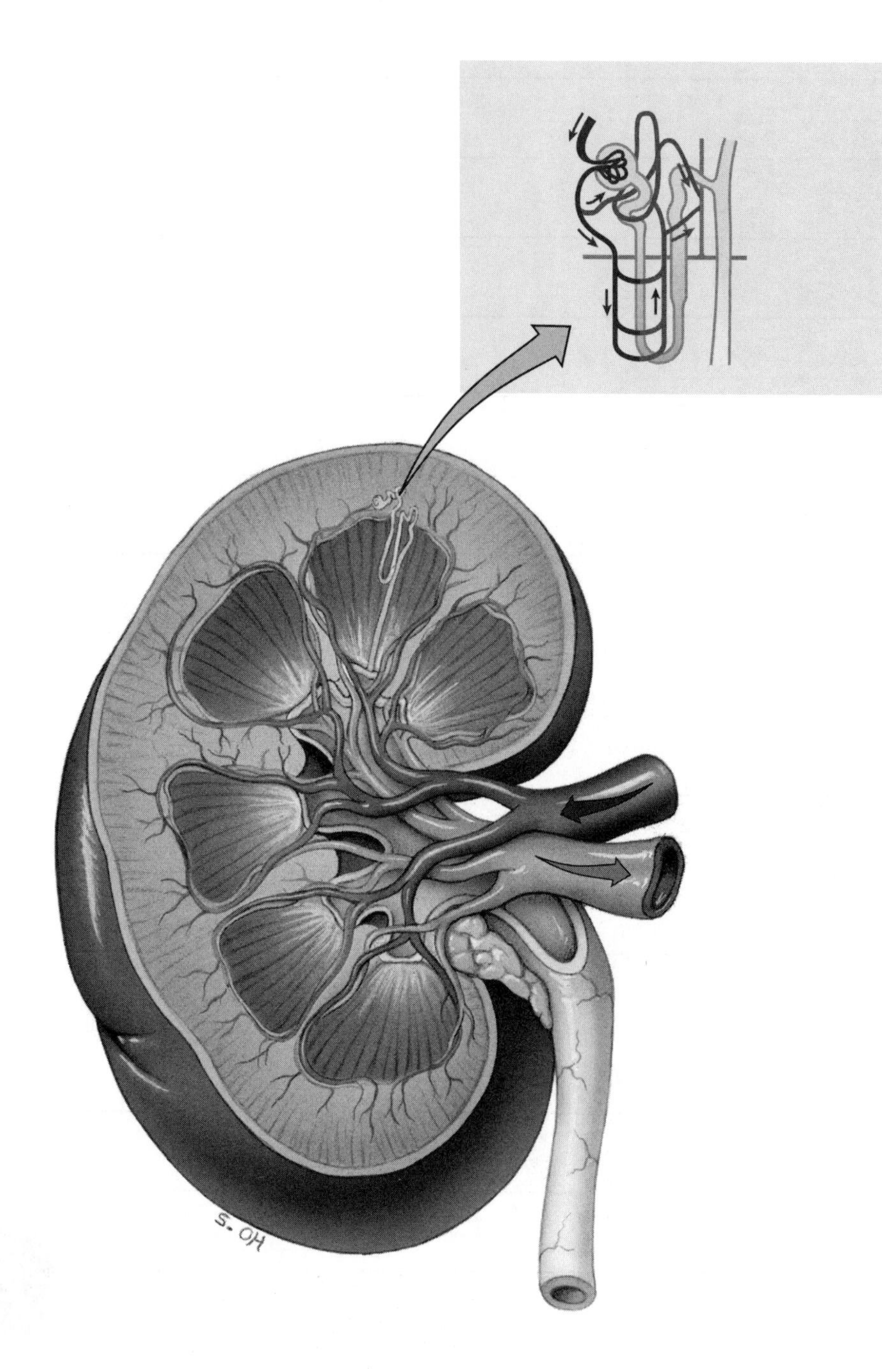

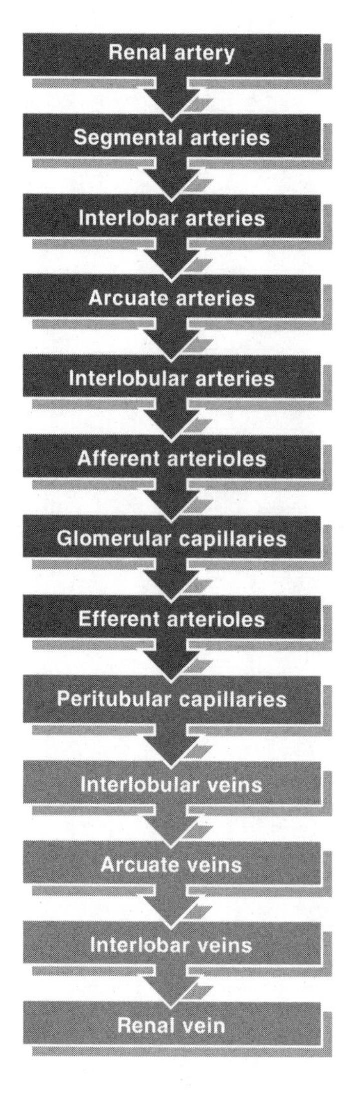

NOTES

25

**Figure 25.5a** The structure of nephrons and associated blood vessels (page 764).

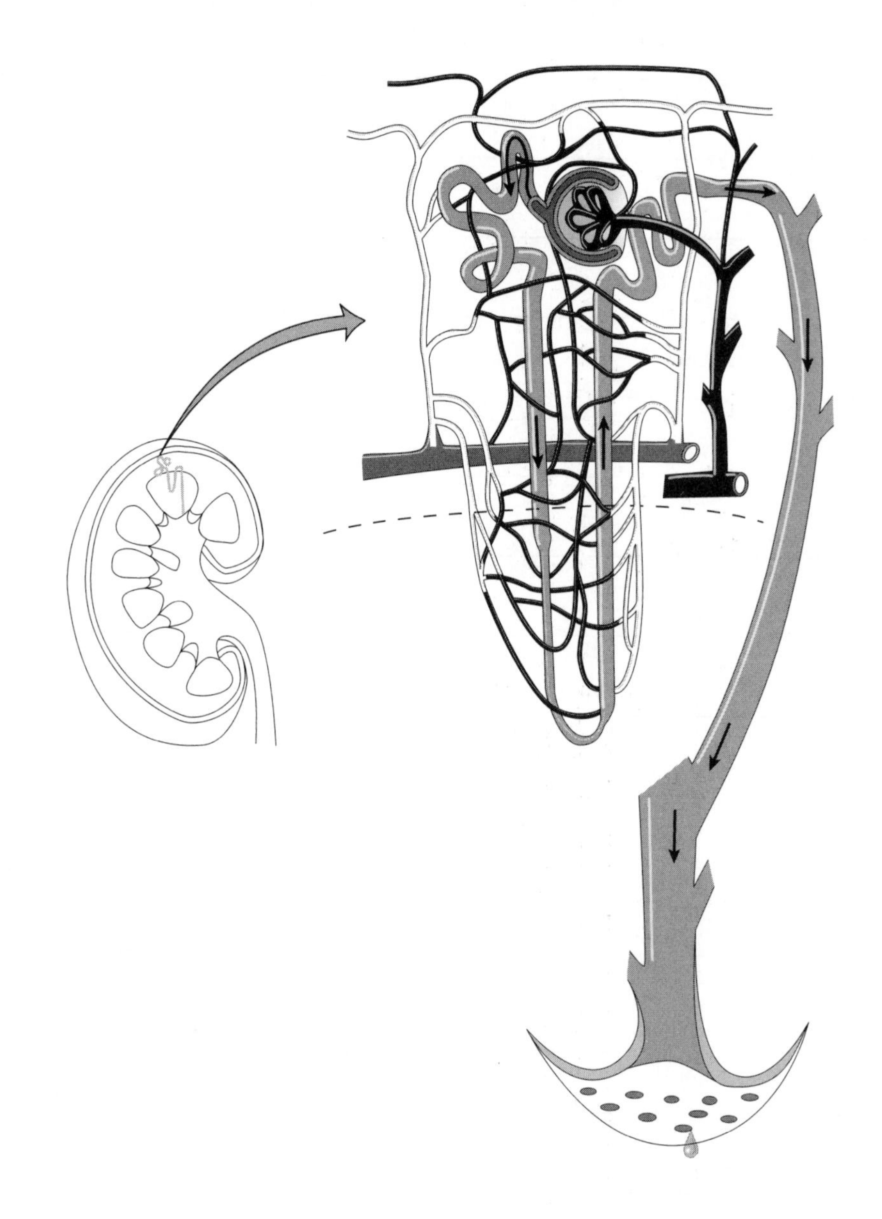

NOTES

25

**Figure 25.5b** The structure of nephrons and associated blood vessels (page 765).

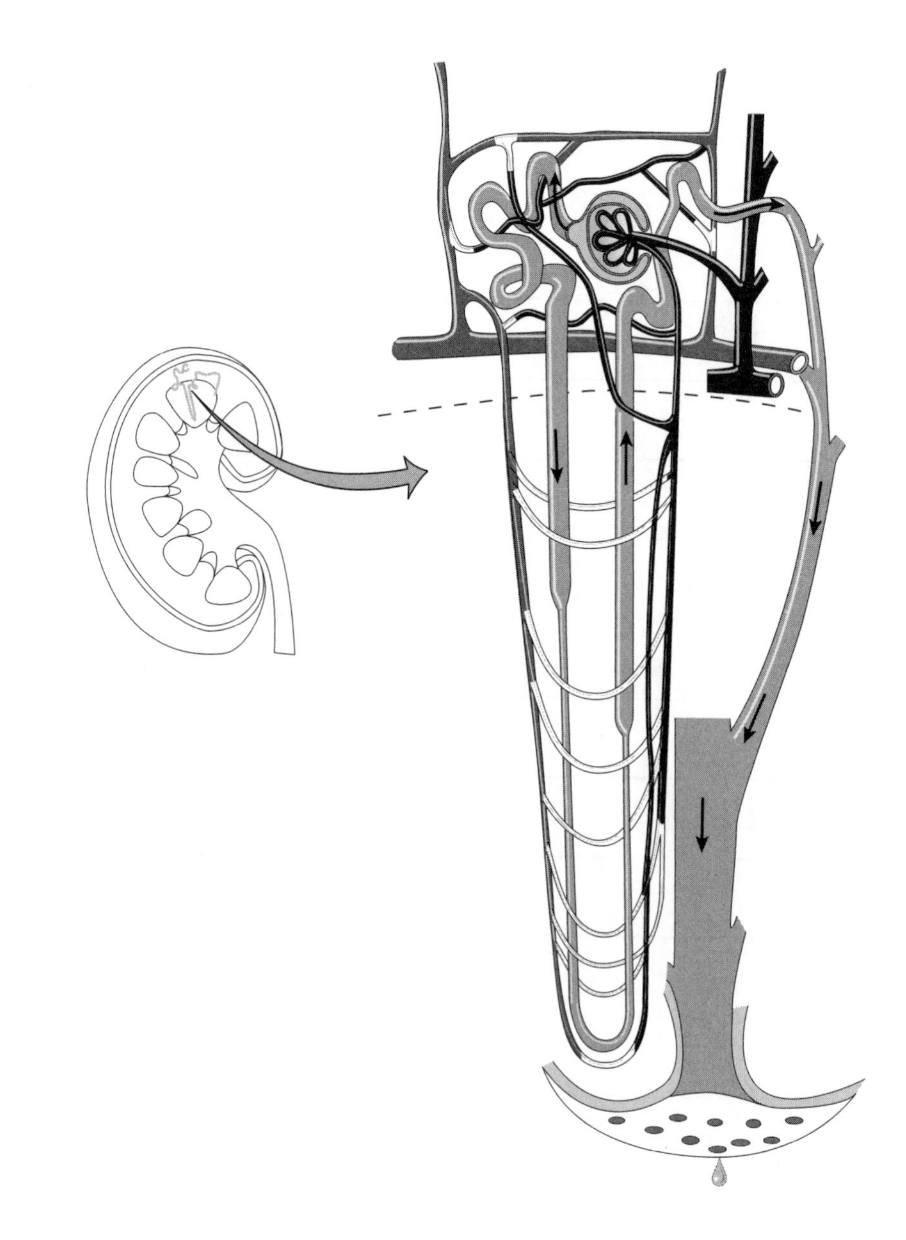

## NOTES

25

**Figure 25.6a** Histology of a renal corpuscle (page 766).

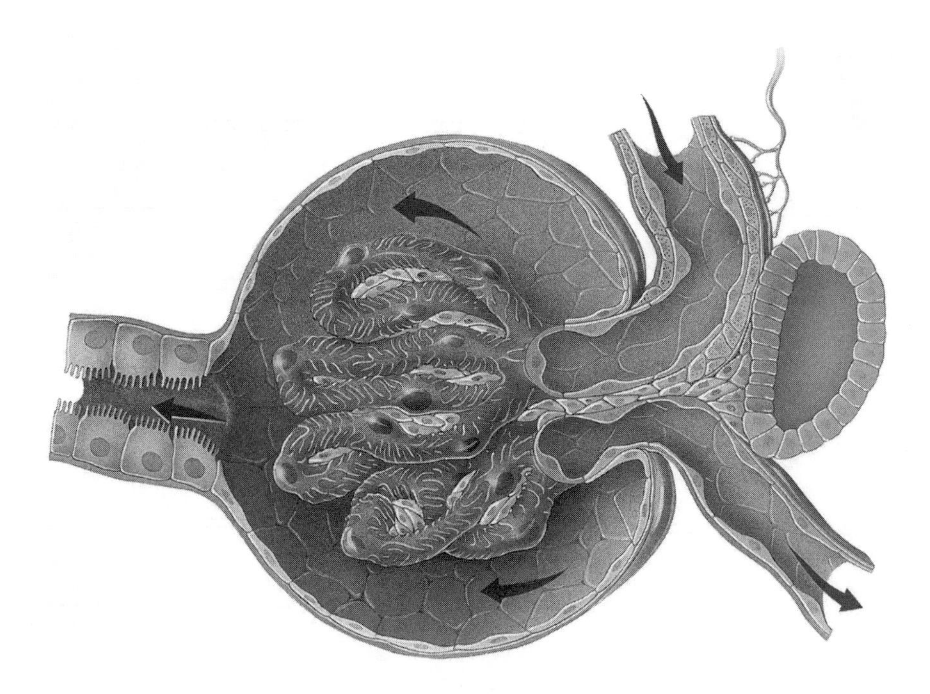

**Figure 25.7** Relation of a nephron's structure to its three basic functions (page 768).

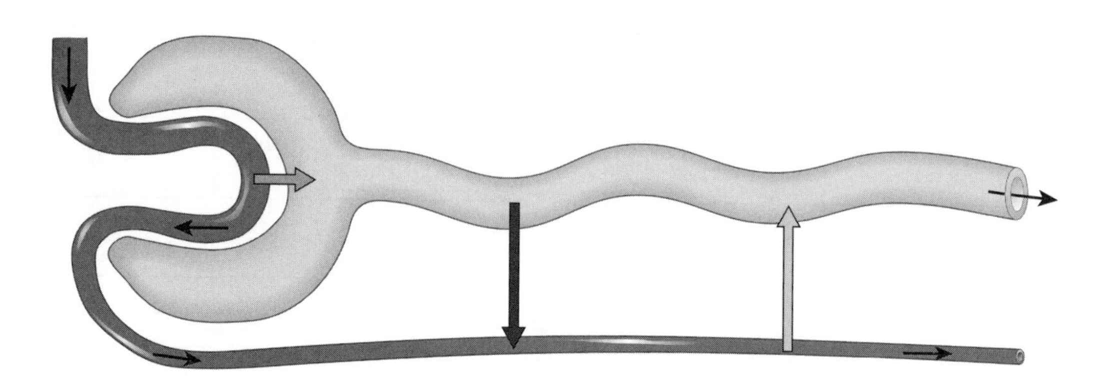

NOTES

25

**Figure 25.8a** The filtration (endothelial–capsular) membrane (page 769).

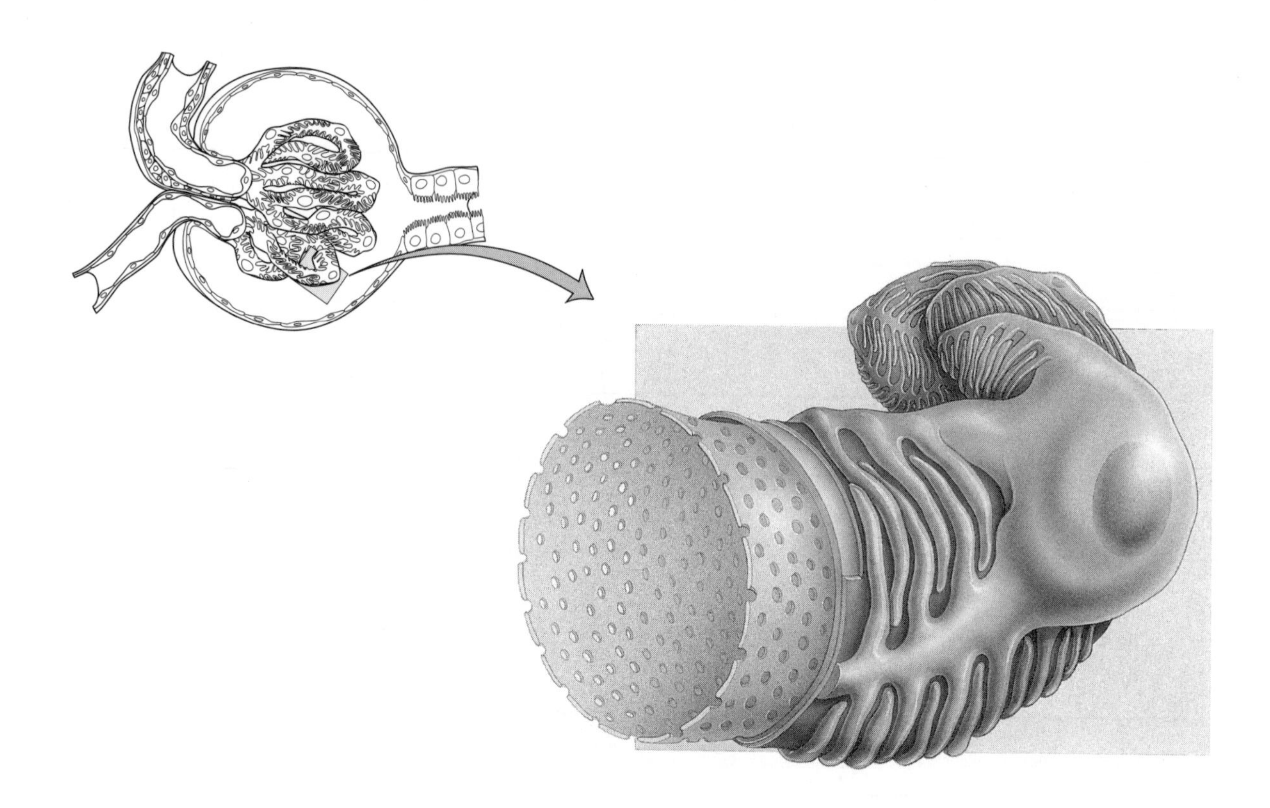

**Figure 25.9a** Ureters, urinary bladder, and urethra (shown in a female) (page 771).

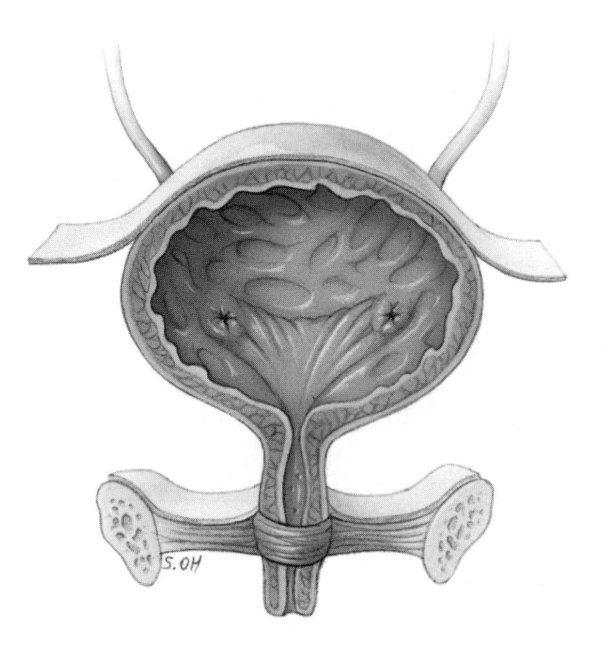

NOTES

25

**Figure 25.12a–d** Development of the urinary system (page 775, 776).

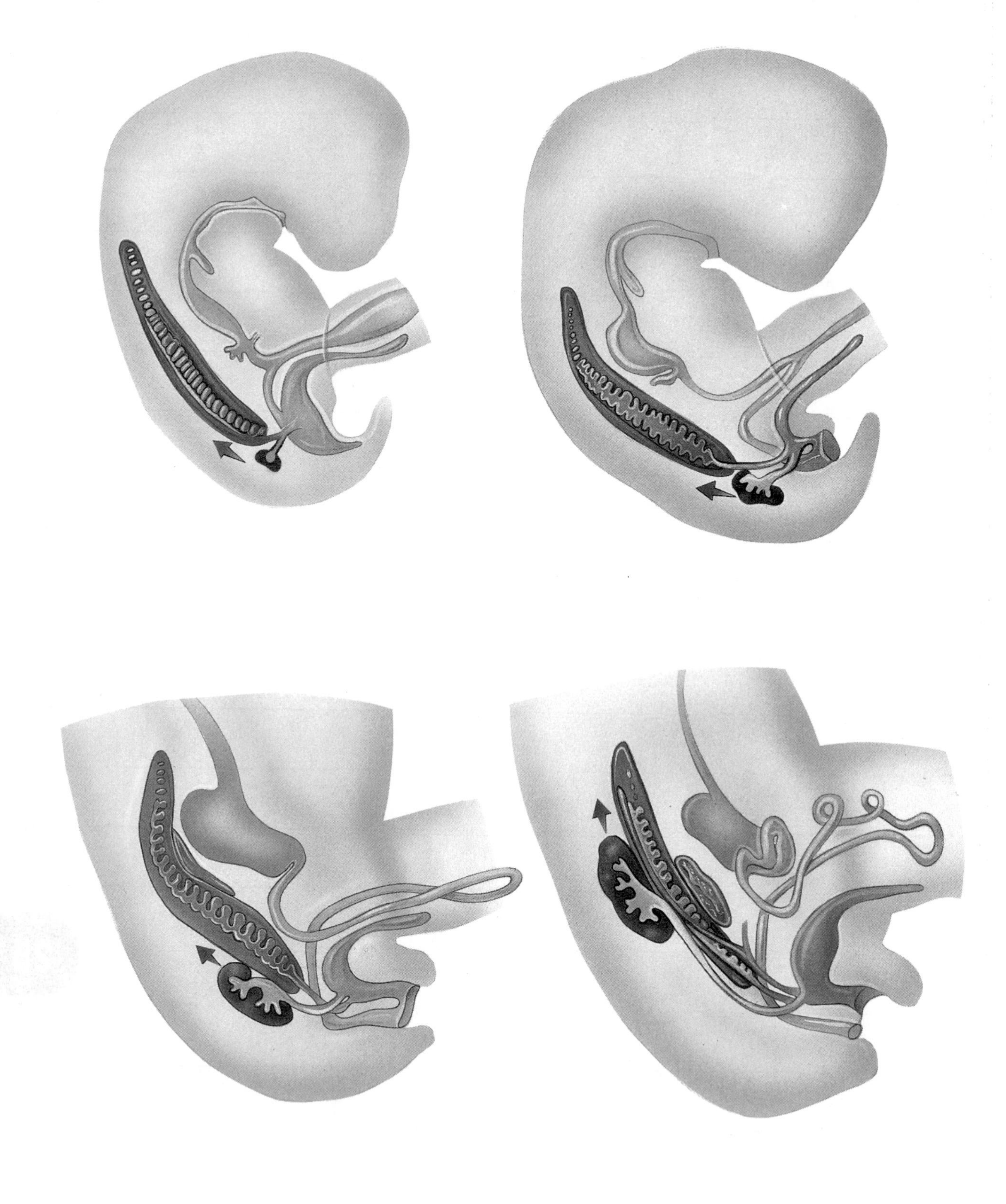

NOTES

25

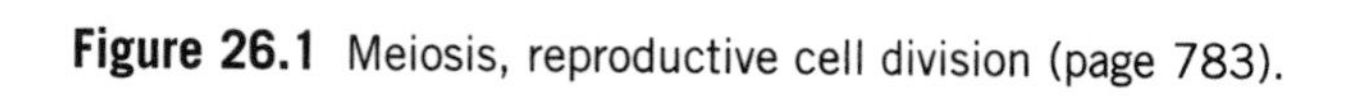

**Figure 26.1** Meiosis, reproductive cell division (page 783).

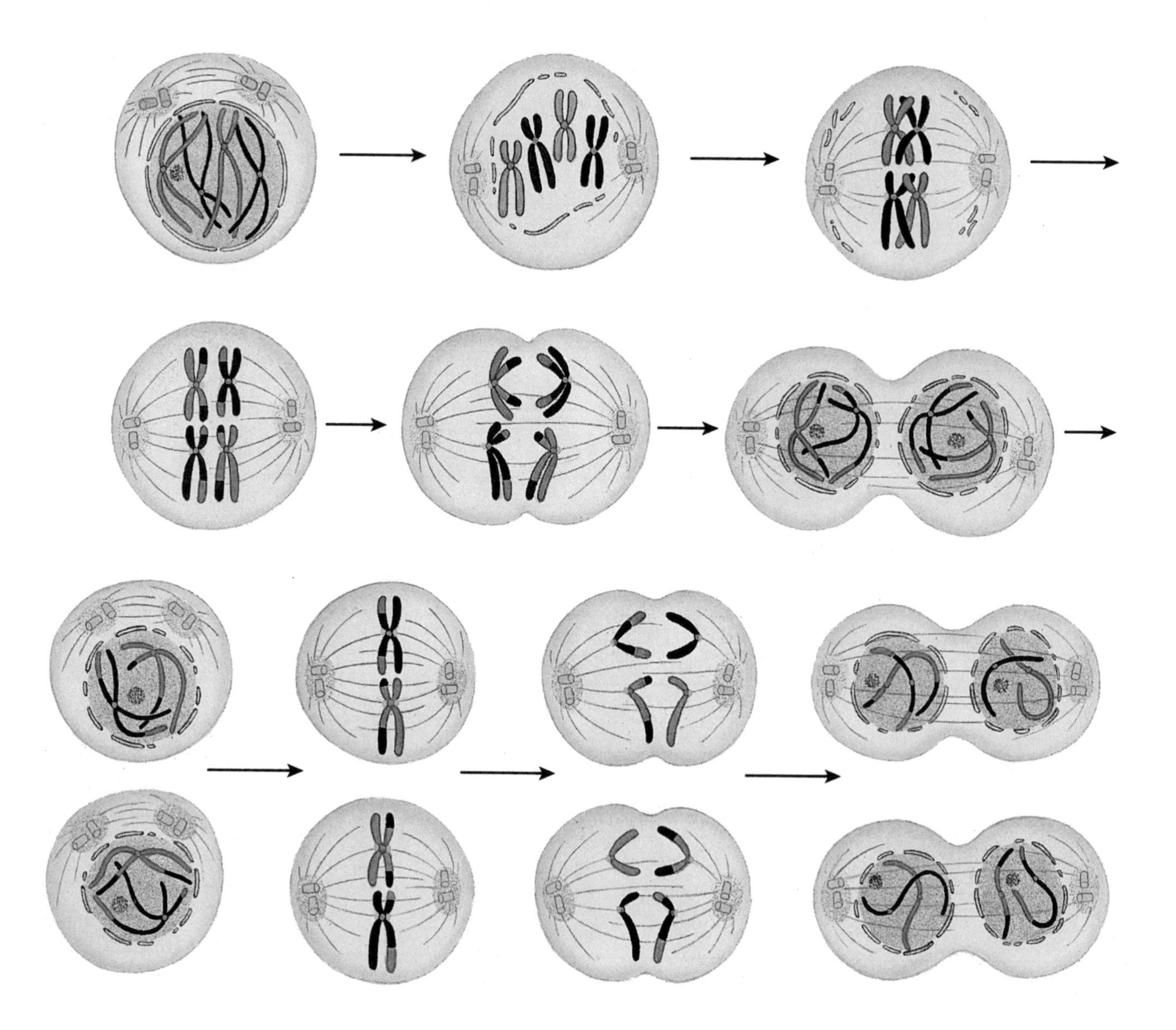

**Figure 26.2** Crossing-over within a tetrad during prophase I of meiosis (page 783).

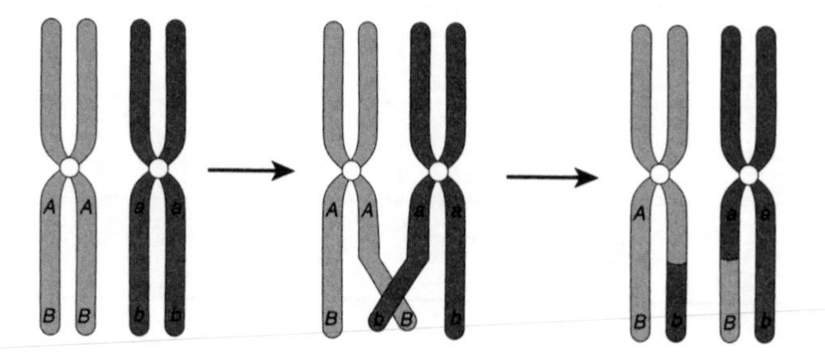

## NOTES

26

**Figure 26.3a** Male organs of reproduction and surrounding structures (page 784).

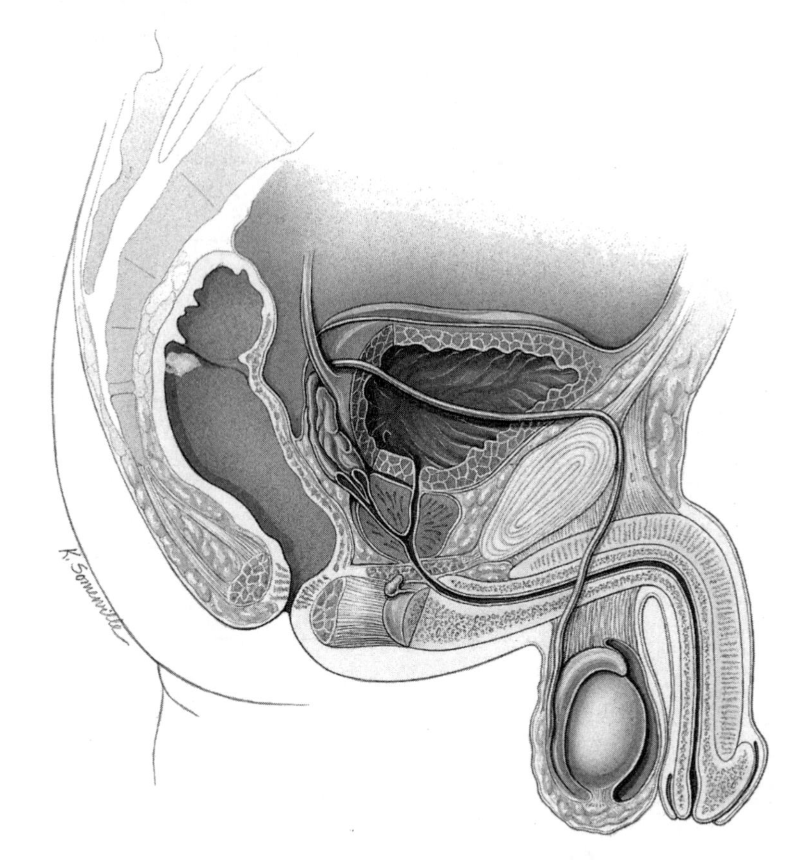

**Figure 26.4** The scrotum, the supporting structure for the testes (page 786).

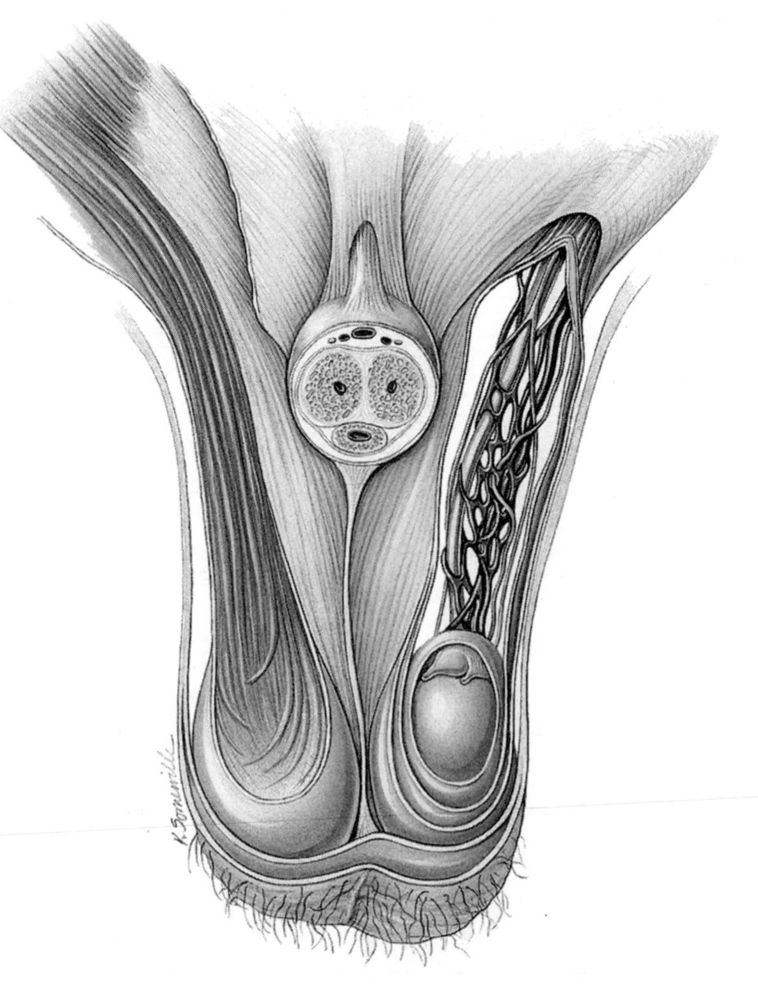

NOTES

26

**Figure 26.5a** Internal and external anatomy of a testis (page 787).

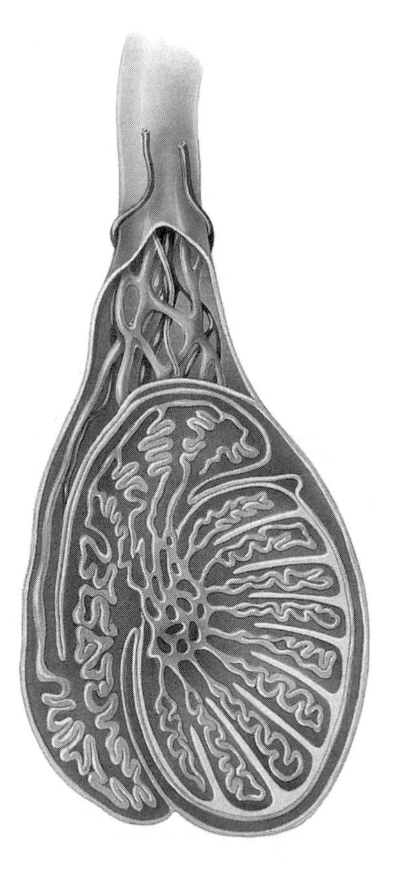

**Figure 26.6b** Microscopic anatomy of the seminiferous tubules (page 788).

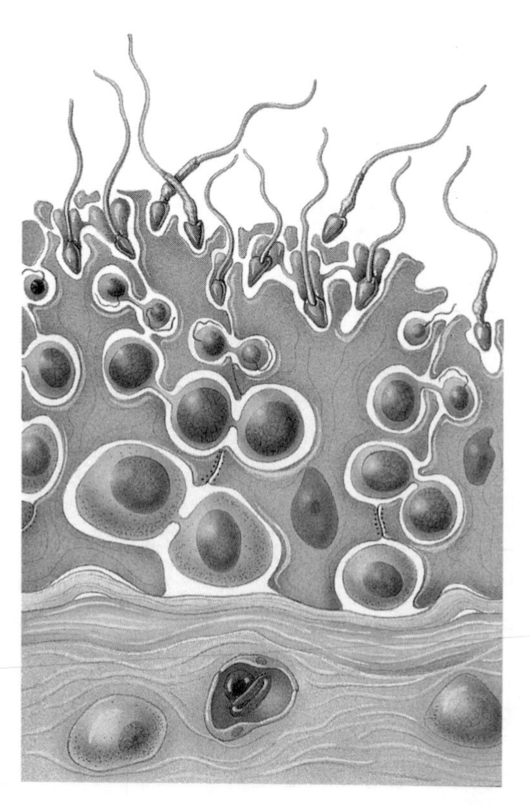

## NOTES

26

**Figure 26.7** Events in spermatogenesis (page 789).

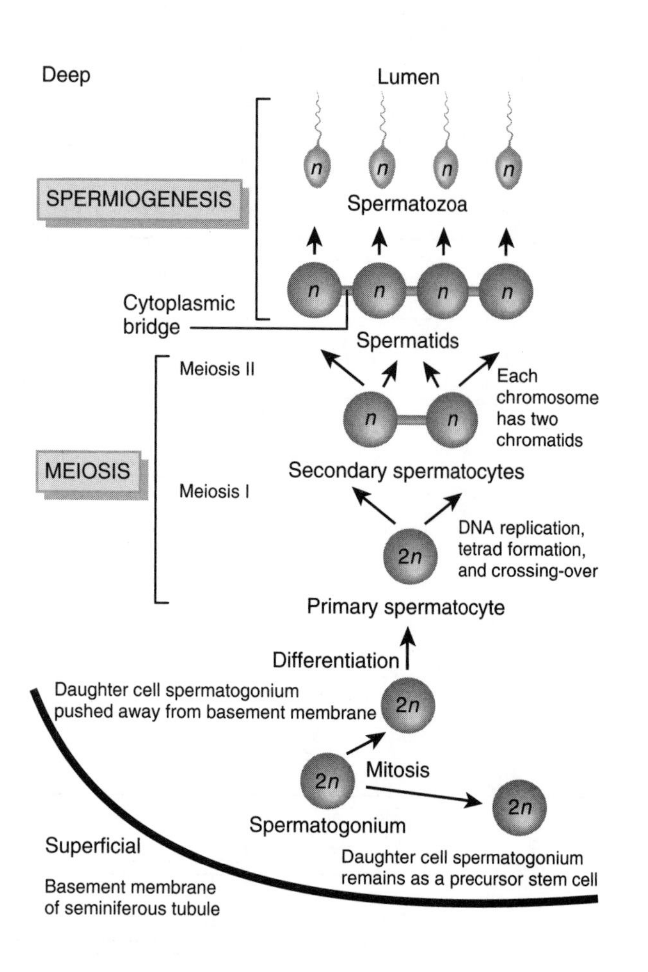


**Figure 26.8** A sperm cell (spermatozoon) (page 790).

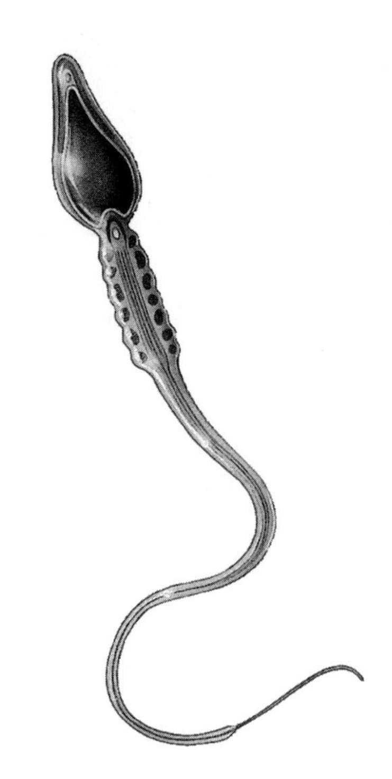

NOTES

26

**Figure 26.11** Locations of several accessory reproductive organs in males (page 793).

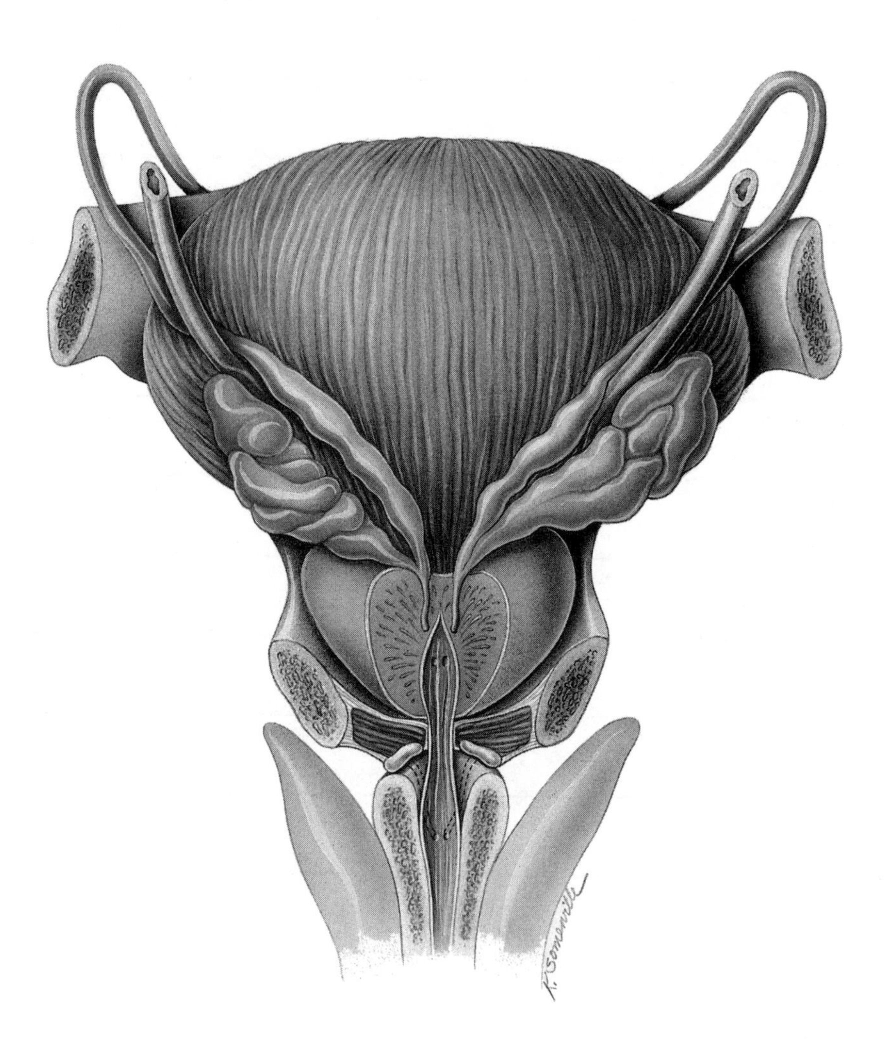

## NOTES

26

**Figure 26.12** Internal structure of the penis (page 795).

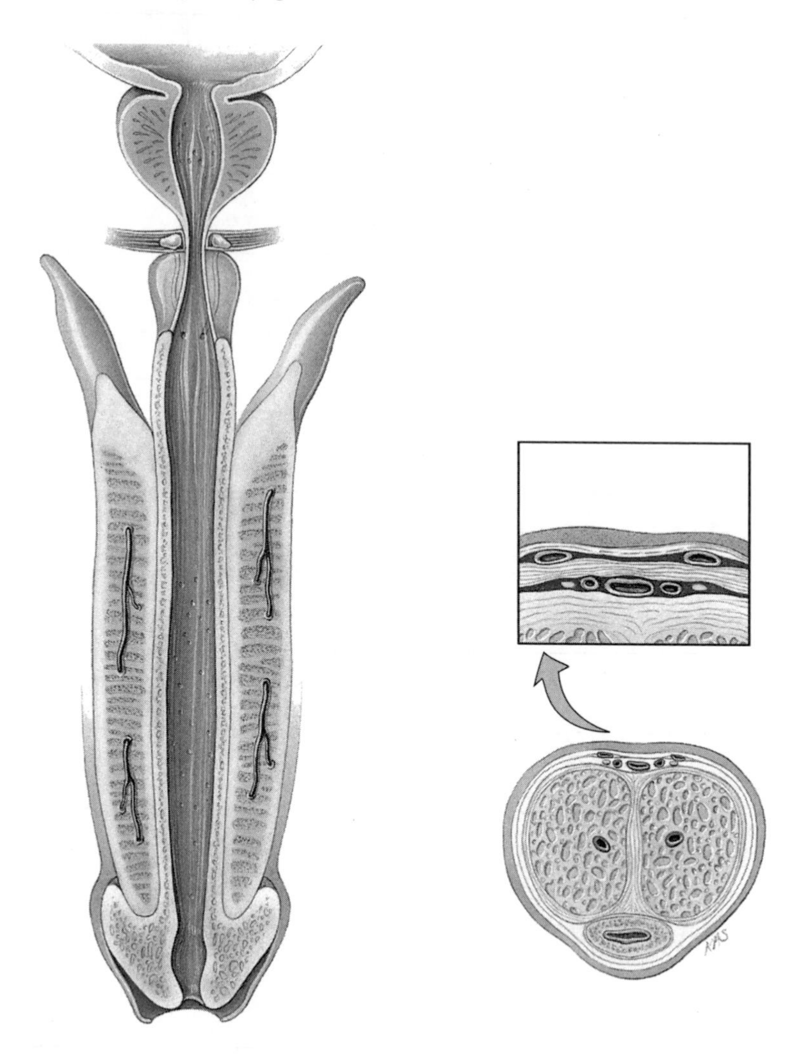

**Figure 26.13a** Organs of reproduction and surrounding structures in females (page 796).

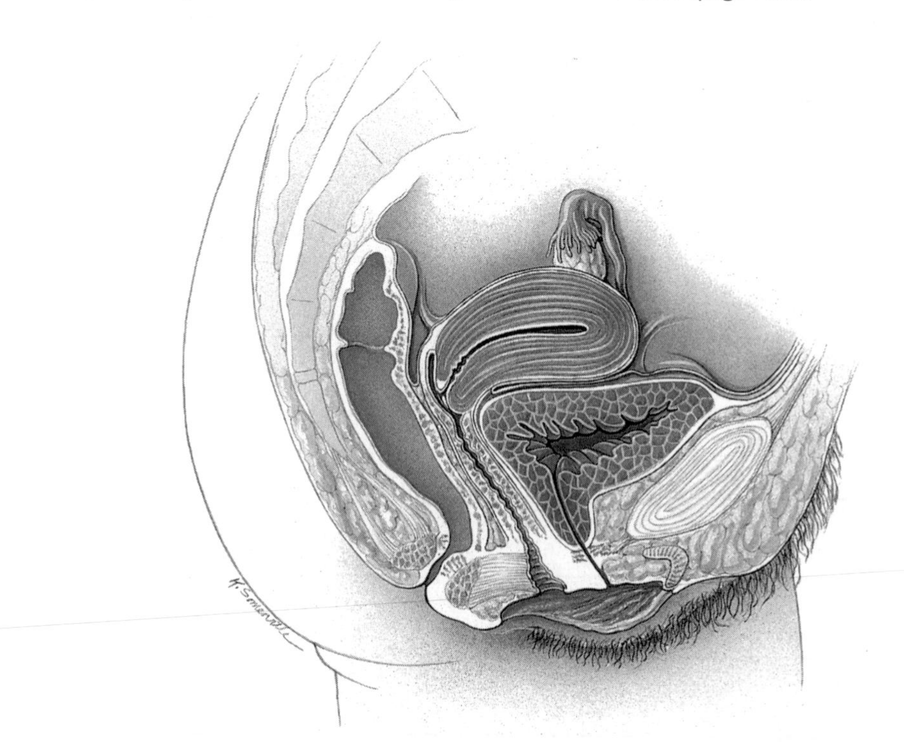

NOTES

**Figure 26.14** Relative positions of the ovaries, the uterus, and the ligaments that support them (page 798).

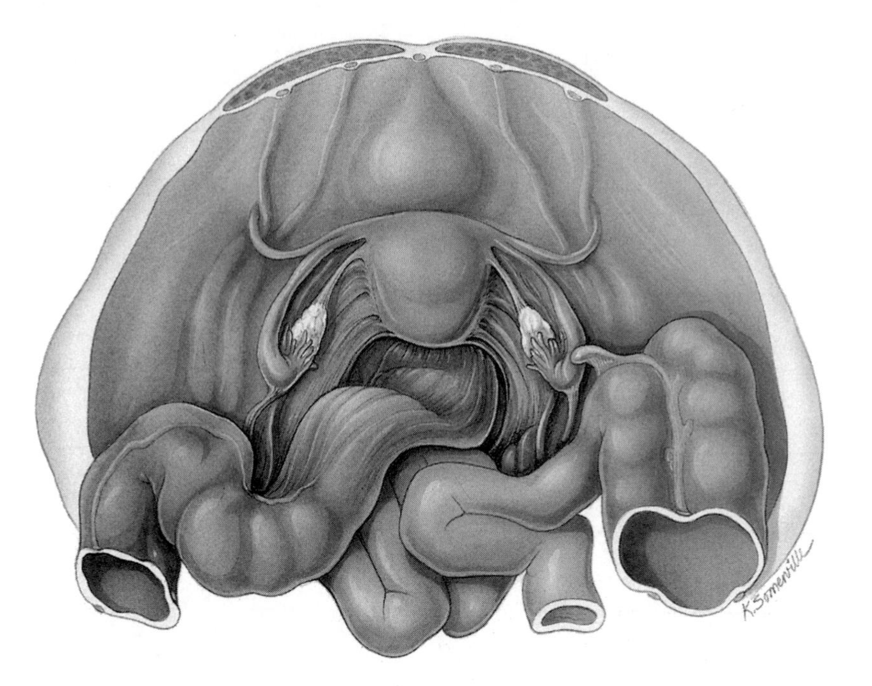

**Figure 26.15** Histology of the ovary (page 799).

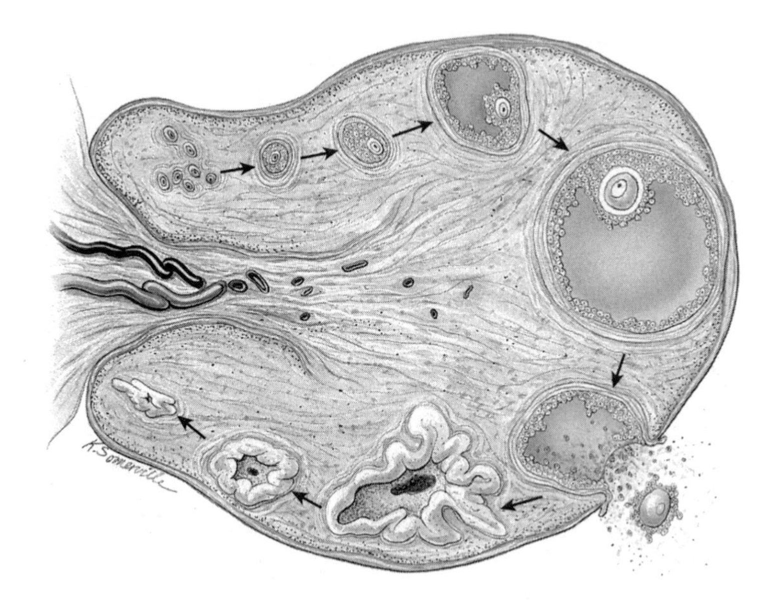

NOTES

26

**Figure 26.17** Oogenesis (page 800).

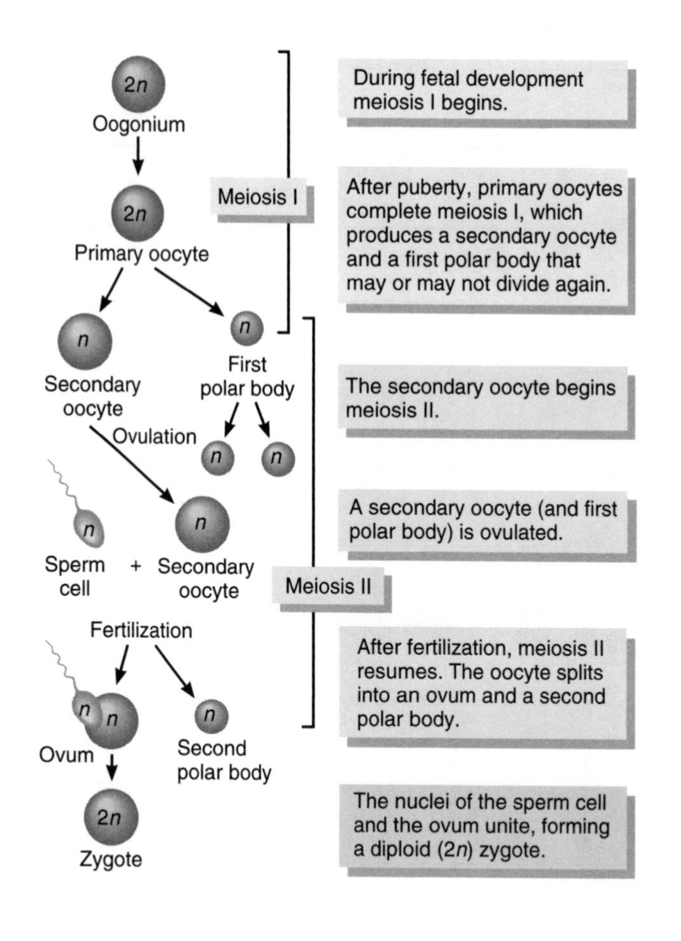

NOTES

26

**Figure 26.18a** Relationship of the uterine (Fallopian) tubes to the ovaries, uterus, and associated structures (page 801).

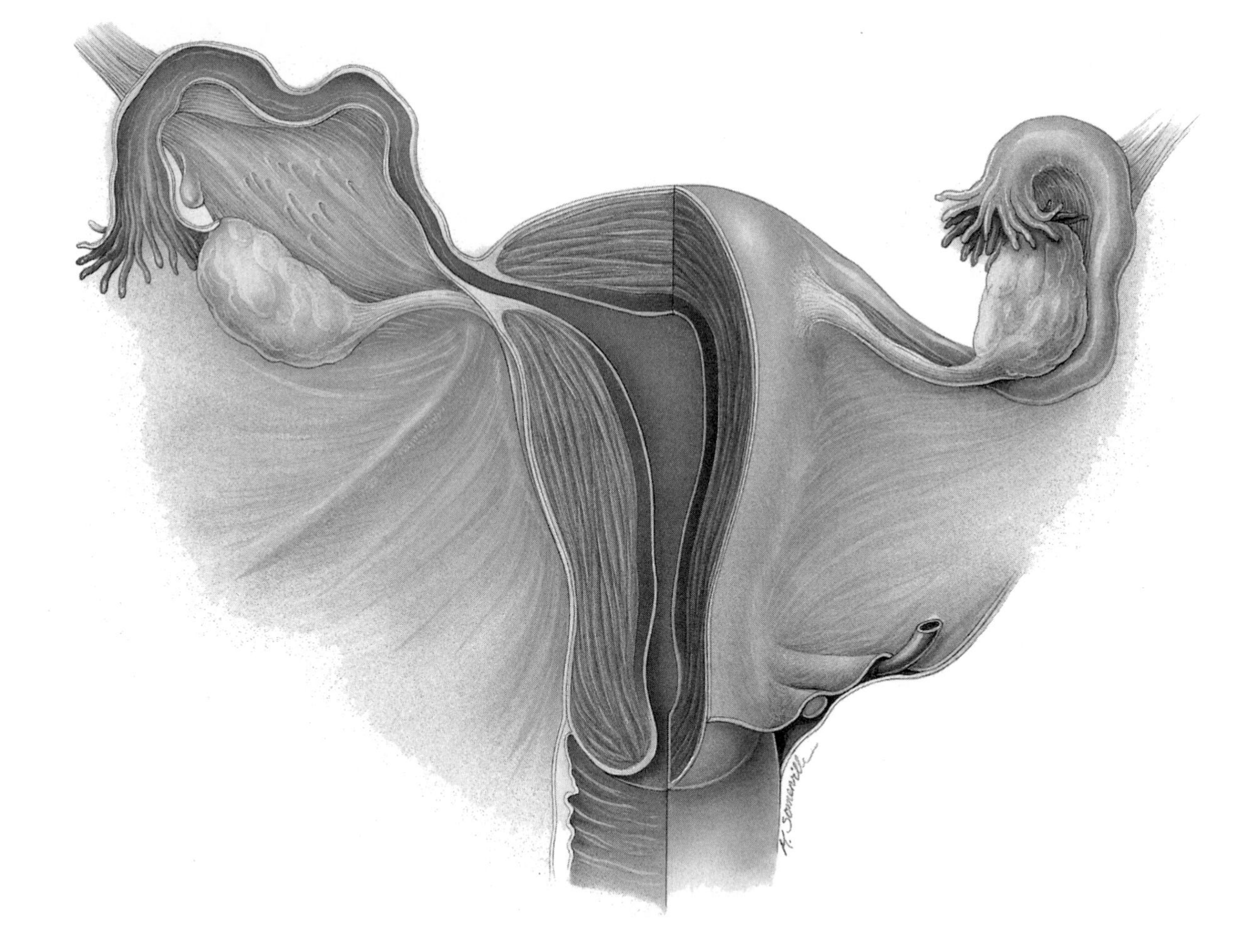

**Figure 26.21** Blood supply of the uterus (page 804).

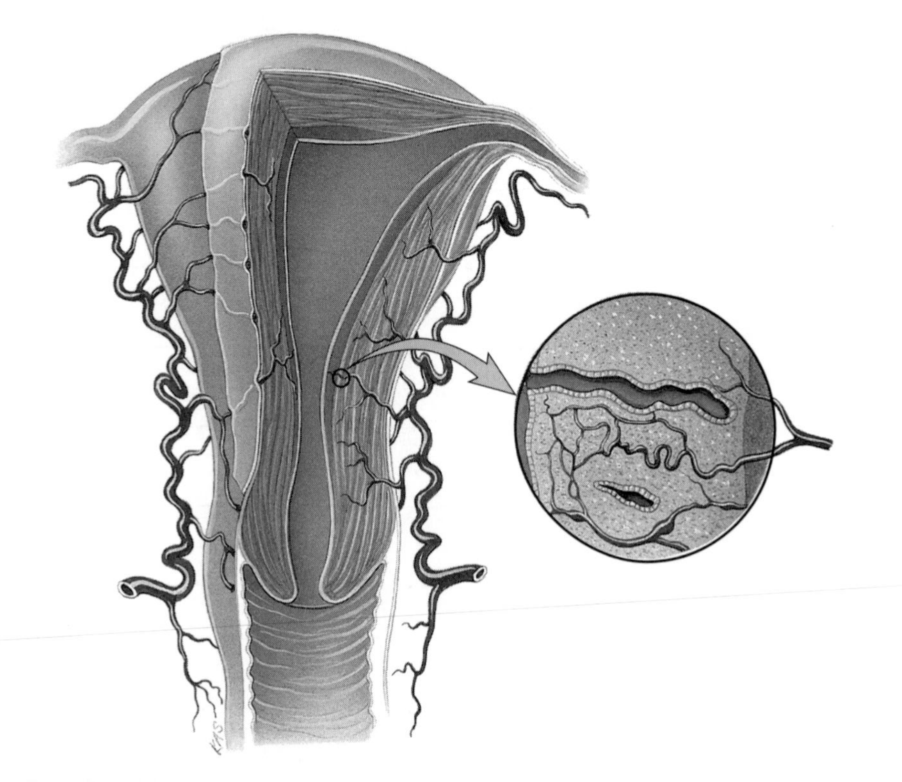

NOTES

26

**Figure 26.22** Components of the vulva (pudendum) (page 806).

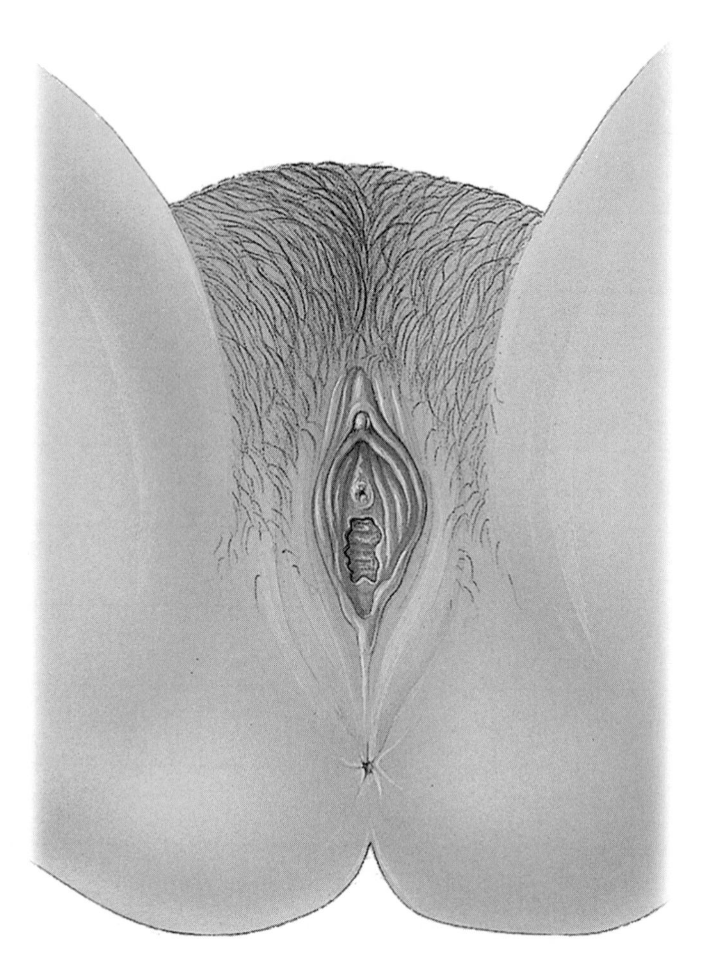

**Figure 26.23** Perineum of a female (page 807).

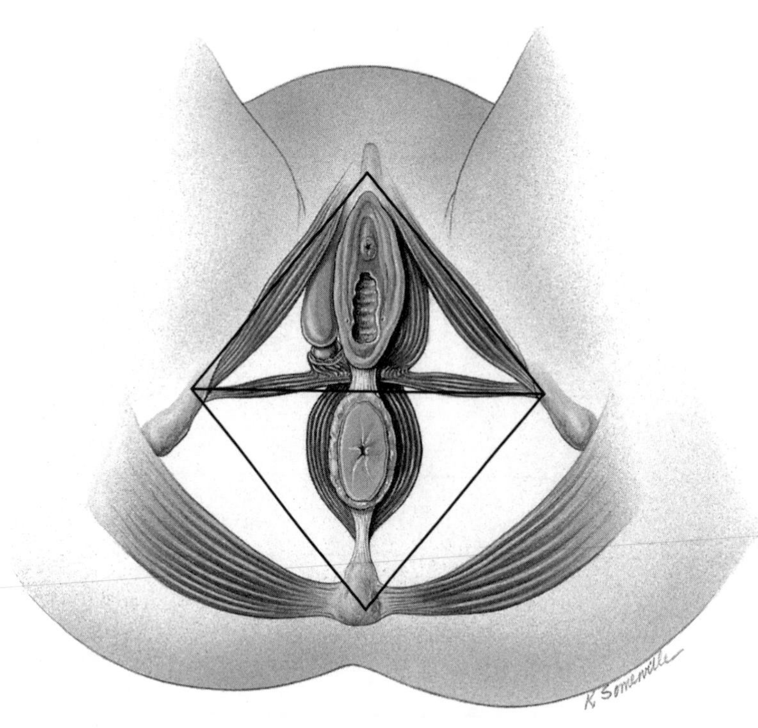

NOTES

26

**Figure 26.24a,b** Mammary glands (page 808).

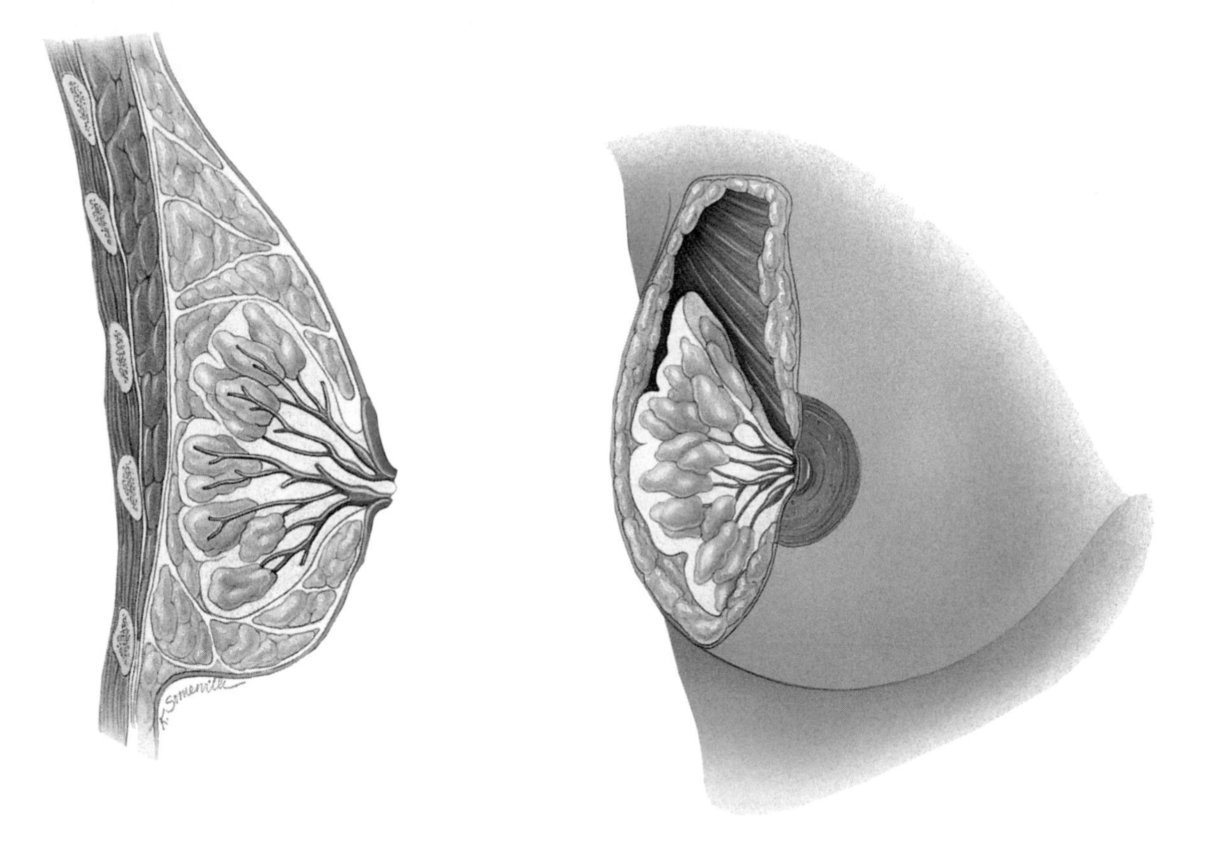

**Figure 26.26** The female reproductive cycle (page 811).

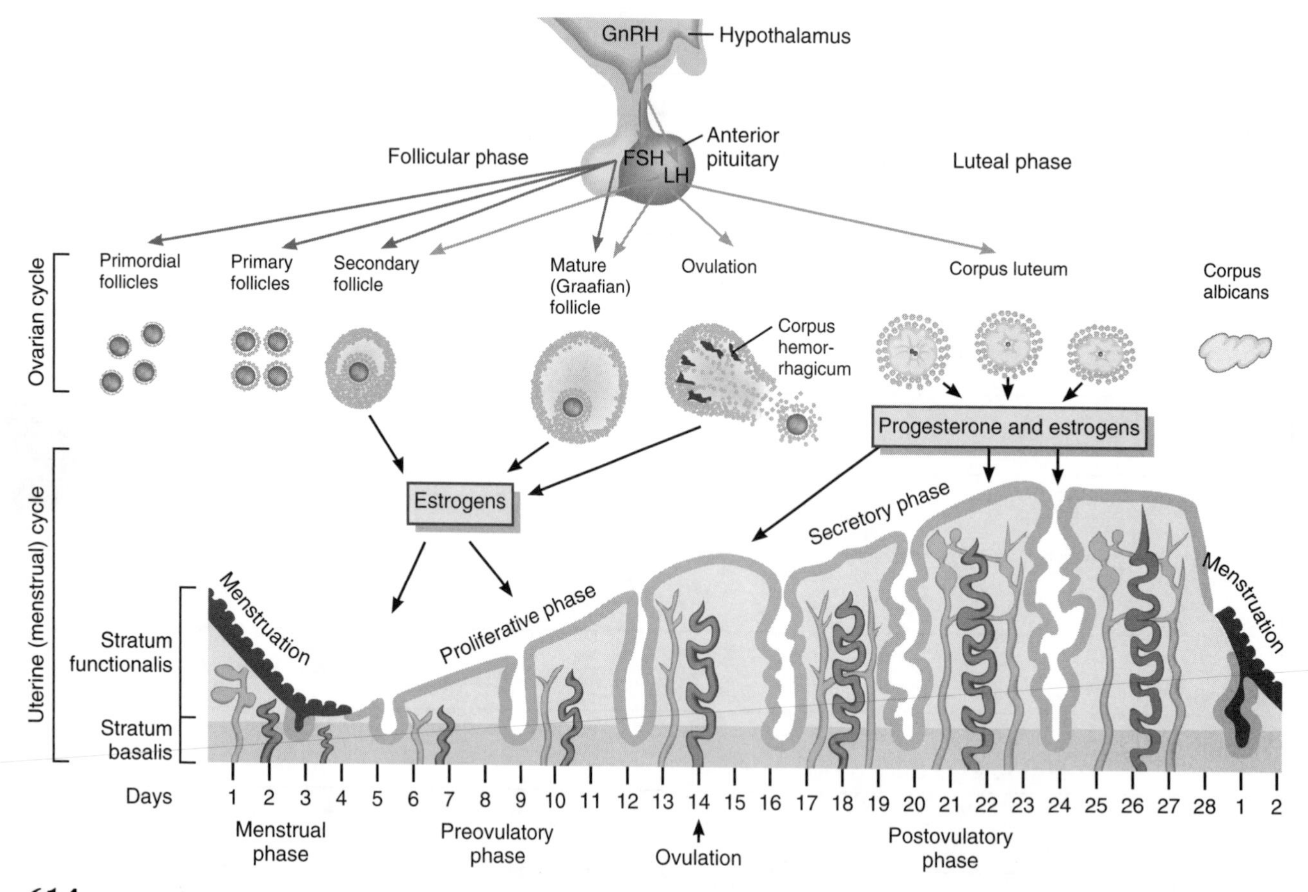

NOTES

26

**Figure 26.27** Development of the internal reproductive systems (page 815).

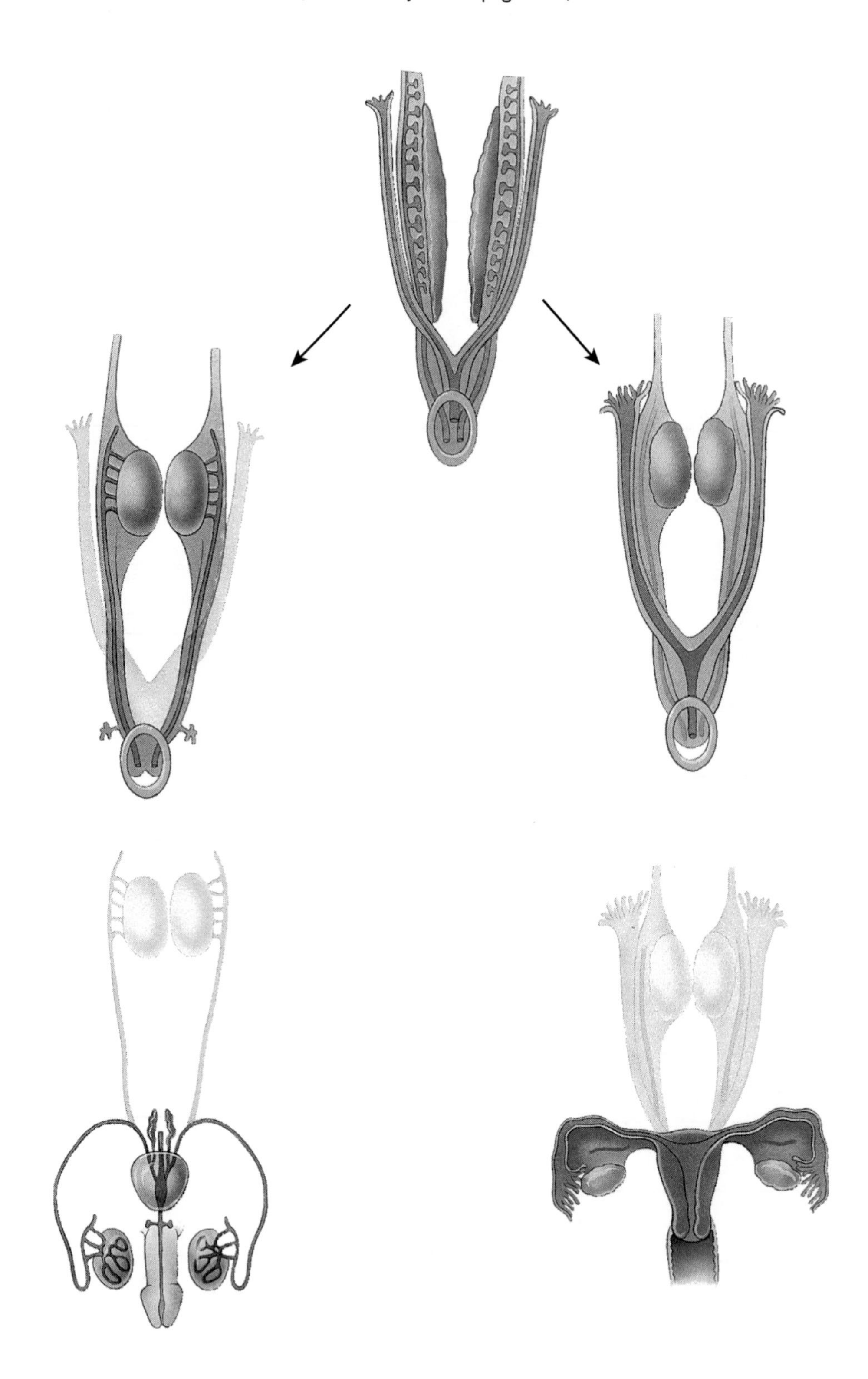

NOTES

26

**Figure 26.28** Development of the external genitals (page 816).

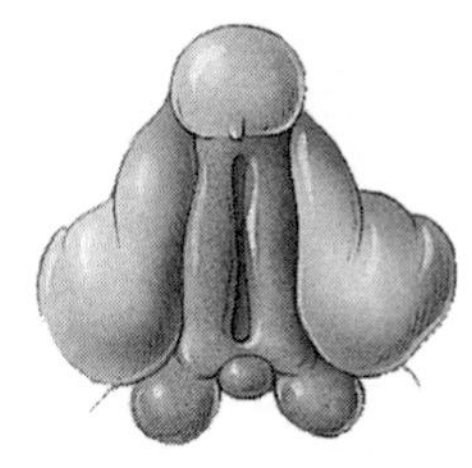

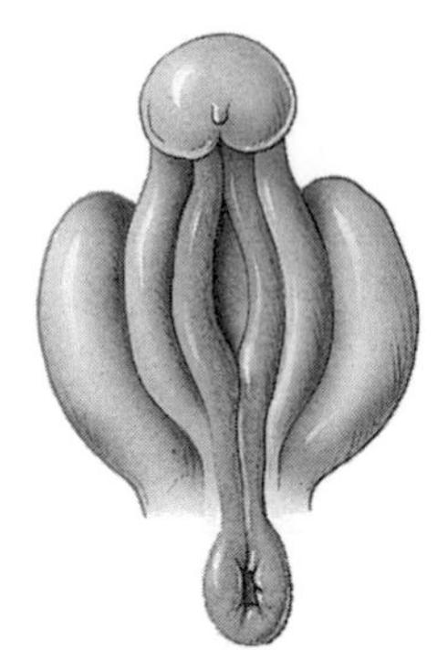

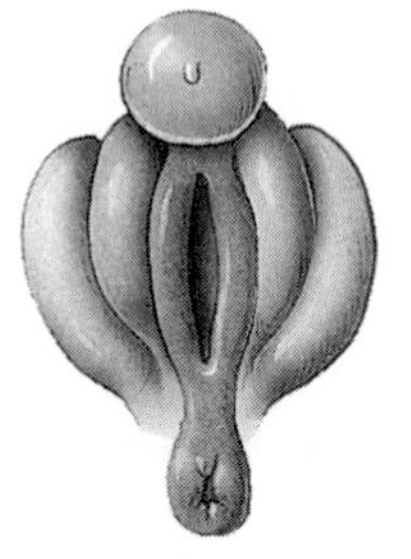

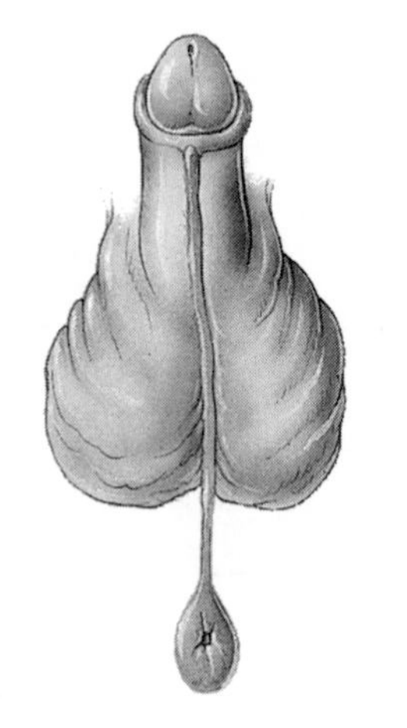

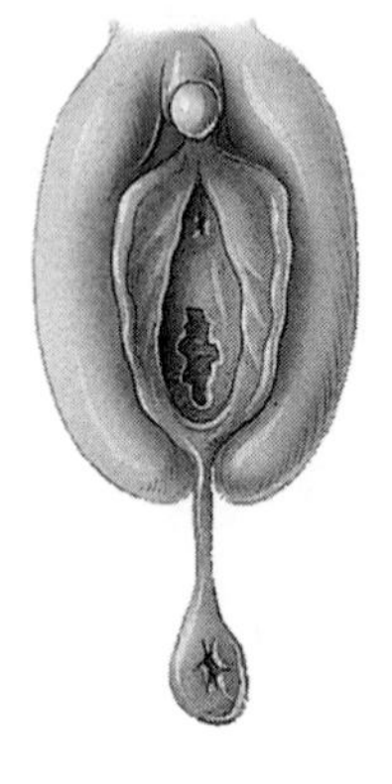

NOTES

26

**Figure 27.1a** Selected structures and events in fertilization (page 827).

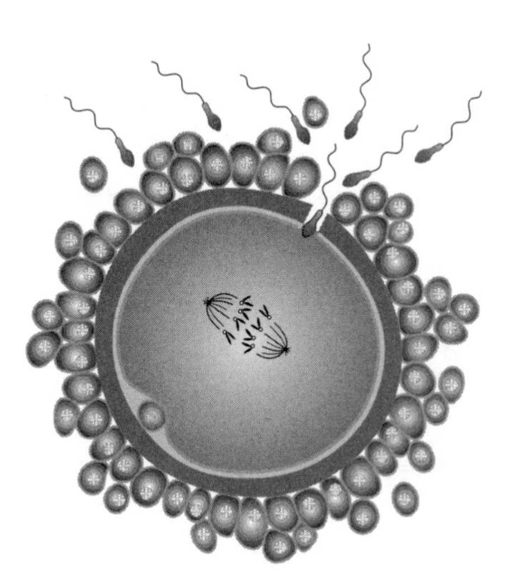

**Figure 27.2** Cleavage and the formation of the morula and blastocyst (page 828).

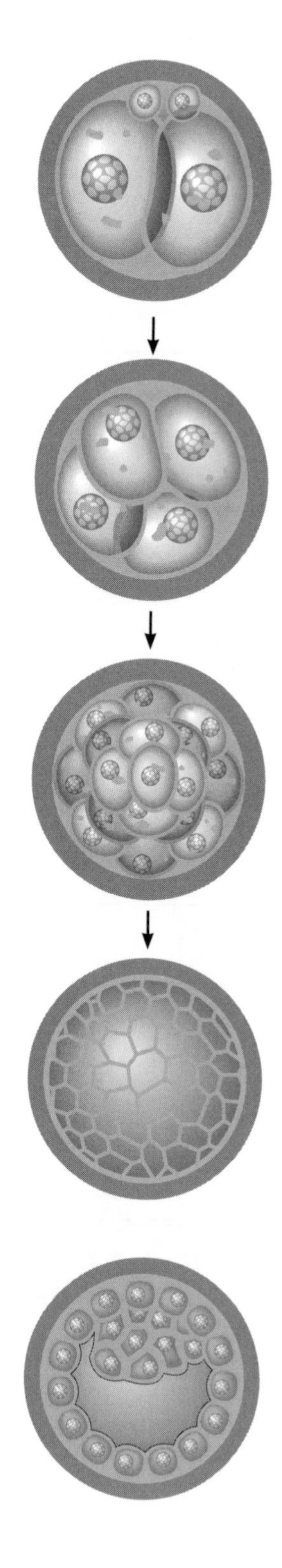

NOTES

27

**Figure 27.3** Relation of a blastocyst to the endometrium of the uterus at the time of implantation (page 829).

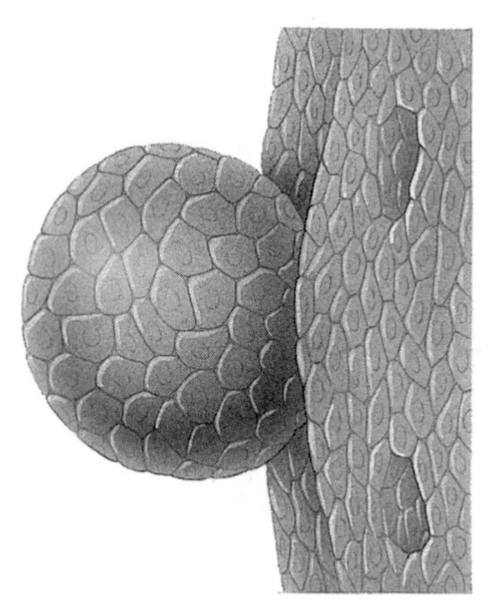

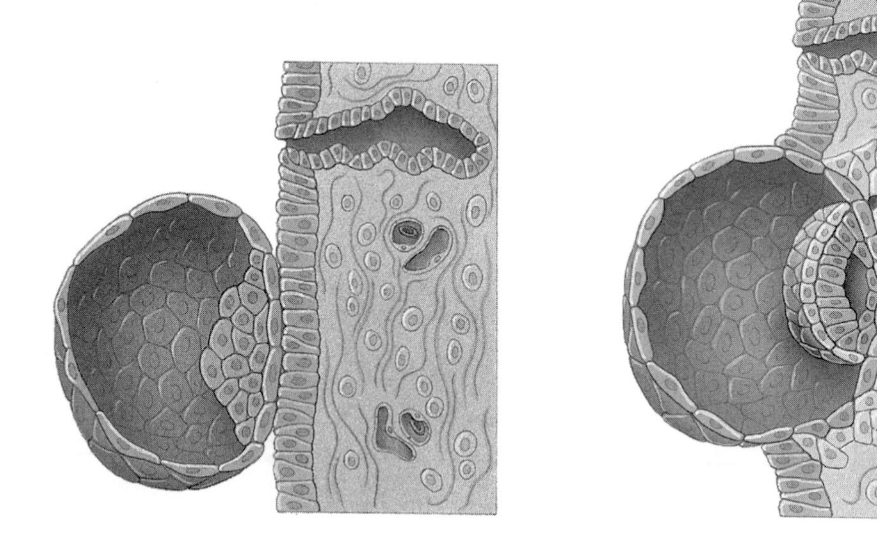

NOTES

27

**Figure 27.4** Summary of events associated with fertilization and implantation (page 830).

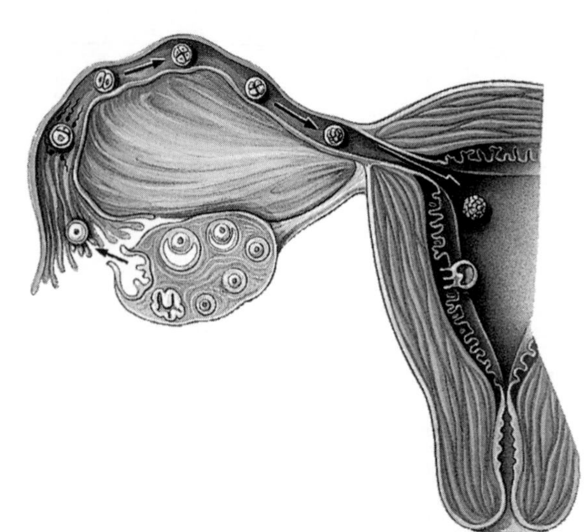

NOTES

27

**Figure 27.5a,b** Formation of the primary germ layers and associated structures (page 832).

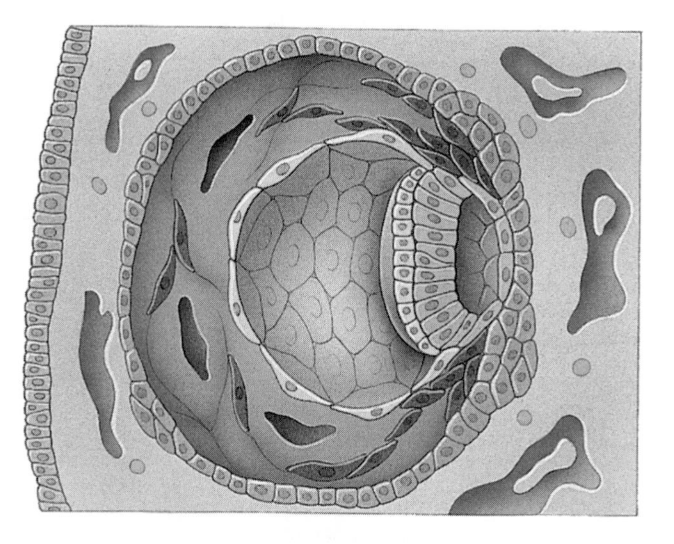

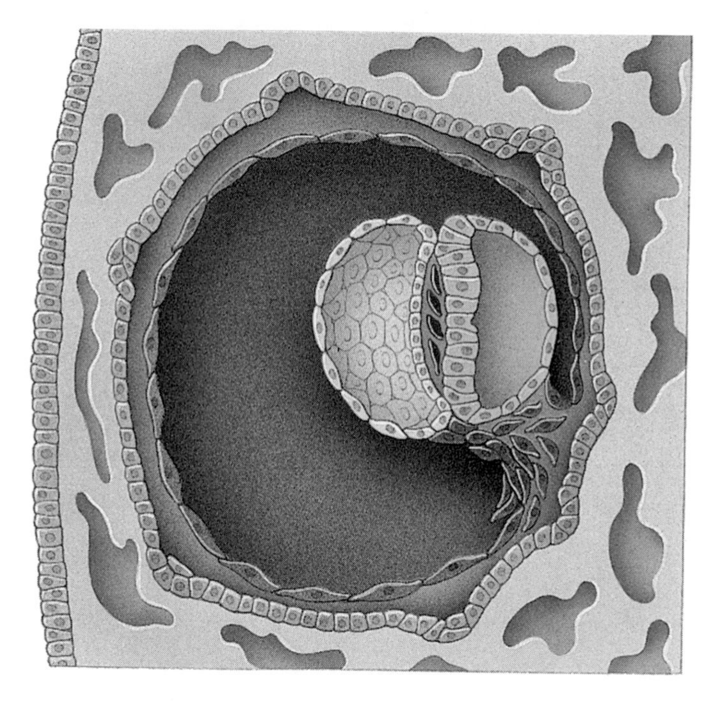

NOTES

27

**Figure 27.5c** Formation of the primary germ layers and associated structures (page 833).

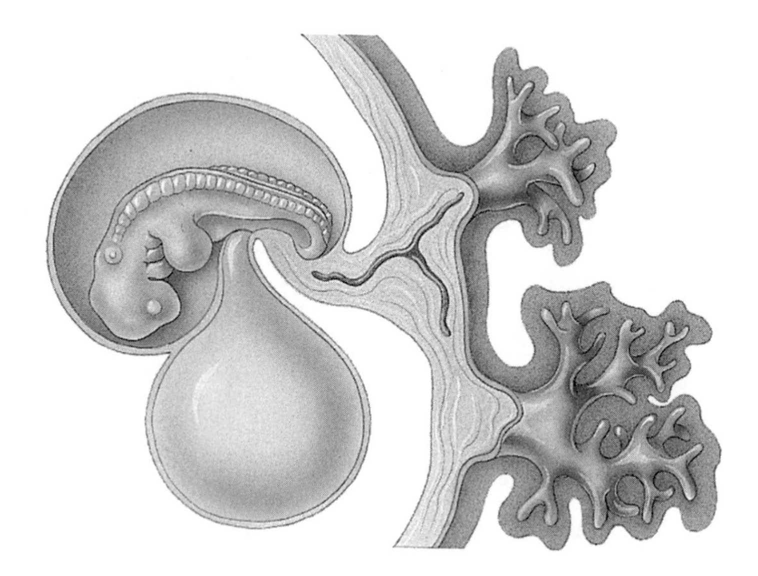

**Figure 27.6a** Embryonic membranes (page 834).

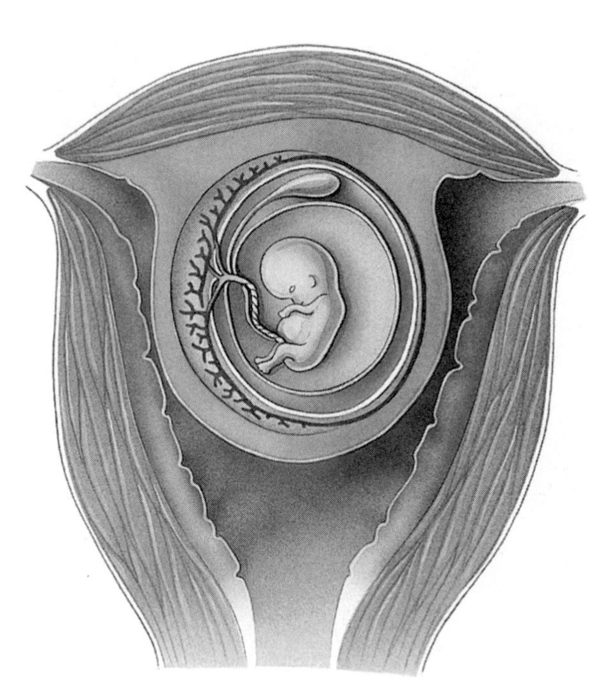

NOTES

27

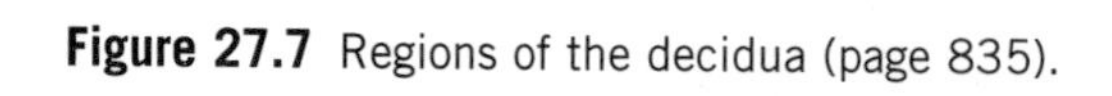

**Figure 27.7** Regions of the decidua (page 835).

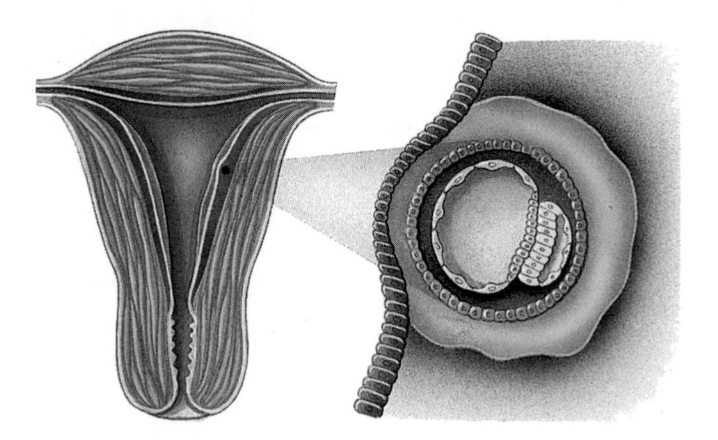

**Figure 27.8a** Placenta and umbilical cord (page 836).

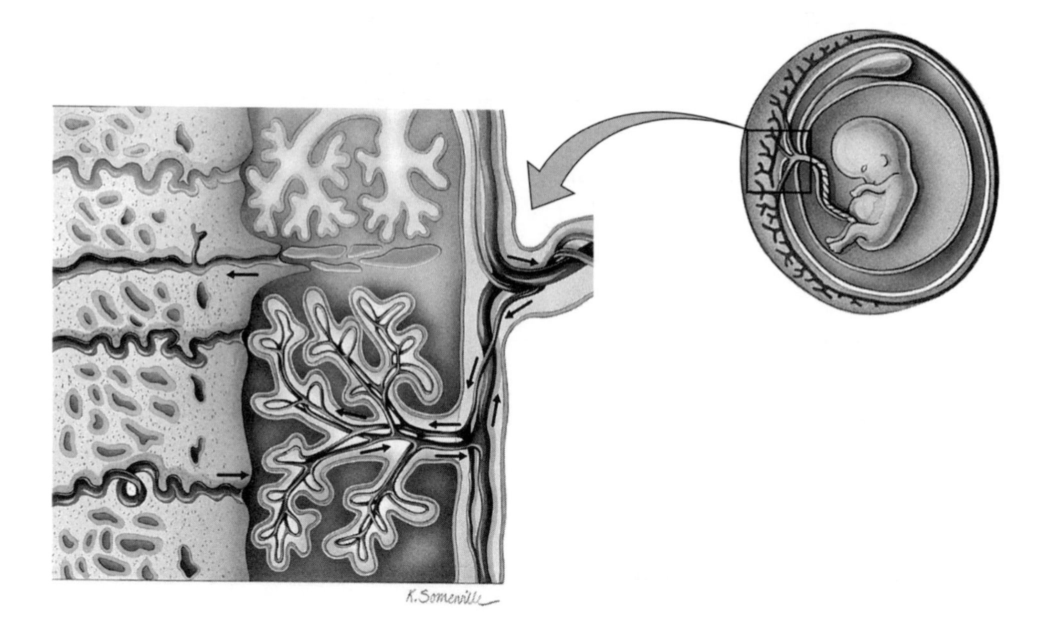

NOTES

27

**Figure 27.9** Amniocentesis and chorionic villi sampling (page 839).

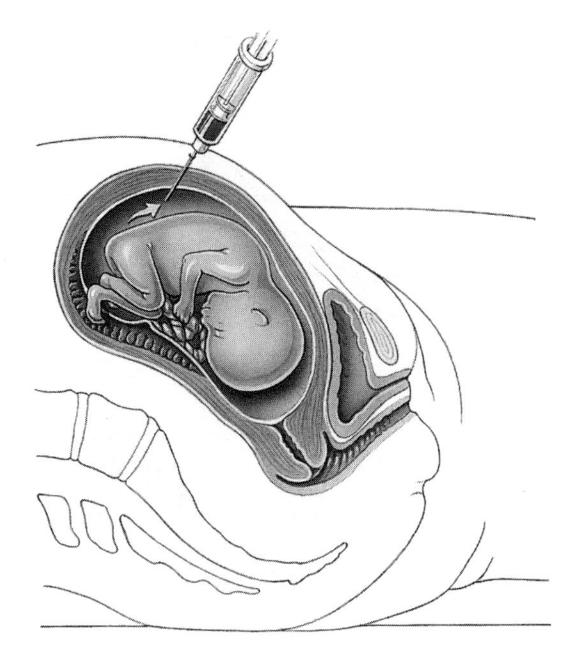

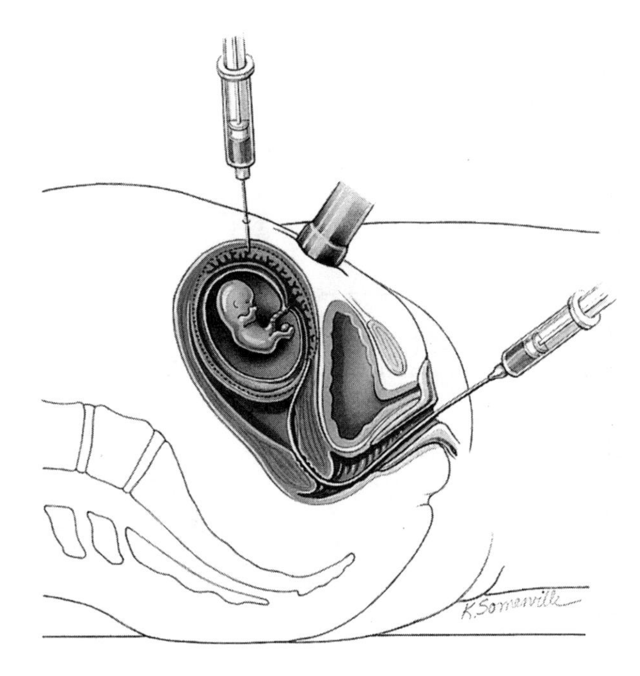

NOTES

27

**Figure 27.10** Normal fetal position at end of a full-term pregnancy (page 841).

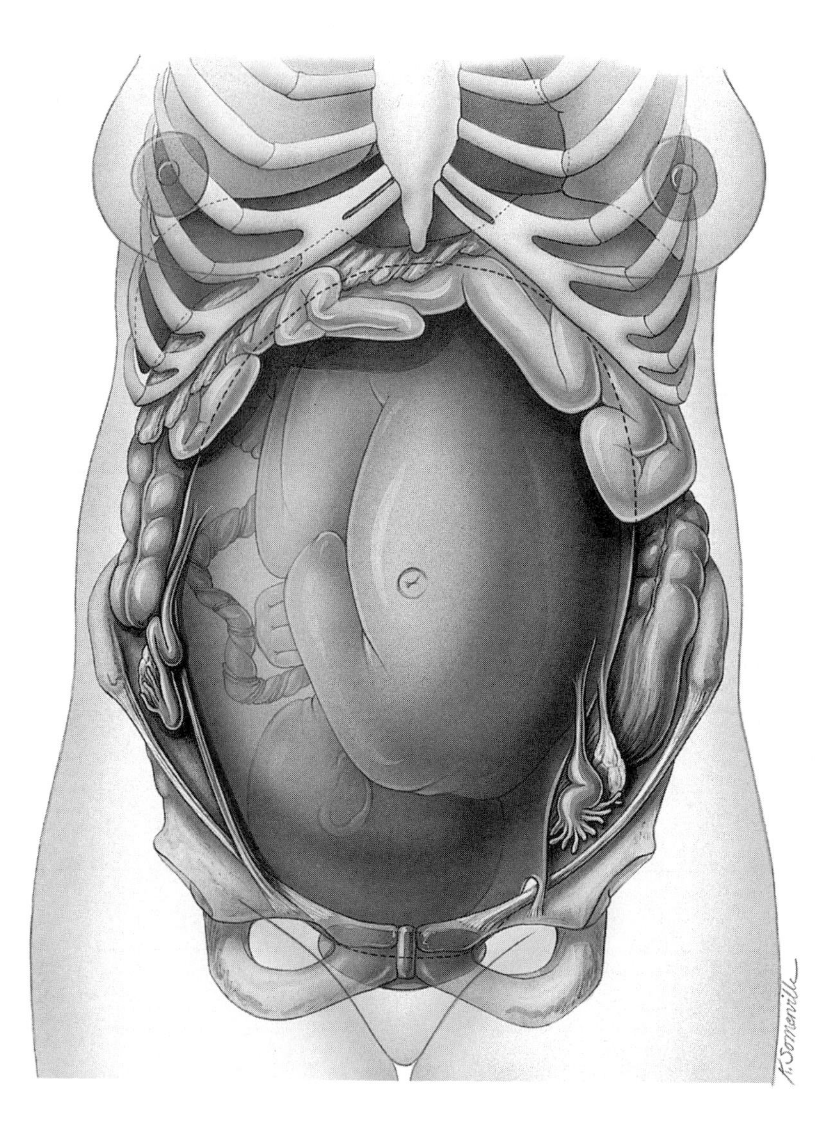

NOTES

27

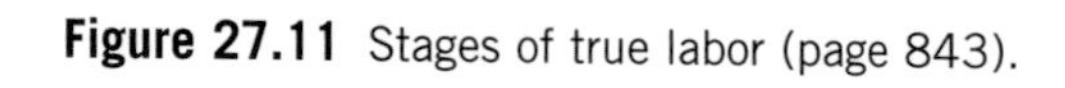

**Figure 27.11** Stages of true labor (page 843).

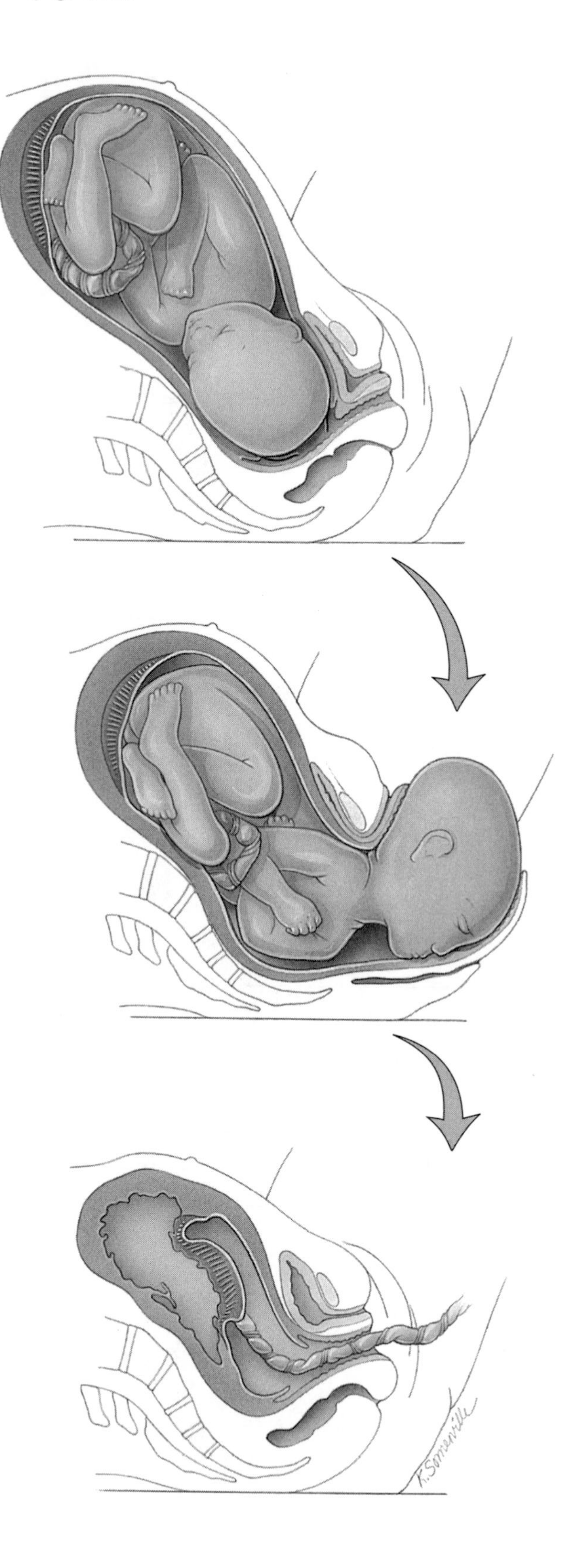

NOTES

# PHOTO CREDITS

**Chapter 1** Page 2: John Wilson White. Page 6 (top right): Stephen A. Keffier and E. Robert Heitzman, *An Atlas of Cross-Sectional Anatomy*, Harper & Row, Publishers, New York, 1979. Page 6 (center right): Lester V. Bergman/Project Masters, Inc. Page 6 (bottom right): Martin Rotker. Pages 12 & 14: Courtesy Mark Nielson.

**Chapter 2** Page 40: Courtesy Michael H. Ross.

**Chapter 3** Pages 46, 50 (bottom), 70 (top): Biophoto Associates/Photo Researchers. Pages 48, 50 (top), 52, 54, 56 (bottom), 62, 64, 66, 70 (bottom), 72, 74, 76: Courtesy Michael H. Ross. Page 56 (top): Lester V. Bergman/The Bergman Collection. Page 68 (top): Ed Reschke. Page 68 (bottom): Andrew J. Kuntzman. Page 78 (top): R. Calentine/Visuals Unlimited. Page 78 (bottom): ©Ed Reschke.

**Chapter 4** Page 82: ©L. V. Bergman/Bergman Collection. Page 86: Science Photo Library/Photo Researchers.

**Chapter 5** Page. 90: Courtesy Mark Nielsen. Page 102 (top): The Bergman Collection. Page 102 (bottom): Biophoto Associates/Photo Researchers. Page 110: National Library of Medicine/Photo Researchers.

**Chapter 6** Page 142: Courtesy Mark Nielsen.

**Chapter 7** Page 160: Courtesy Mark Nielsen.

**Chapter 9** Page 198: Courtesy Denah Appelt and Clara Franzini-Armstrong.

**Chapter 10** Pages 220, 222, 238: Courtesy Mark Nielsen.

**Chapter 11** Pages 278. 280, 282 (right), 284, 286, 288, 290, 292, 294, 296, 298: John Wiley & Sons. Page 282 (left): Courtesy Lynne Marie Barghesi.

**Chapter 12** Page 304: John D. Cunningham/Visuals Unlimited.

**Chapter 13** Page 306: John Wiley & Sons. Pages 310, 314, 316, 320: Courtesy Mark Nielsen.

**Chapter 14** Page 332 (center): Courtesy Dennis Strete. Page 332 (bottom): Courtesy Michael H. Ross. Page.336: Courtesy Mark Nielson.

**Chapter 15** Page 374: Courtesy Michael H. Ross.

**Chapter 22** Page 510: Courtesy Michael H. Ross.

**Chapter 24** Page 550: Courtesy Michael H. Ross. Page 566: Courtesy Mark Nielson.